ADVANCED DIGITAL SIGNAL PROCESSING
OF SEISMIC DATA

Seismic data must be interpreted using digital signal processing techniques in order to create accurate representations of petroleum reservoirs and the interior structure of the Earth. This book provides an advanced overview of digital signal processing (DSP) and its applications to exploration seismology using real world examples. The book begins by introducing seismic theory, describing how to identify seismic events in terms of signals and noise, and how to convert seismic data into the language of DSP. Deterministic DSP is then covered, together with non-conventional sampling techniques. The final part covers statistical seismic signal processing via Wiener optimum filtering, deconvolution, linear-prediction filtering, and seismic wavelet processing. With over 60 end-of-chapter exercises, seismic data sets, and data processing MATLAB codes included, this is an ideal resource for electrical engineering students unfamiliar with seismic data, and for Earth scientists and petroleum professionals interested in DSP techniques.

WAIL A. MOUSA is Associate Professor of Electrical Engineering at King Fahd University of Petroleum and Minerals (KFUPM), Saudi Arabia. His research interests include digital signal processing and its application in geophysics, with emphasis on seismic data processing. He has taught courses on "Geo-signal data processing" to graduate and undergraduate students in electrical engineering since 2007; and previously worked as a research scientist for Schlumberger. He was also Chair of the World Petroleum Council-Youth Committee (WPC-YC) for three years.

ADVANCED DIGITAL SIGNAL PROCESSING OF SEISMIC DATA

WAIL A. MOUSA

King Fahd University of Petroleum and Minerals

CAMBRIDGE
UNIVERSITY PRESS

CAMBRIDGE
UNIVERSITY PRESS

University Printing House, Cambridge CB2 8BS, United Kingdom

One Liberty Plaza, 20th Floor, New York, NY 10006, USA

477 Williamstown Road, Port Melbourne, VIC 3207, Australia

314–321, 3rd Floor, Plot 3, Splendor Forum, Jasola District Centre, New Delhi – 110025, India

79 Anson Road, #06–04/06, Singapore 079906

Cambridge University Press is part of the University of Cambridge.

It furthers the University's mission by disseminating knowledge in the pursuit of
education, learning, and research at the highest international levels of excellence.

www.cambridge.org
Information on this title: www.cambridge.org/9781107039650
DOI: 10.1017/9781139626286

First published 2020

Printed in the United Kingdom by TJ International Ltd. Padstow Cornwall

A catalogue record for this publication is available from the British Library.

Library of Congress Cataloging-in-Publication Data
Names: Mousa, Wail Abdul-Hakim, author.
Title: Advanced digital signal processing of seismic data / Wail Mousa.
Description: Cambridge, United Kingdom ; New York, NY : Cambridge
University Press, 2019. | Includes bibliographical references and index.
Identifiers: LCCN 2018058437 | ISBN 9781107039650 (hardback)
Subjects: LCSH: Seismic reflection method. | Signal processing–Digital techniques. |
Petroleum–Prospecting–Data processing.
Classification: LCC TN269.84 .M68 2019 | DDC 511.22028/7–dc23
LC record available at https://lccn.loc.gov/2018058437

ISBN 978-1-107-03965-0 Hardback

Additional resources for this publication at www.cambridge.org/mousa

To my dear parents and family.

To KFUPM and my great country.

Contents

Preface   *page* xi

Part I   Seismic Theory Background   1

1   Introduction   3
   1.1   Introduction   3
   1.2   Oil and Gas Formation and Accumulation   5
   1.3   Geological Classification of Petroleum Reservoirs   5
   1.4   Oil and Gas: From Exploration to Production   8
   1.5   Geophysical Surveys   10
   1.6   The Seismic Surveying Method   12
   1.7   Seismic Data Acquisition   14
   1.8   Seismic Data Processing   29
   1.9   Seismic Data Interpretation   39
   1.10 Summary   40
   Exercises   40

2   Seismic Theory and Reflection Surveying: A Necessary Background   42
   2.1   Introduction   42
   2.2   Theory of Elasticity   43
      2.2.1   Stress   43
      2.2.2   Strain   44
   2.3   The Wave Equation and d'Alembert's Solution   48
   2.4   Seismic Waves   58
      2.4.1   Body Waves   60
      2.4.2   Surface Waves   63
   2.5   Seismic Wavefronts and Raypaths   64
   2.6   Seismic Wave Velocity of Rocks   65
   2.7   Propagation Effects on Seismic Waves Amplitudes   67

		2.7.1	Amplitude Attenuation	67
		2.7.2	Dispersion	70
	2.8	Raypaths in Layered Media		73
		2.8.1	Reflection and Transmission of Normally Incident Seismic Wave Rays	73
		2.8.2	Reflection and Refraction of Obliquely Incident Rays	78
		2.8.3	Critical Refraction	80
		2.8.4	The Phenomena of Diffraction	82
	2.9	Seismic Events Geometry in Layered Media		82
		2.9.1	Geometry of Direct Raypaths	83
		2.9.2	Single Layer Reflector	83
		2.9.3	Single Layer Refractions	89
		2.9.4	Single Layer Dipping Reflectors	89
		2.9.5	Geometry of Diffractions	92
		2.9.6	Sequences of Horizontal Reflectors	92
	2.10	Characteristics of Seismic Events and Accompanying Noise		97
		2.10.1	Linear Events	98
		2.10.2	Hyperbolic Events	101
		2.10.3	Noise	103
		2.10.4	Examples of Real Seismic Data	107
	2.11	Summary		107
	Exercises			111
Part II	**Deterministic Digital Signal Processing for Seismic Data**			**115**
3	Spectral Analysis of Seismic Data and Useful Transforms			117
	3.1	Introduction		117
	3.2	Discrete-Time(Space) Signals and Systems: A Review		117
		3.2.1	Discrete-Time(Space) Signals	117
		3.2.2	Discrete-Time(Space) Systems	122
	3.3	The z-Transform		132
		3.3.1	The Forward z-Transform	132
		3.3.2	Rational z-Transforms	139
		3.3.3	The Inverse z-Transform	142
		3.3.4	Analysis of LSI Systems in the z-Domain	148
	3.4	The Fourier Transform		150
		3.4.1	The Discrete Fourier Transform	155
	3.5	Spectral Analysis of 2-D Seismic Data		163
	3.6	Radon Transform		172
		3.6.1	Radon Transform and Seismic Data Processing	177
		3.6.2	Linear Radon Transform	178

		3.6.3 Parabolic Radon Transform	182
3.7		Summary	186
		Exercises	189

4		Sampling Theorem for Seismic Data	193
	4.1	Introduction	193
	4.2	Sampling Theorem for Time and Spatial Continuous Functions	194
		4.2.1 1-D Sampling	194
		4.2.2 2-D Sampling	198
		4.2.3 Alias Effects on Seismic Data	202
	4.3	Overview of Compressive Sensing Applications in Seismic Data Processing	206
		4.3.1 Properties of Compressive Sensing	209
		4.3.2 Mathematical Theory	209
		4.3.3 Missing Seismic Trace Interpolation	211
		4.3.4 Primary Arrivals and Multiples Separation	214
	4.4	Summary	219
		Exercises	219

5		Seismic Applications of Digital Filtering Theory	221
	5.1	Introduction	221
	5.2	Filter Design	222
	5.3	Design of FIR Digital Filters	225
		5.3.1 Design of FIR Digital Filter Based on Windowing Techniques	226
		5.3.2 Design of FIR Filters Using Frequency Sampling	235
		5.3.3 FIR Projections onto Convex Sets Based Digital Filters	238
	5.4	Design of IIR Digital Filters	248
	5.5	Seismic Wavefield Extrapolation 1-D FIR and IIR Filters	255
	5.6	Two-dimensional Filters for Seismic Data	261
	5.7	Summary	269
		Exercises	269

| | **Part III Statistical Digital Signal Processing for Seismic Data** | | 271 |

6		Fundamentals of Digital Optimal Filtering	273
	6.1	Introduction	273
	6.2	The Wiener Optimum Filter	274
	6.3	Application of Optimum Filters to Reflection Seismology	279
	6.4	Summary	284
		Exercises	284

7   Seismic Deconvolution   285
     7.1   Introduction   285
     7.2   The Seismic Deconvolution Model   286
     7.3   Seismic Deconvolution Based on Wiener Optimum Filtering   289
            7.3.1   Spiking Deconvolution   289
            7.3.2   The Linear Prediction Filter   290
     7.4   FX Deconvolution   295
     7.5   Summary   295
     Exercises   296

8   Seismic Wavelet Processing   298
     8.1   Introduction   298
     8.2   Seismic Wavelets   299
            8.2.1   Two-Length Wavelets or Minimum-Delay Wavelets   301
            8.2.2   Zero-Phase and Symmetric Wavelets   306
     8.3   Seismic Wavelet Processing   309
     8.4   Summary   309
     Exercises   310

     References   313
     Index   323

Preface

It is well known that Digital Signal Processing (DSP) plays an important role in many applications of science and engineering disciplines, including seismology, sonar, radar, medical, and communications. In the case of seismology, the application of signal processing theory began with the work of the Geophysical Analysis Group at the Massachusetts Institute of Technology (MIT) between 1960 and 1965, where it was one of the great historical milestones in seismic data processing.

Interestingly, oil and gas are still considered to be extremely important natural resources for human beings, with many beneficial applications. Such precious resources are buried in deep land or marine subsurface geological structures. In order to produce oil, we need first to estimate as accurately as possible an image of the subsurface. This can be done by listening to the echo caused by artificial earthquakes via a surveying method known as seismic exploration. The process generally requires acquisition, processing, and interpretation of seismic data, where DSP plays an important role in estimating seismic subsurface images.

It is known that the energy industry already faces critical shortages across the entire spectrum of skilled positions and our future needs will only grow. This requires education and training of tens of thousands of new scientists and engineers, in addition to geologists, geophysicists, and other vital workers. This is all the more important given the low number of students joining geosciences-related disciplines. Therefore, with the increased demand for oil and gas production, seismic data acquisition, processing, and interpretation will require more manpower and innovative thinking. Part of this innovation will rely on advancing seismic signal analysis and processing techniques and algorithms. And this book, in essence, discusses deterministic and statistical DSP theories but with various examples based on seismic exploration data. It provides a blended mix between the theoretical as well as the practical aspects of DSP and its application to the processing of seismic data.

For electrical engineers, particularly those in the area of DSP, the book provides the seismic surveying theory background that is necessary for understanding seismic data and, hence, for preparing them to properly process seismic data in academia or industry. It also covers deterministic and statistical DSP theory, with a focus on practical DSP for seismic data processing in a language that electrical engineers will understand. The book also serves as an important reference for geosciences researchers and professionals, who are interested in digging deeper into the theory of DSP and achieving a more precise and in-depth understanding of their applications to seismic data, which enables them to develop advanced seismic data processing algorithms with a DSP center. The main book features are as follows:

- Suggested senior level undergraduate and graduate Electrical Engineering course syllabi.
- Sufficient examples, illustrations, and figures for each chapter as applicable.
- Exercises, at the end of each chapter, including computer assignments for various topics using MATLAB.
- Synthetic and real seismic data short gathers for illustration figures, and computer assignments.

Chapter 1 is an introduction with the main aim of motivating readers regarding the subject of seismic signal processing. It also focuses on general seismic data acquisition, processing workflow, the seismic convolution model, and seismic interpretation. On the other hand, Chapter 2 mainly gives an intensive overview of the fundamentals and physical principles on which seismic methods are based. It provides the geophysical background that is necessary for understanding seismic data and, hence, the reader will obtain a clearer understanding of how to properly process the data in order to ultimately obtain better seismic images that are used for accurate interpretation.

Signal analysis in the spectral or other domains is very important and assists in obtaining a better understanding of signals. When dealing with seismic data sets, it becomes almost standard to analyze them in the 2-D frequency-wavenumber domain. Also, other discrete transforms are very useful for processing seismic data sets such as the Radon transform, which can be used for seismic wavefield decomposition as well as seismic multiple removal. Of course, this would also require a brief review of the z-transform and the various usages in seismic applications. Hence, in Chapter 3, spectral analysis of seismic data and useful transforms are discussed.

Chapter 4 presents sampling theory for seismic data, where whenever we talk generally about digital data and their acquisition systems, we must explain Shannon sampling theory for sampling continuous time (space) signals. Also, the chapter explains the aliasing effects due to under-sampling of seismic data sets. This, of course, includes some discussion of how to choose the best parameters for sampling seismic data given the opportunity to do so. Moreover, the theory of compressive sensing (CS) is currently considered the state of the art theory of DSP, with many applications related to signal and image compression, signal recovery, and many other applications. CS is currently used for various seismic data processing problems. Hence, in this chapter we introduce CS principles and provide a few seismic data processing-related applications.

Seismic applications of digital filtering theory are presented in Chapter 5. 1-D FIR and/or IIR digital filters, such as low-pass or band-pass, are used heavily to enhance the signal-to-noise (SNR) ratio of acquired seismic data. Furthermore, 2-D digital filters such as fan filters have become standard in removal of surface waves accompanying seismic data records. Solving the wave equation numerically may also require using FIR or IIR digital filters such as explicit depth wavefield extrapolation filters.

When using low-pass or band-pass filters to attenuate unwanted seismic energy from seismic data records, so-called seismic vertical resolution will be affected, i.e., the high frequency components of seismic waves are attenuated and some processes such as deconvolution must, therefore, be used. Also, multiple or ghost noise types can be greatly attenuated when using predication error filters. Seismic wavelets and wavelet processing are essential to understand, since wavelets are a main component of the seismic convolution model. All of these filters are various applications of Wiener optimum filtering processes. Hence, in Chapters 6–8 we provide the fundamentals of optimum filtering and show different applications of this important theory in seismic data processing.

I would like to take this opportunity to thank my protégé H.E. Professor Khaled S. Al-Sultan, Previous Rector, King Fahd University of Petroleum & Minerals (KFUPM), and the university for their support. I would also like to thank the Electrical Engineering Department, KFUPM, and its chairman, Dr. Ali A. Al-Shaikhi, for the opportunity to teach, over the last 10 years, many special topics courses around the book subject, which greatly helped to shape the book's content and exercises. I cannot forget my dear colleague, Professor Abdullatif A. Al-Shuhail, Earth Sciences Department, who has been my research partner since 2008, and has supported me in the course and its seismic theory background over the last 10 years. Without his assistance, neither the book nor the course would exist. Special thanks are due to my colleagues from Saudi Aramco, including Dr. Saleh Al-Dossary, for being welcoming and supportive on my visits with students to their Seismic Data Processing facilities for allowing me access. Also, I sincerely thank Professor Stewart Greenhalgh, Saudi Aramco Chair Professor, Geosciences Department, College of Petroleum Engineering & Geosciences, KFUPM, for taking the time to review the book. I thank also Dr. Arbab Latif, Mr. Abdullah F. Al-Battal,

Dr. Naveed Iqbal, and Mr. Ammar Jamie, KFUPM, who helped in the final compilation of this book. Many thanks are due to the Cambridge University Press team, who have been supportive and patient with me since we signed the book contract: Phil Meyler, Susan Francis, Emma Kiddle, Sarah Lambert, and Esther Migueliz Obanos. Finally, very special and sincere appreciation is due to my family, friends, and colleagues, who have constantly encouraged me to pursue this book: my father Professor Abdul-Hakim M. Mousa, Dr. Jacquelin E. Elliott, Dr. Ahmad A. Masoud, Dr. Tariq Y. Al-Naffouri, Mr. Ayman F. Al-Lehyani and Dr. Abdulaziz O. Al-Kaabi.

Figures taken from "Seismic Data Analysis" (Yilmaz, 2001) carry the CC BY-SA 3.0 Unported License and appear in the print and e-book versions of the title (see https://library.seg.org/doi/book/10.1190/1.9781560801580) as well as the SEG Wiki, operated by the Society of Exploration Geophysicists (see https://wiki .seg.org/wiki/Seismic_Data_Analysis).

Part I

Seismic Theory Background

1

Introduction

1.1 Introduction

Oil and gas are important natural energy resources, since they are used in nearly every aspect of human life from fuelling cars and generating power to producing plastics and fertilizers. Their impact on human civilization is such that continuing and improving the success of the overall production process is vital. Even with the forecast decreasing use of, for example, gasoline in cars, oil and gas products will still be extensively used. Oil and gas exist in reservoirs located thousands of metres below the Earth's surface and ocean floors. Such reservoirs only exist in certain locations, dependent on the geological history of the area. An increasingly essential precondition for successful extraction of gas and oil is the accurate determination of the subterranean geophysical structure. The knowledge obtained is not simply that of knowing where to look for oil and gas; it is also relevant to the management of the production process after the crude oil is found and extracted. But first, the geological strata need to be imaged. Imaging of subsurface layers is done by a method known as reflection seismology. This geophysical technique relies on the generation of artificial seismic waves and the recording of their reflections from different geological layers. However, the acquired seismic data do not reveal an accurate image of what lies beneath the surface, unless appropriate signal processing methods are carefully employed. In fact, the seismic signals acquired are among the most challenging geo-signal data to process and interpret, since more than 80% of them have extrinsic sources, or are simply noise. In the case of land-based seismic data, for example, the primary seismic reflection signals sought are overwhelmed by surface wave noise (known as ground roll), wave scatter, and so on. The signals of interest in marine exploration seismology are usually masked by multipath waves (multiples).

Digital signal processing (DSP) has long been widely accepted as the primary tool for data conditioning and information extraction. It plays an important role in many applications in science and engineering, including seismology. The actual application of DSP theory to the exploration of subterranean layers began with the work of the Geophysical Analysis Group at the Massachusetts Institute of Technology (MIT) between 1960 and 1965, which proved to be a significant milestone in the history of seismic data processing. This book deals with the application of DSP to the processing of reflection seismic data that are used for hydrocarbon exploration (otherwise known as Seismic Exploration). This type of seismic data has been successfully used for hydrocarbon exploration for approximately 100 years, where measurements can reach a depth of up to 10,000 m. This geophysical method does not assist in showing oil and/or gas (sometimes gas can be seen in time-migrated data) as such or even in terms of quantity. This is due to the low resolution of the seismic reflection data, where in order for the seismic waves to travel deeper in the subsurface, fewer frequency components are acquired. The normal seismic data frequency bandwidth is only between 15 and 60 Hz. However, the idea of processing seismic waves that are reflected from various geological interface layers of the strata will result in clearly identifying such interface boundaries between the layers. In addition, as will be explained in this chapter, oil and gas are usually trapped and accumulated in certain geological structures. Hence, petroleum reservoirs are most likely to be found in such potential geological structures, and reflection seismic will certainly identify those structures.

Geophysics students, particularly applied geophysics students, would normally have covered the subject matters that are related to oil and gas formation, types of petroleum reservoirs based on their geological classification, seismic theory and seismic reflection theory, seismic data acquisitions and processing essentials, and so on. DSP students reading this book, however, would be lacking such important knowledge and background. Therefore, this chapter will provide brief background information for DSP students about how oil and gas were formed and accumulated. This is followed by a basic description of the geological classification of petroleum reservoirs. Exploration geophysics and the basics of geophysical surveying methods will then be briefly explained. The seismic surveying method, the seismic convolution model, and an overview of seismic data processing are then given. At the end of this chapter, readers will have the opportunity to meet seismic data, probably for the first time, and, as an overview, see various effects of processing on data. Note that the necessary seismic reflection theory background (that will hopefully lead to the understanding of the physics behind seismic events and the accompanying noise) will be covered in Chapter 2.

1.2 Oil and Gas Formation and Accumulation

It might be interesting to note that some evidence exists that millions of years ago ancient seas covered much of the present land surface of the Earth including the Arabian peninsula and the Gulf of Mexico. Rivers flowing down to these seas carried large amounts of mud and sedimentary materials into the sea for many years. The sea floors slowly sunk and were squeezed due to the creation of accumulated mud and sedimentary materials layers. Such layers were the direct result of the distribution and deposition of mud and sedimentary material layer upon layer. Eventually, these formed rocks (where oil and gas are found) such as sandstone, shale, and carbonate rocks, all of which belong to the sedimentary category. The source of petroleum was constituted mainly by two organism resources: (a) a very large amount of small plants and animal life, which came into the sea from the land with the river mud and sedimentary materials, and (b) small marine life that existed in the sea already. These organisms died but were buried by the depositing silt, which kept them protected from ordinary decay. Many reactions such as pressure and temperature have over the years caused the dead organisms to change into oil or gas, depending on the temperature conditions.

Rock structures contain pore spaces that oil, gas, or even water can occupy. For example, oil, gas, and water can occupy the pore spaces between the grains of sandstone rocks or the pore spaces in addition to the cracks and vugs of other rocks like limestone and dolomite. These rocks where oil and gas were formed are known as *source rocks*. If they were sealed by a layer of rocks that is impermeable (known as the cap rock), then the accumulating petroleum within their pore spaces will be trapped and form the petroleum reservoir. When the pore spaces in the source rocks are not trapped, then under the effects of pressure and gravity, oil and gas will migrate from the source rocks until they are trapped in other capped rocks. Within the trap rock, oil and gas as well as water are segregated because of their density differences, with oil coming between gas and water. A *petroleum reservoir* can then be defined as the geological structure in which petroleum has been trapped and has accumulated. A petroleum reservoir can be a source rock or a rock to which petroleum has migrated.

1.3 Geological Classification of Petroleum Reservoirs

Petroleum reservoirs are usually classified according to their geological structure and their production mechanism. However, for an applied geophysicist and, hence, a DSP reader, it is sufficient to know about these reservoirs' geological structures. The reservoirs come in various sizes and shapes, and they are geologically classified according to their formation conditions. Structural-based reservoirs exist, which are physical arrangements of rocks, where traps result from deformation of the rocks by outside forces (Sheriff, 2006). Examples include but are not limited to:

1. Anticline- and dome-shaped reservoirs[1]: These are formed by the folding of the rock as shown in Figure 1.1. An anticline is considered as a fold in stratified rocks, where the rocks dip in opposite directions from a crest and the corresponding layers are, therefore, convex upward. A dome, on the other hand, is a structure where all of the rock beds dip away from a central area. While the anticline is long and narrow, the dome is circular in outline. Oil and gas are migrated upward through the pore spaces in the rocks and they are trapped by the cap rocks and the shape of the structure. Examples of such structures are Dammam and Bahrain domes.

2. Faulted reservoirs: Shearing and offsetting will cause faulting of the strata. Movement of impermeable rocks opposite rocks with pore space formations, which contain oil and gas, will create the trap. Both the tilt of the petroleum-bearing rock and the created fault trap the oil and gas in the reservoir. Figure 1.1 shows an example of a normal faulted reservoir. In this case, the faults drop one side down and push the other side up to place the reservoir rock against impermeable sealing rocks, forming a structural fault trap. An example is the Dunvegan gas field in northwestern Alberta. Another example of a faulted reservoir is the thrust fault. In the foothills of Western Canada, east of the Rockies, the original limestone layer was first folded and then thrust-faulted over itself. An overlying seal of impermeable rock completes the structural trap. Examples include the Turner Valley oil and gas field and Jumping Pound gas field, both in south-western Alberta.

3. Salt-dome reservoirs:[2] Such reservoirs are shaped like a dome and are formed due to the upward movement of a large and impermeable salt dome that has deformed and lifted the overlying layers of rock. Oil and gas can be trapped between the cap rock and the underlying impermeable rock layer or in between two impermeable layers of rock and the salt dome, as can be seen in Figure 1.1.

On the other hand, stratigraphic reservoirs exist, where traps are formed by variations within the sedimentary rocks themselves. Rock classification is organized according to different schemes, which do not have one-to-one correspondence. In other words, lithostratigraphic and geochronologic or chronostratigraphic

---

[1] In 1885, I. C. White, an American geologist, was the first to associate the existence of organic matter, the presence of reservoir rocks, and an anticlinal trap with the successful location of oil and gas (Shoup et al., 2003).

[2] In 1900, P. Higgins (a businessman and self-taught geologist) started drilling on Spindletop Hill, Texas, which is located on a salt dome, with the salt capped by a fractured dissolved limestone, which acts as a reservoir, producing in 1901 100 K barrels/day. This has led to exploration of salt structures along the US gulf coast region (Shoup et al., 2003).

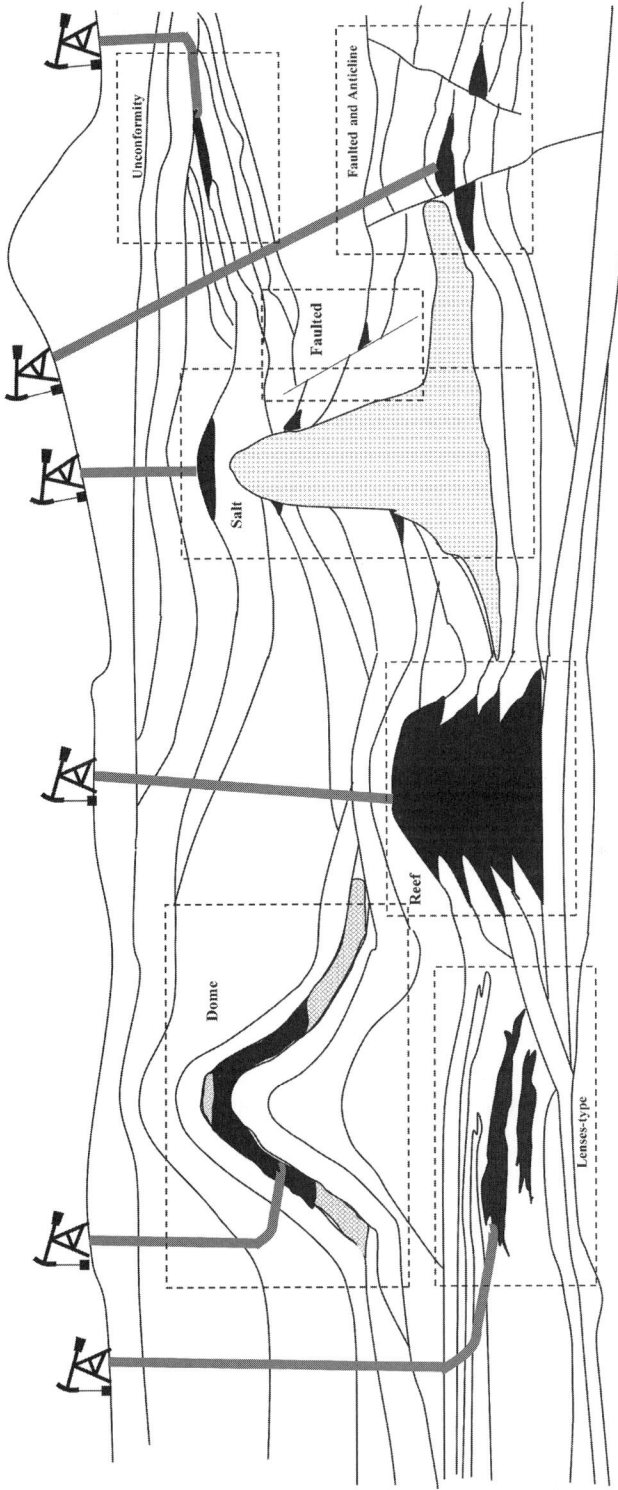

Figure 1.1 Common structural and stratigraphic petroleum reservoirs.

subdivisions sometimes cross each other (Sheriff, 2006). Examples of stratigraphic reservoir types are listed as follows:

1. Unconformity reservoirs[3]: Figure 1.1 shows an example of this type of reservoirs. Such reservoir types are formed as the result of an unconformity, where the impermeable cap rocks lay down across cutoff surfaces of lower geological beds.
2. Lenses-type reservoirs: Pore space rocks that contain oil and/or gas are surrounded in these reservoir types by impermeable rock formations, probably due to irregular deposition of sediments, leading to the creation of shale rock types. An example of a sandstone lenses-type reservoir is shown in Figure 1.1.
3. Reef-type reservoirs: in this type, hydrocarbon is trapped in the core of the reef, with fore-reef talus and back-reef lagoonal muds acting as lateral seals, and basinal mudstones as top seals.

Note that combination reservoirs exist, which can be formed based on a combination of folding, faulting, abrupt changes in the pore spaces, and so on. Such types are common. For more details on petroleum reservoirs, the reader is recommended to see, for example, Onajite (2013); Gluyas and Swarbrick (2014); Dow and Magoon (1994); Selley and Sonnenberg (2014); Miall (2008).

1.4 Oil and Gas: From Exploration to Production

It is important to understand that the process of extracting oil and/or gas is not straightforward. It basically requires a great deal of effort emerging from different areas of expertise from both engineering and science disciplines, and consumes a lot of time and money. The process of extracting oil and/or gas involves, in principle, three main stages:

1. Geological and Geophysical Surveys: Such surveys are the most important step toward extracting oil and gas. The aim of geological surveys is to examine the surface geology, formation outcrops, and surface rock samples. On the other hand, geophysical surveys aim to generate approximate maps of the subsurface structures by measuring various physical properties of rocks such as their densities, velocities, resistivity values, and many others. The data collected from both survey types are usually combined together, and without them, particularly the geophysical ones, it is difficult to claim the existence of oil and/or gas. More light will be shed on geophysical surveying methods in Section 1.5.
2. Drilling Exploratory Wells: The data collected from the geological and geophysical surveys are extremely useful in formulating probable definitions and

---

[3] In 1934, the American geologist A. I. Levorsen suggested that unconformities and changes in sedimentary environments control the trapping potential for hydrocarbons (Miall, 2008).

realizations for geological structures that might contain oil and/or gas. That is, they greatly assist in the determination of possible rock caps (traps) and, hence, potential petroleum reservoirs. The question that comes next is, does oil and/or gas exist? If so, then is it economical to develop oil and/or gas fields based on their quantity? The answer to such questions is determined only by drilling exploratory wells, called *wildcat wells*, in locations that are identified by geologists and geophysicists. During the drilling of wildcat wells, rock samples are extracted and used for further analysis. Once drilling is completed, various well logs are taken using logging techniques. Such tools are attached to electric cables know as *wirelines*, and as they are lowered they transmit and receive signals that are recorded, processed digitally, and then further analyzed to produce maps of the rocks and fluid properties versus depths, initial reservoir pressures, reservoir productivity, and many other important data to petroleum engineers.

3. Development of Oil and Gas Fields: The data collected from the previous steps, under the condition that the discovered reservoirs are economically productive, are combined and used to generate many important subsurface maps that will be of great importance to petroleum engineers. Examples of such maps are porosity maps, permeability maps, contour maps, and all are used to construct, through computer simulations, conceptual models that describe details of the subsurface structure and the oil/gas location within the subsurface structures. This is followed by integrated efforts of geologists, geophysicists, and petroleum engineers as one team to plan for developing oil/gas fields and the remaining operations that should ultimately lead into commercial production of fossil fuel. This includes drilling production wells, casing and cementing them, reservoir management, and so on.

Figure 1.2 summarizes the process described previously.

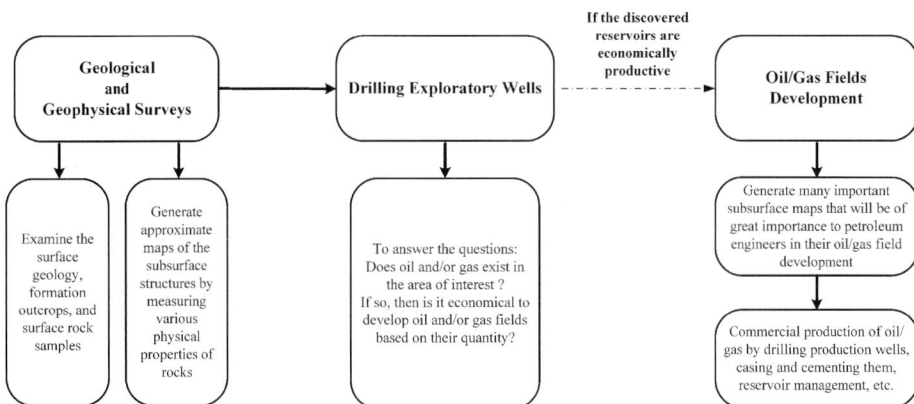

Figure 1.2 Oil and/or gas: From exploration to production.

1.5 Geophysical Surveys

In simple terms, exploration geophysics can be defined as the branch of geophysics that uses geophysical surveying methods to measure various subsurface rock properties in order to somehow infer the presence of oil and gas using various analysis and processing techniques (Sheriff, 2006; Telford et al., 1990). Rocks come with many physical properties and, hence, various interesting geophysical surveying methods exist that ultimately infer some of these properties that can assist in identifying the rock types accurately. Geophysical surveying methods can be applied to a wide range of investigations at various scales, such as studying the Earth. The main aim of geophysical surveying methods is to perform measurements within geographically restricted areas, seeking knowledge of the distribution of physical properties at depths that reflect the geology of the subsurface location of interest. For the sake of completeness, the following commonly used geophysical surveying techniques for hydrocarbon exploration are briefly described:

- **Gravity survey**: This method is considered among the least expensive surveying methods for locating a possible petroleum reservoir. This method involves the use of an instrument called a gravimeter that picks up a reflection of the density of the subsurface rock. The gravimeter can detect, for example, the presence of salt domes, which may indicate the presence of an anticline structure. In places where the gravimeter detects stronger or weaker than normal gravity forces, a geologic structure containing hydrocarbons may exist. The depth of investigation using this method is almost unlimited.

- **Electrical and electromagnetic surveys**: Such surveys can be defined as measurements that are made at or near the surface of natural or induced electric fields and can reveal information of up to 2 km of depth. They aim to map mineral concentrations and/or to map geological basements. These measurements can be made in many ways to determine various results. Electrical conductivity between different rocks makes such measurements feasible. Examples of electric measurements include self potential, magnetotellurics, and resistivity.

- **Magnetic survey**: This surveying method involves measurement of magnetic pull, which is affected by the type and depth of the subsurface rocks. It is usually used for determining the existence and depth of subsurface volcanic formations and basement rocks, which contain high concentrations of magnetite. This information is useful in identifying the presence of sedimentary formations above the basement rocks. The measurements are usually made more easily and cheaply than most geophysical measurements, where corrections are practically unnecessary. It basically can cover depth locations from the surface to the Curie isotherm. It basically infers magnetic susceptibility and remanent magnetization. However, they lack uniqueness of interpretation.

- **Seismic reflection survey**: Such a surveying method is the most widely used and well-known geophysical technique among many geophysical surveying techniques. Seismic reflection data can be processed to reveal details of geological structures on scales from the top tens of metres of the Earth's crust to its inner core (Kearey et al., 2002; Sheriff and Geldart, 1995; Yilmaz, 2001). Part of its success lies in the fact that the raw seismic data is processed to produce images of the subsurface structure. A geologist can then make an informed interpretation by understanding how the reflection method is used and seismic sections (images) are created. The analysis of seismic data is performed for many applications such as petroleum exploration, determination of the Earth's core structure, monitoring earthquakes (Spanias et al., 1991; Yilmaz, 2001). If one is considering, for example, land seismic surveys, then seismic signals are generated by a source (transmitter), such as an explosion. These signals propagate through Earth layers. Some of these signals will be reflected, refracted, and lost due to attenuation). At the surface, the reflected signals are then recorded by a receiver. The strength of the reflected signal depends on the impedance contrast between adjacent layers. More discussion on this surveying method is presented in subsequent sections.

A summary of these geophysical methods is given in Table 1.1. Readers seeking further information on geophysical surveys are encouraged to see Telford et al. (1990).

Table 1.1 *Commonly used geophysical surveying methods for hydrocarbon exploration*

Geophysical surveying method	Depth	Measured parameters	Extracted rock physical quantity
Gravity	All	Spatial variations in the strength of the gravitational field of the Earth	Rock density contrasts
Electrical and Electromagnetic	About 2 km	Earth resistance and response to electromagnetic radiation	Lateral/vertical changes in resistivity, conductivity, and inductance of rocks
Magnetic	Surface to Curie isotherm	Spatial variations in the strength of the geomagnetic field of the Earth	Magnetic susceptibility and remanent magnetization contrasts
Seismic	All	Travel times and amplitudes of reflected seismic waves	Velocity layer contrasts

1.6 The Seismic Surveying Method

In 1913, the Canadian inventor Reginald Fessenden conceived the basic ideas for the seismic reflection technique but it was not until the year 1923 that geophysical measurements, in general, were starting to be used by oil companies on an industrial scale. The use of seismic data began to gain industrial prominence in 1929 (mainly seismic refraction), when it was used to map salt domes and their related petroleum traps in various places in the world, including Germany and the United States. Alongside the advances achieved in electronics and computer sciences, by which digital recording was made possible during the 1950s and 1960s, seismic reflection made considerable progress. Since then all seismic data are recorded, processed, and stored digitally (Sheriff and Geldart, 1995).

At a large-scale level, the seismic survey analysis scenario involves three main stages (see Figure 1.3). The first stage is the acquisition stage, where the seismic data are collected by an array of receivers (geophones for land and hydrophones for marine), transmission over a narrow band channel (using electric cables [Sheriff and Geldart, 1995] or wireless [Freed, 2008]), and storage of data for processing. One can acquire two-dimensional (2-D) or three-dimensional (3-D) seismic data. Both 2-D and 3-D acquisition receivers can be of a single component, recording seismic primary waves, or multi-component, recording additional vertical and horizontal seismic shear waves. The acquisition is commonly based on the so-called Common Midpoint (CMP) technique. A recording-processing method where each source is recorded at a number of receiver locations and each receiver location is used to record from a number of source locations. After performing some data correction, the data are combined in a way that provides what is known as a CMP seismic section. Such a seismic section approximates seismic signals that would be recorded by a coincident source and receiver at each location, but with improved discrimination against noise. The objective is to attenuate random noise and events whose dependence on offset is different from that of the main primary reflected seismic signals (reflections) (Yilmaz, 2001). Other acquisition methods include Cross-spread and Narrow Azimuth (NAZ). More advanced acquisition techniques such as multi-azimuth (MAZ), wide-azimuth (WAZ), and rich-azimuth (RAZ) are recent innovations that aim to address the illumination problems inherent in traditional NAZ seismic acquisition. These types of advanced reflection seismic surveying acquisitions improve signal-to-noise ratio and illumination in complex geology (Liner, 1999a; Pecholcs et al., 2012; Vermeer, 2012; Campman et al., 2016).

Next, comes a crucial stage, which is the processing of the acquired seismic data sets. Seismic data processing can be considered an unusual concoction of highly esoteric and objective mathematical techniques (including signal processing techniques). This is blended with the subjective approach of geophysicists' interpretation. Seismic data processing includes steps like geometrical spreading

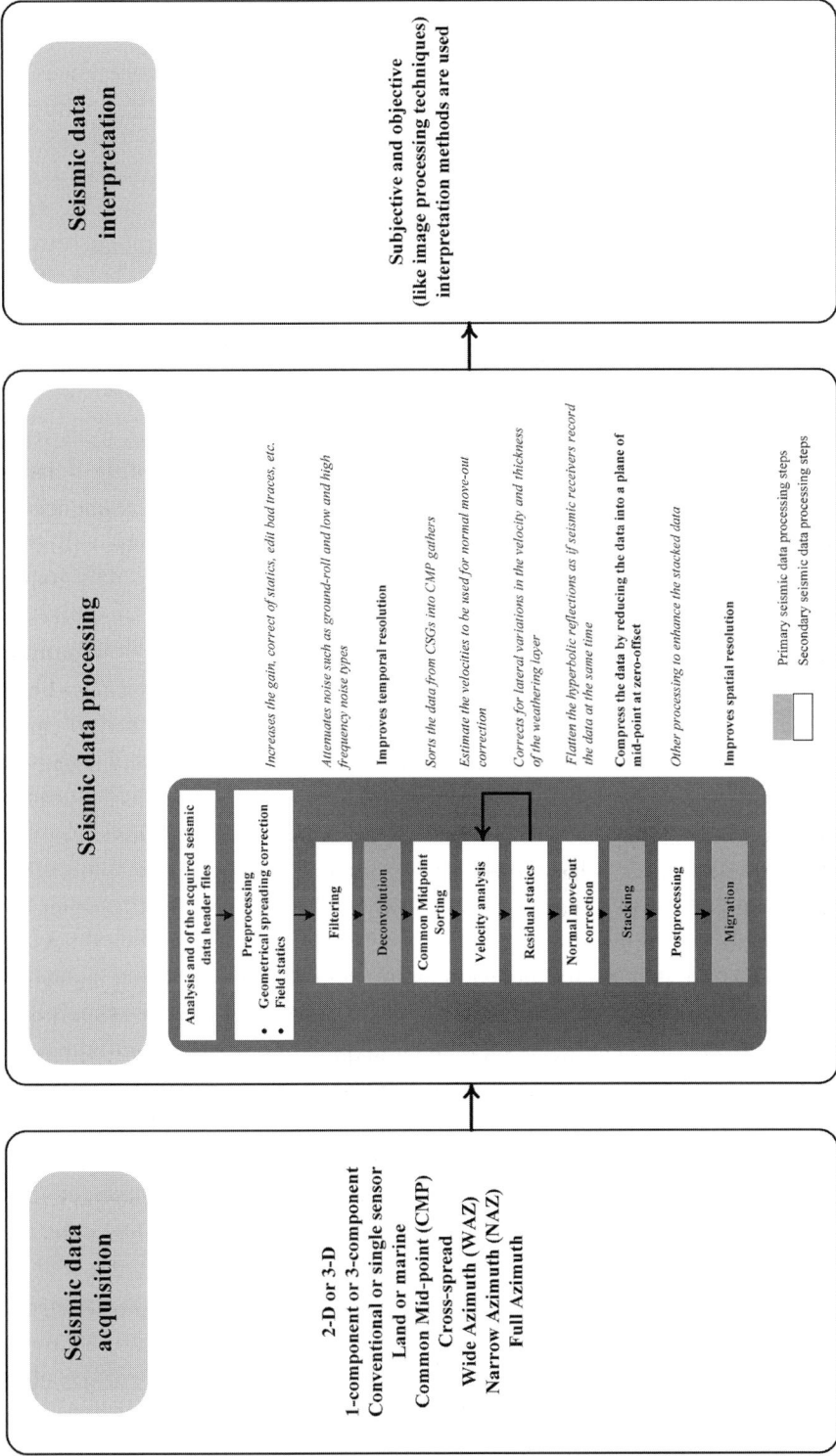

Figure 1.3 The seismic survey analysis scenario, which includes acquisition, processing, and interpretation.

correction, frequency filtering, deconvolution, velocity analysis, static correction, and so on (Yilmaz, 2001).

Finally, the processed data goes to the last stage, which is seismic data interpretation. This stage mainly aims to derive a simple, plausible geologic model that is compatible with the observed data (Al-Shuhail et al., 2017; Yilmaz, 2001). It can utilize subjective and, more recently, objective ways to interpret the seismic data. Many image processing techniques are used for the sake of interpreting seismic images (Al-Shuhail et al., 2017; Avseth et al., 2010; Bacon et al., 2003; Hart, 2011; Herron, 2012; Iske and Randen, 2000; Liner, 2004; Yilmaz, 2001).

1.7 Seismic Data Acquisition

Seismic data acquisition is concerned with data gathering, in the field, of sufficient quality. The process of seismic data acquisition involves various steps, where careful selection of source and receiver arrays at the acquisition step will help to eliminate a large amount of noise. In order to collect the seismic data, many things are taken into consideration that are related to the physics of the problem, local situations and conditions, and the economy. All of these will determine which sources and receivers to be used, and which configurations of arrays to be put in place. To be able to detect seismic reflected waves, the acquisition team should use sources to generate seismic waves of sufficient power and adequate frequency content. The seismic waves within the subsurface are detected and the Earth's motion will be converted into electrical signals using seismic detectors (geophones and hydrophones). Such electrical signals will typically be transported to the recording instruments via electric cables. At the recording instruments, the data will be sampled and quantized and then stored on digital discs. Such discs can be used to retrieve the data at anytime and will be sent for data processing. Note that during the recording process, the signals are monitored, where they are checked to primarily assure that the seismic acquisition equipment is properly functioning.

Considering land seismic surveys, the acquired data are massive, whether considering the conventional or the single-sensor seismic acquisition techniques (Blacquiere and Ongkiehong, 2000). In 2-D, a conventional recording is taken as an example to survey a line of length 10 km, and this simple 2-D acquisition requires 7,200 geophones, 18 shots, and, probably, a terabyte of recorded seismic data. On the other hand, for the same line, the number of single-sensor geophones will be 30,000 and with 18 shots, 3 terabyte of data. The recorded seismic data size will increase when running 3-D acquisition instead of 2-D and will even be tripled when acquiring three-components (3-C), which is gaining more attention nowadays. The location and the type of surface geology and topology may reduce or increase the cost of the acquisition.

In terms of seismic sources, they generate mechanical disturbances that cause a seismic wave motion, with a signal of a certain signature, to travel through the sub-surface from the sources to the seismic detectors. Seismic sources have dominant influences on the signal response that are due to the response of: (a) the sources, (b) detectors, and (c) the seismic recording system (which will be discussed later in this section). Some of the commonly used land seismic sources are dynamite and vibroseis.

Dynamite, as a land seismic source, is very cheap and usually used in nonurban areas. Its signature resembles an impulse. The burning of dynamite takes place in a very short time, resulting in the generation of sudden high pressures and tem-peratures. Nearby the explosive, a nonlinear zone is created, where the rock and soil will have undergone some permanent change due to the explosion. A nice characteristic of dynamite is that it resembles a (bandlimited) form of the delta pulse. Based on the field experimental work of Ziolkowski and Bokhorst (1993), the dynamite signatures are shown in Figure 1.4. Particularly, Figure 1.4a shows that the dynamite source signature is a pulse with a sharp peak at the beginning in the time domain, resulting in a minimum phase wavelet source, while the amplitude and phase spectrum are given in Figure 1.4b and c, respectively.

The use of a vibrator as a land seismic source (known as vibroseis) has become very popular for so many reasons, but it can mainly be used in urban areas. The vibrator is a surface source, and emits seismic waves by forcing vibrations of the vibrator baseplate, which is kept in tight contact with the Earth, through a pull-down weight. The direction in which the plate vibrates can result in generating compressional vibrators (the motion of the plate is in the vertical direction). Also, it can result in generating shear vibrations, where the motion of the plate is in the horizontal direction. The signal emitted by the vibrator is band-limited and expanded impulse. It typically has a duration of some 10–15 seconds. The band limitation has two aspects: At the low frequency end it is dictated by the mechanical limitations of the system and the size of the baseplate. The high frequency limit is determined by the mass and stiffness of the baseplate. The source signal, unlike dynamite, is a sweep and not impulsive. Hence, the cross-correlation of the sweep along with the recorded seismic signals are used to obtain the source signature and the seismic signal. Note that the vibroseis source's wavelet signature is mixed or maximum phase, which affects the time arrivals of seismic data, and, therefore, incorrect interpretations. An example of a vibroseis sweep signal, its spectra, and its autocorrelation is presented in Figure 1.5.

The detectors used for land seismic exploration work are called geophones, since they are used to hear echoes from the Earth underneath. They are used for both land (onshore) and seabed (offshore) surfaces. They detect ground velocity produced by seismic waves and transform motion into electrical signals. Although several

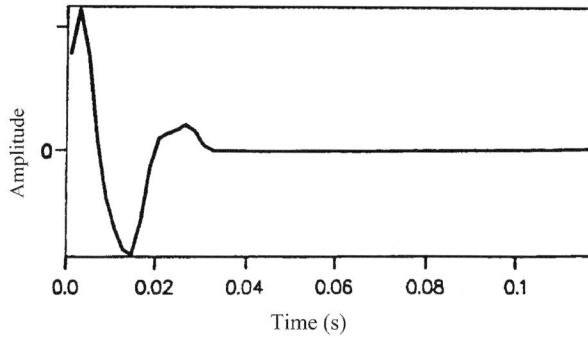

Time (s)

(a)

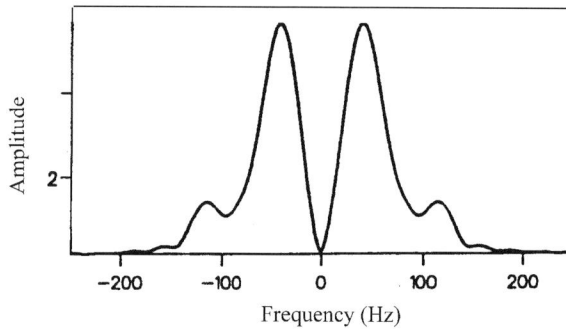

Frequency (Hz)

(b)

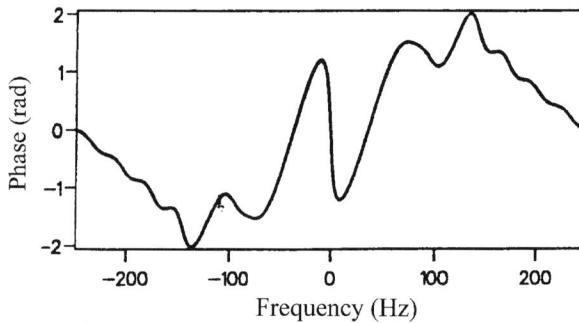

Frequency (Hz)

(c)

Figure 1.4 (a) Seismic dynamite signature signal in the time domain. (b) The amplitude spectrum of the signal in (a) with a bandwidth of about 100 Hz, while (c) shows the phase spectrum of the signal in (a) being nonlinear within the bandwidth (courtesy of Ziolkowski and Bokhorst, 1993.)

types of geophones exist, the moving-coil geophone is the most commonly used in seismic exploration. It is composed mainly of two systems. Let $x_1(t)$ denote the analog seismic signal (that resulted from the convolution between the embedded wavelet and the subsurface reflectivity function to be discussed later) and denote

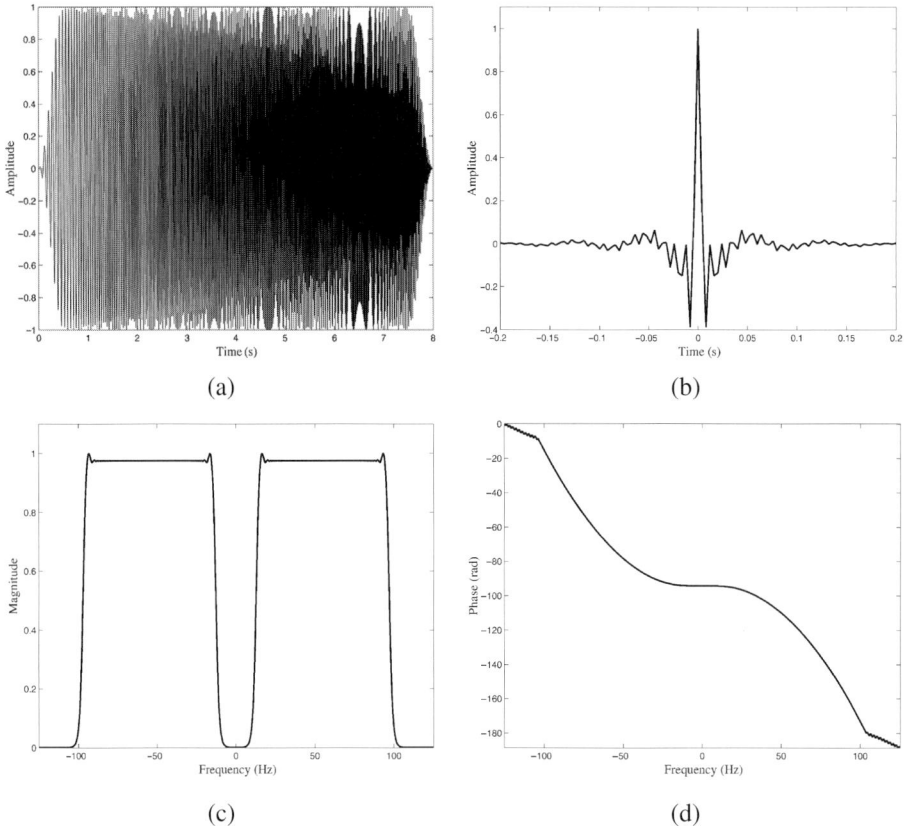

(a)

(b)

(c)

(d)

Figure 1.5 (a) An 8-second seismic vibroseis signature signal in the time domain with a taper of 500 ms. (b) Its autocorrelation. (c) The amplitude spectrum of the signal in (a) with a frequency range between 10 and 100 Hz, while (d) shows the phase spectrum of the signal in (a).

the output from the geophone system by $r_g(t)$. The geophone system is composed of two Linear Time Invariant (LTI) subsystems (see Figure 1.6). The first is a mechanical system that is characterized using the following differential equation:

$$\frac{d^2x_2(t)}{dt^2} + 2h_0\omega_0\frac{dx_2(t)}{dt} + \omega_0^2 x_2(t) = x_1(t). \tag{1.1}$$

The second is an electrical LTI system, which is characterized by:

$$r_g(t) = \Lambda\frac{dx_2(t)}{dt}, \tag{1.2}$$

where h_0 is the geophone damping constant, ω_0 is the natural frequency, and Λ is the transduction constant of the geophone system. The analog transfer function

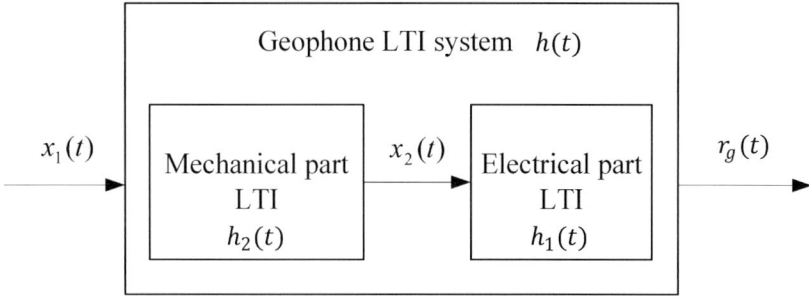

Figure 1.6 Block diagram showing the geophone LTI system. The system is a cascade of a mechanical system followed by an electrical system.

(in the Laplace domain with s being a complex variable) of a geophone can be obtained based on both Equations (1.1) and (1.2) to be:

$$H(s) = \frac{\Lambda s}{s^2 + 2h_o\omega_o s + \omega_o^2}. \tag{1.3}$$

Figure 1.7a displays the impulse response of a typical geophone system with $\Lambda = 6 \times 10^6$ Volts/m/s, $\omega_0 = 20\sqrt{3}$ rad/s and $h_0 = 20/\omega_0$ s/(kg × rad). Figure 1.7b and c shows its geophone system frequency response, where clearly the geophone operates as a low pass filter. Note that one can solve the differential equations for the given values of Λ, ω_0, and h_0 to obtain the following impulse and frequency responses, respectively:

$$h(t) = 7.5443 \times 10^5 e^{-21t} \cos(27.55t + 37.32°), \text{ for } t > 0, \tag{1.4}$$

$$H(f) = \frac{j12\pi \times 10^5}{-4\pi^2 f^2 + j84\pi f + 1,200}. \tag{1.5}$$

Exploration land and marine seismic reflection surveys are commonly acquired using the Common Midpoint (CMP) method, sometimes called the CMP fold acquisition method. Other names for this method include common reflection point (CRP) and common depth point. Seismic waves reflected from the same common depth or midpoint between a group of sources and detectors (receivers) form a CMP gather. The number of recorded seismic waves on seismic detectors (known as *traces*) in a gather is called the fold of that gather. It is noted here that the CMP spacing is half of the trace (receiver) spacing in a survey. The basic idea of this method is that the point on the surface halfway between the seismic source and seismic receiver is shared by numerous pairs of sources and receivers, following Snell's first law (Kearey et al., 2002). The redundancy among the pairs of sources and receivers enhances the acquired seismic data at the seismic data processing stage

(a)

(b)

(c)

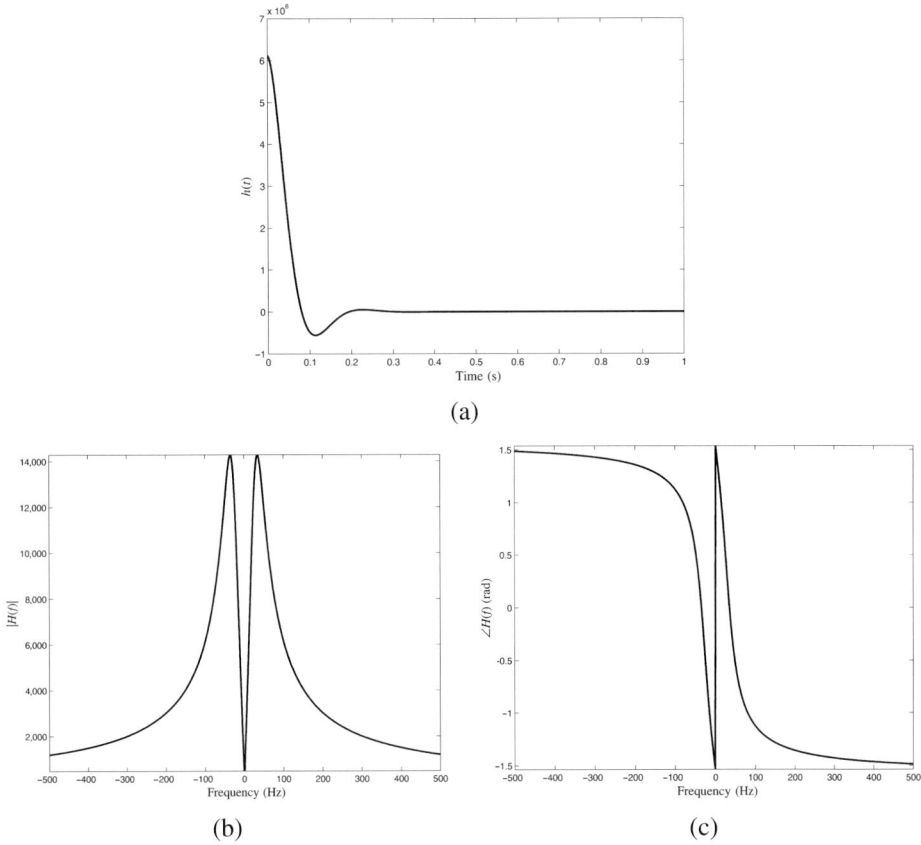

Figure 1.7 The impulse response of the geophone system with $G = 6 \times 10^6$ Volts/m/s, $\omega_0 = 20\sqrt{3}$ rad/s and $h_0 = 20/\omega_0$ s/(kg $\times$ rad) is presented in (a). (b and c) The geophone system frequency response.

known as Stacking (that will be explained in Section 1.8). For each pair of sources and receivers, the common midpoint is vertically above a common depth point of interest.

When geophysicists meet with their seismic data for the first time, they carefully examine the acquired seismic data and the acquisition parameters. Usually, seismic data are stored in hard disks and saved in various standard data formats such as the well-known SEG-Y or Seismic Unix formats (Forel et al., 2005; Yilmaz, 2001). All stored data on a disk comes with its header file. The header file contains information and tabulation of parameters used to acquire the data. For 2-D seismic data acquisition, a very instructive chart that geophysicists usually can obtain from the seismic header files is the so-called *Stacking Chart* (Milkereit, 1989). A stacking chart is a chart in which the x-axis indicates the receiver's location and the y-axis indicates the source's locations. The stacking chart is also used to sort seismic traces into various gathers or profiles such as shot gathers, receiver gathers, offset gathers, or CMP gathers. Typically, on a stacking chart:

- Points along one diagonal have a common midpoint; the resulting gather is called a common midpoint (CMP) gather.
- Points along the other diagonal have a common offset. The resulting gather is called a common offset gather (COG).
- Points along a horizontal line have a common source, and the gather is called a common shot or source gather (CSG).
- Points along a vertical line have a common receiver, where this is called a common receiver gather (CRG).

Figure 1.8 shows a schematic for a typical stacking chart, while Figure 1.9a and b shows various gather configurations. The stacking chart in Figure 1.8a shows the case for a 10-trace cable (labeled by squares) with 25 m receiver group spacing and 16 shots (labeled by starts) with spacing of 12.5 m. Each recording channel will provide a seismic trace from consecutive shots to a CMP gather, and 10-fold coverage. This is true for all the CMP gathers on the stacking chart except at the beginning of each line, where the fold of coverage gradually builds up (this zone is called the roll-on or taper on). Similarly, at the end of each line the fold coverage gradually drops off (this is called the roll-off or taper off zone). This will be better addressed with some examples in Section 1.8, when examining the header of a 2-D seismic data set.

More recently, seismic acquisition has started to look into the acquisition from a different perspective to that of the CMP technique to another perspective called Wavefield recording. This is due to the recent developments in seismic data acquisition such as wide azimuth surveys and symmetric sampling methods. Interested readers can look into (Liner, 1999a; Pecholcs et al., 2012; Vermeer, 2012; Campman et al., 2016). However, and without loss of generality, the focus here will continue on the CMP acquisition method.

Additionally, the acquisition process can be done based on a technique known as Analogue Receiver Arrays (see Figure 1.10a), where each of these arrays is composed of n receivers. In this technique, also known as conventional acquisition, the a set of these arrays is placed. Each detector records a single trace, resulting in n recorded traces per an analogue array. Finally, the arithmetic mean of each analogue array (average of all traces recorded in such array) is given as an output of that array and considered a single trace (Vermeer, 2012; Yilmaz, 2001). This helps in reducing noise during acquisition (Kearey et al., 2002; Yilmaz, 2001). The single-sensor technique, on the other hand, replaces the analogue receiver array with one sensor. Such a technique provides superior data quality through the efficient recording and

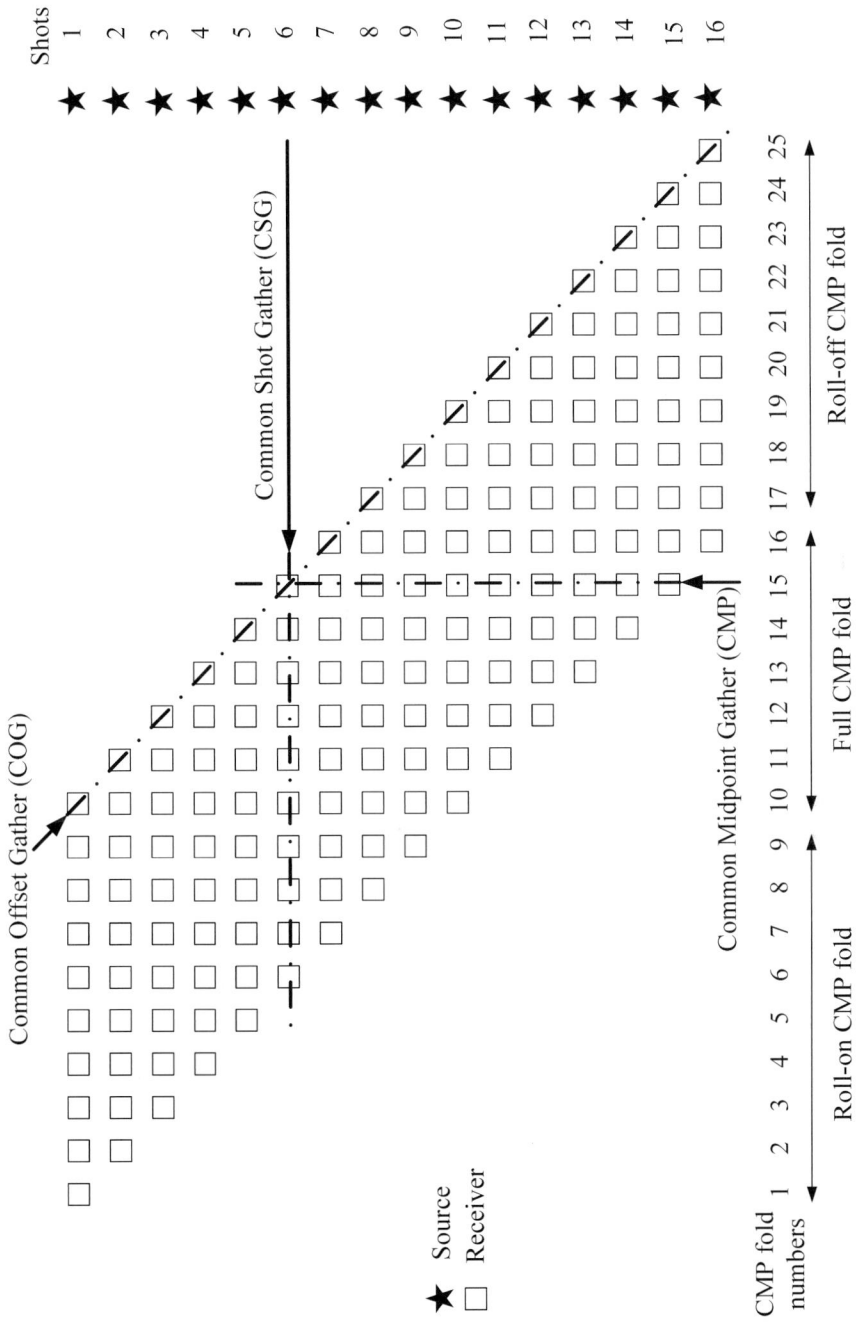

Figure 1.8 A seismic stacking chart for the CMP coverage acquisition method.

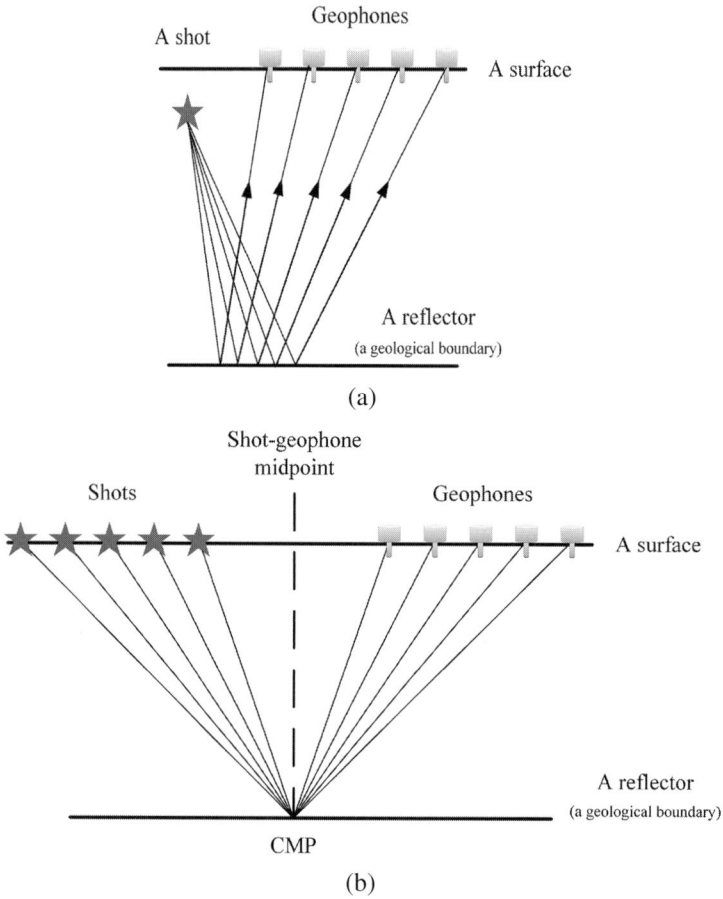

Figure 1.9 (a) A common shot gather configuration. (b) A common midpoint gather configuration.

processing of thousands of live single-sensor channels. The system yields denser digital spatial sampling of the seismic data. Hence, the composition of the single receiver records from digital array forming provides better noise attenuation and a high resolution per seismic trace (see Figure 1.10b) (Baeten et al., 2000; Blacquiere and Ongkiehong, 2000). Single receiver recordings have the advantage that better noise-rejection filters can be designed. The filters can be either time-invariant or adaptive (Özbek, 2000a,b).

Finally, the recording systems can be divided into analog and digital, where the analog recorded data are almost obsolete and seismic data are rarely recorded on magnetic tapes. Digital recording, which was first introduced in seismic work in the 1960s, is used to convert the seismic data into discrete-time-space/digital signals. The digital system is known as the digital seismic amplifier, where such a system

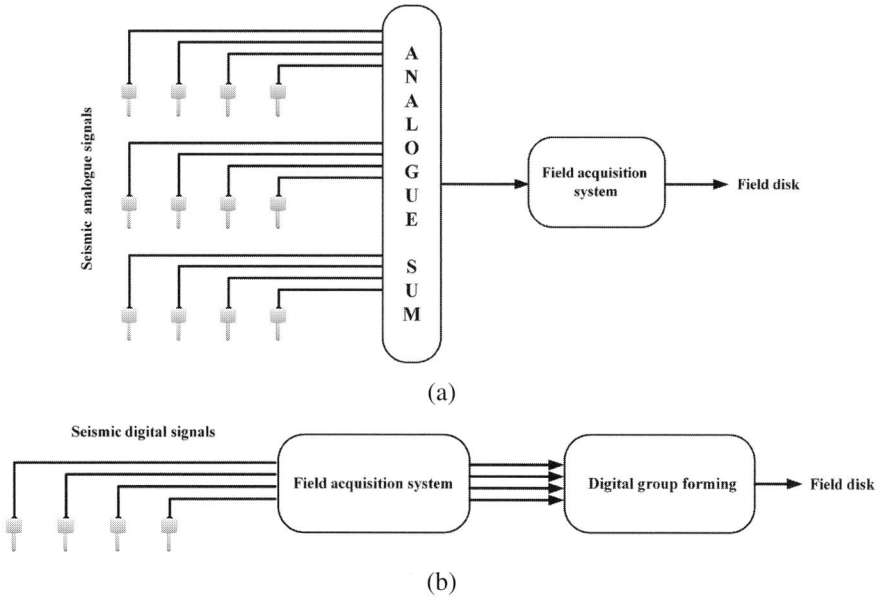

(a)

(b)

Figure 1.10 (a) Analogue receiver array acquisition technique. To obtain a single trace, an array of seismic traces is required. Here, a 3 by 4 array is provided. The analogue sum (arithmetic mean) is used to obtain one single seismic trace at the output of the system array. (b) Point-receiver or Single-sensor array acquisition technique. Unlike the analogue receiver array technique in (a), each receiver records only one seismic trace. Here, four receivers will record four seismic traces.

usually has enough dynamic range so that no pre-filtering is required prior to the first amplification stage. A typical digital seismic system is mainly composed of the following elements:

- Pre-amplifiers: increase the gain of the signal by a constant amount before multiplexing and converting the analog signal into digital.
- Anti-alias filters or pre-filters: are lowpass filters that are used before sampling to attenuate frequency components of the seismic signals that are above the Nyquist frequency to avoid aliasing when sampling.
- Multiplexers: are high speed switches that are used to take each output of the anti-aliasing stage and connect each seismic channel to a floating-point amplifier.
- Analog-to-digital (A/D) convertors: each A/D samples in time and quantizes the amplitudes of the seismic signals. This results in a series output of pulses, which provide the polarity and signal amplitude value in bits at the sampling instant.

Figure 1.11 shows a schematic diagram of this process.

At a small-scale level, a recorded seismic signal (seismic trace) represents a combined response of a layered ground and a recording system to a seismic source

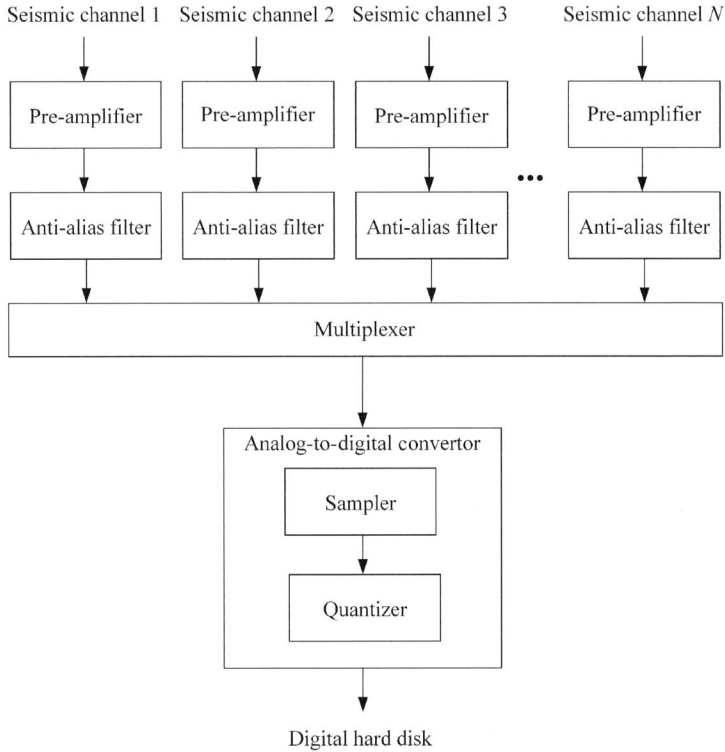

Figure 1.11 A schematic diagram of a typical digital seismic recording system. Before seismic traces are recorded on a hard disk, each seismic channel (from an analogue group array or single-sensor array as seen in Figure 1.10) will go through a pre-amplifier, an anti-alias filter, the multiplexer, and, finally, an A/D convertor.

signal (known as a seismic impulse or wavelet). Any display of a collection of one or more seismic traces is termed a seismic section or a *seismogram* (e.g. a CGS or a CMP). Assuming that the seismic wavelet shape remains unchanged as it propagates through such a layered ground, the resultant seismic trace may be regarded as the convolution of the input impulse with a time series known as a *reflectivity function*. The reflectivity function is composed of a series of unit sample/delta functions (or spikes as geophysicists prefer to call them). Each unit sample has an amplitude related to the reflection coefficient at a layer boundary and a traveltime[4] equivalent to the two-way reflection time from the surface to that boundary. All of this is based on the assumption that the subsurface can be regarded as a Linear-Time (Shift) Invariant (LTI or LSI) System (Robinson, 1967; Robinson and Treitel, 2000).

[4] Traveltime is the time difference between zero time and the arrival time of a seismic event. It can be a one-way time such as for direct waves or a two-way time such as for reflected waves.

Since the source wavelet has a finite length, individual reflections from closely spaced boundaries may overlap in time on the resultant seismic section. Assuming that one is directly dealing with a discrete-time seismic signal, then the seismic convolution model can mathematically be written as:

$$g[n] = r[n] * w[n], \tag{1.6}$$

where $r[n]$ denotes the subsurface impulse response (reflectivity) function and $w[n]$ is called the embedded or equivalent wavelet, which would be affected from a single isolated interface. Wavelets will be discussed in more detail in Chapter 8. It is assumed throughout the book that seismic signals are discrete-time and/or space signals. These are functions of an independent variable n that is an integer, i.e., $n \in \mathbb{Z}$, where $\mathbb{Z}$ is the set of integer numbers. In principle, $w[n]$ represents the overall convolution of many involved systems' impulse responses (along with other unwanted signals that will be discussed in detail in Chapter 6) such as the source pulse or signature $s_s[n]$, the receiver impulse response $r_g[n]$, the A/D converter system impulse response $g_{A/D}[n]$. That is (for the time being):

$$w[n] = s_s[n] * r_g[n] * g_{A/D}[n]. \tag{1.7}$$

The convolution model in Equation (1.6) often includes additive noise $n_r[n]$, which is usually but not necessarily assumed to be random. Therefore, the convolution model of a noisy seismic trace is modeled as follows:

$$g[n] = r[n] * w[n] + n_r[n]. \tag{1.8}$$

Figure 1.12 represents an ideal seismic convolution model (Kearey et al., 2002; Robinson and Treitel, 2000; Yilmaz, 2001), where the seismic input wavelet is convolved with the Earth's reflectivity function and the output is contaminated with noise. Note that ideally, $r[n]$ is the response of the subsurface LSI system due to a unit sample impulse $\delta[n]$, i.e., $r[n]$ is the impulse response of the subsurface LSI system.

Many unwanted wave signals corrupt seismic records with noise. This includes random or incoherent noise, such as instrument signals, and coherent noise like the ground-roll noise and multiples (as will be explained later in Chapter 2). As a consequence of these effects, seismic traces generally have a complex appearance and $r[n]$ cannot often be recognized without the application of suitable signal/data processing techniques. The purpose of processing such data can, in general, be viewed as the process of attenuating noise, improving both the vertical and horizontal resolution of the data and, ultimately, estimating the input wavelet $w[n]$ in order to estimate the reflectivity function $r[n]$. Hence, the acoustic impedances (or other related properties) of the subsurface layers could be estimated, with or without the help of other geophysical measurements.

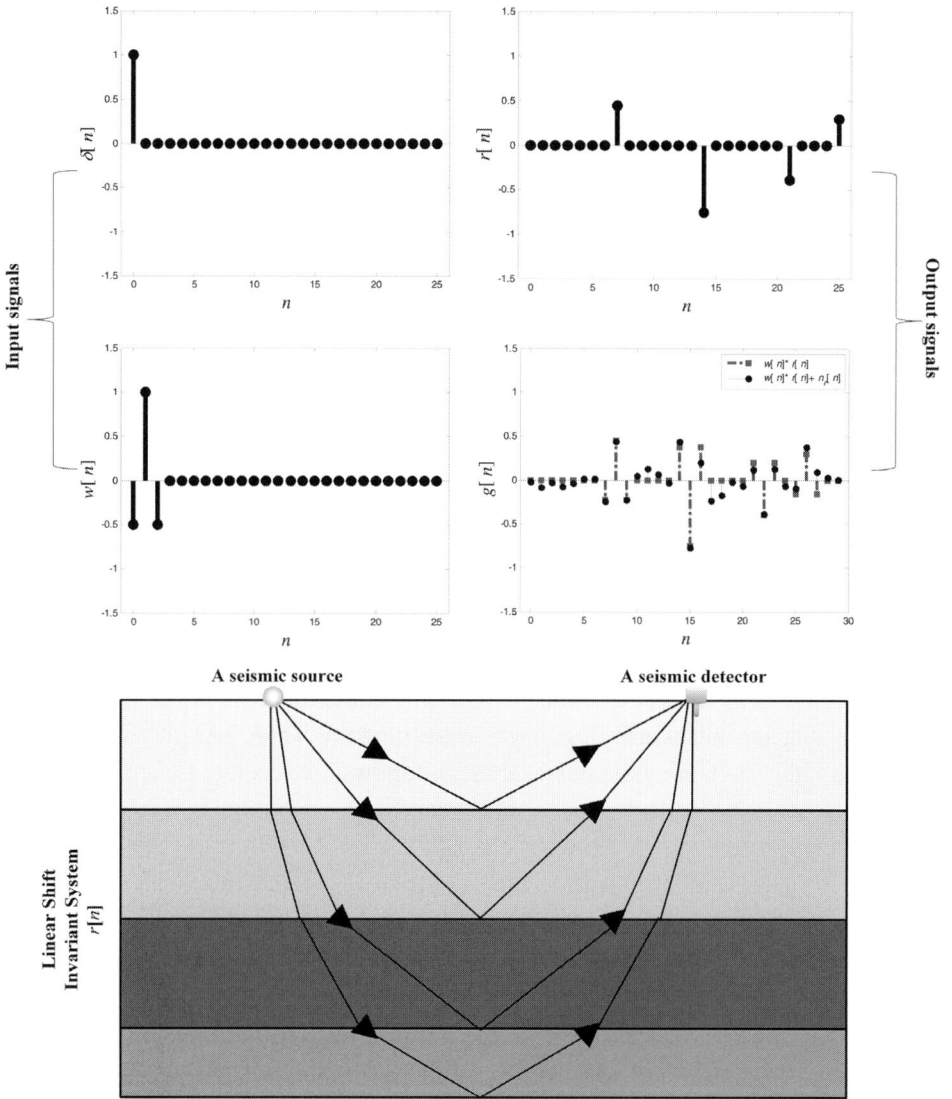

Figure 1.12 Convolution seismic data model. A seismic wavelet $w[n]$ is convolved with the reflectivity function $r[n]$ to obtain a seismic trace $g[n]$. Note that $r[n]$ is related to the geological section of the subsurface through the reflection coefficient of each geological boundary and the two-way travel time of the seismic wave.

Using the Discrete-Time Fourier Transform (DTFT),[5] the convolution in time, seen in Equation (1.6) while considering Equation (1.7), is equivalent to a multiplication in the Fourier domain. Therefore, the seismic trace signal is represented in the frequency domain by:

$$G(e^{j\omega}) = R(e^{j\omega})S_s(e^{j\omega})R_g(e^{j\omega})g_{A/D}(e^{j\omega}), \tag{1.9}$$

where $j = \sqrt{-1}$. In such a domain, the seismic trace consists of (complex) multiplications of the individual transfer functions, where $G(e^{j\omega})$ can be written as a multiplication of magnitudes and addition of phases as follows:

$$G(e^{j\omega}) = |G(e^{j\omega})|e^{j\angle G(e^{j\omega})}, \tag{1.10}$$

where:

$$|G(e^{j\omega})| = |R(e^{j\omega})||S_s(e^{j\omega})||R_g(e^{j\omega})||g_{A/D}(e^{j\omega})|, \tag{1.11}$$

and,

$$\angle G(e^{j\omega}) = \angle R(e^{j\omega}) + \angle S_s(e^{j\omega}) + \angle R_g(e^{j\omega}) + \angle g_{A/D}(e^{j\omega}). \tag{1.12}$$

Before moving to the next section, it is probably time for the reader to look into raw acquired seismic data sections displayed in different forms. There exist several ways of displaying seismic data. For the sake of the book, 1-D and 2-D displays of seismic data will be considered. In the case of a 2-D seismic section (e.g., CSG, CMP, a seismic stacked section, seismic images), some of the most commonly used displays are, for example, the wiggle display, as seen in Figure 1.13a. It basically plots seismic trace amplitudes as a function of time, where the positive peaks are to the right of every trace and the negative ones are to the left of every trace. Another common example to display seismic data is the so called variable area display, which shades the area under the wiggle trace to make coherent seismic events (see Figure 1.13b, rendering the positive peaks)[6] evident. The combination of variable area and wiggle is known as a wiggle-variable area display (Figure 1.13c). Additionally, the variable density display represents amplitude values by the intensity of shades of gray (black is for maximum amplitude value and white is for the minimum amplitude value) like the one shown in Figure 1.13d. Also, colors (blue is for maximum amplitude value and red is for the minimum amplitude value) could be used as well (see, for example, Al-Shuhail et al. [2017]). For 1-D seismic data, one can simply plot the seismic trace signal as a function of time. Figure 1.14 shows an example of various real 1-D seismic signals extracted from a 2-D seismic section.

---

[5] The DTFT is the Fourier transform of discrete-time signals and will be discussed in Chapter 4.
[6] A seismic event is the arrival of a new seismic wave, usually indicated by a phase change and an increase in amplitude on a seismic record. It may be a reflection, refraction, diffraction, surface wave, or random noise.

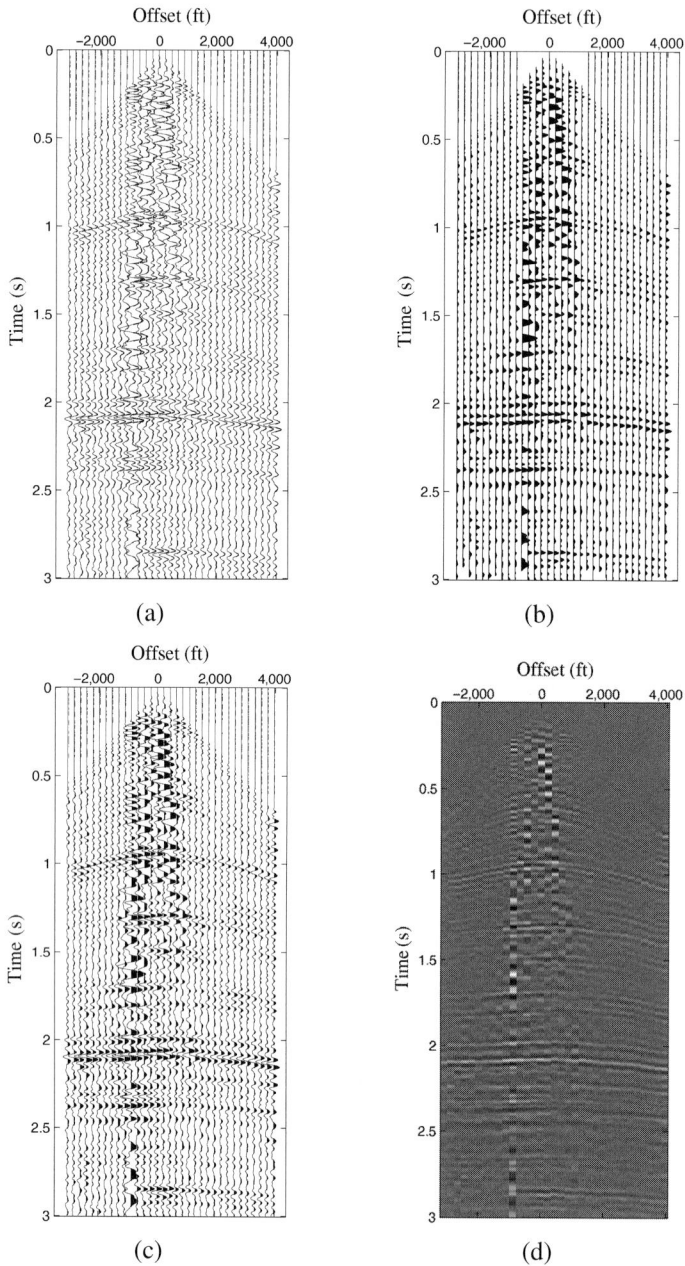

Figure 1.13 Various displays for a seismic shot gather section from (Mousa and Al-Shuhail, 2011): (a) wiggle display, (b) variable area display, (c) wiggle-variable area display, and (d) gray-scaled variable density display. Note that the vertical axis represents the two-way travel time increasing downward, while the horizontal axis represents the distance from left to right.

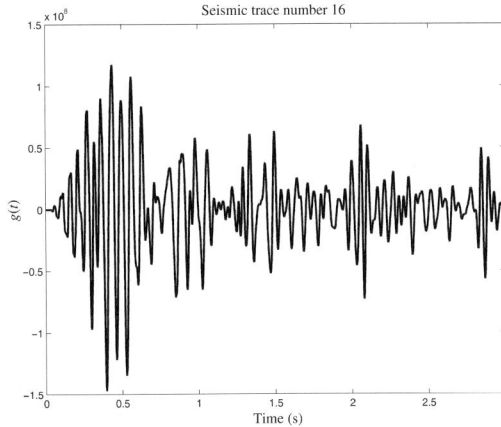

Figure 1.14 Seismic trace signal number 16 from the seismic data shown in Figure 1.13.

1.8 Seismic Data Processing

Seismic data processing can be considered a sequence of cascaded operations that attenuate noise accompanying seismic data as well as making geometrical corrections such that the final image will truly show a subsurface image. Processing of seismic data includes, but is not limited to, filtering, common mid-point (CMP) sorting, velocity analysis, normal move-out (NMO) correction, and stacking. Each seismic trace has three primary geometrical factors that determine its nature: shot position, receiver position, and the position of the subsurface reflection point, which is considered to be the most critical (Kearey et al., 2002). Before processing, this position is unknown, but a good approximation can be made by assuming that this reflection point lies vertically under the position on the surface midway between the shot and the receiver for that particular trace (if one recalls, it is called the midpoint). Traces reflected from the same CMP define a CMP gather. The CMP gather is important for seismic data processing because the subsurface velocity can be derived using it. In general, the seismic reflection energy is very weak and it is essential to increase the signal-to-noise ratio (SNR) of most data. Once the velocity is known, the traces in CMP gathers can be corrected for NMO, which is basically a way of correcting for time differences that occur due to offset[7] in a CMP gather, i.e., to get the equivalent of a *zero-offset* trace. This implies that all traces will have the same reflected pulses at the same time, but with different random and coherent noise. So combining all the traces in a CMP gather will average out noise and the SNR increases. This process is known as stacking.

---

[7] An offset is the distance from the source-point to a geophone or the centre of a geophone group.

In general, the main objectives of seismic data processing are to improve the seismic resolution and increase the SNR of the data. These objectives are achieved through three primary stages. In their usual order of application, they are (Yilmaz, 2001):

1. Deconvolution, which increases the vertical resolution of the seismic data.
2. Stacking, which increases the SNR of the seismic data.
3. Migration, which increases the horizontal resolution of the seismic data.

In addition to these primary stages, secondary processes may be implemented at certain stages to condition the data and improve the performance of these three processes (refer to Figure 1.3). Before the seismic data processing steps are explored, a processed real data is used to explain the steps as an example. These 2-D land seismic data are from Mousa and Al-Shuhail (2011). The steps are listed as follows:

1. Analysis of the acquired seismic data is essential for a seismic data processor before moving forward. This process aims at carefully analyzing the acquired and stored seismic data by carefully looking into the acquisition parameters, the geometry used, the type of sources and receivers, and the status of the data itself. The first step involved is to examine the header and extract useful information about the data. From the examination of the real data header, the following information is extracted:

   - Number of shots = 18.
   - Source type = dynamite in 80–100 ft depth holes.
   - Number of channels per shot = 33.
   - Receiver type = Vertical-component geophones.
   - Array type = 12-element inline.
   - Number of traces in line = 594.
   - Receiver interval = 220 ft.
   - Shot interval is variable.
   - Time sampling interval = 2 milliseconds (ms).
   - Number of time samples per trace = 1501.
   - Data format = SEG-Y.
   - Byte swap type = Big endian.
   - Data file name = data.sgy.
   - Geometry has already been set up and recorded in the trace headers.
   - Uphole times at shot locations have been recorded in the trace headers.
   - An 8–64-Hz bandpass filter has been applied to the data in the field.

   These are different geometrical variables obtained from the seismic header. Recall from Section 1.7 that surface seismic reflection surveys are commonly acquired using the common midpoint (CMP) method, where the data are acquired in the shot gather configuration. The stacking chart for the real data

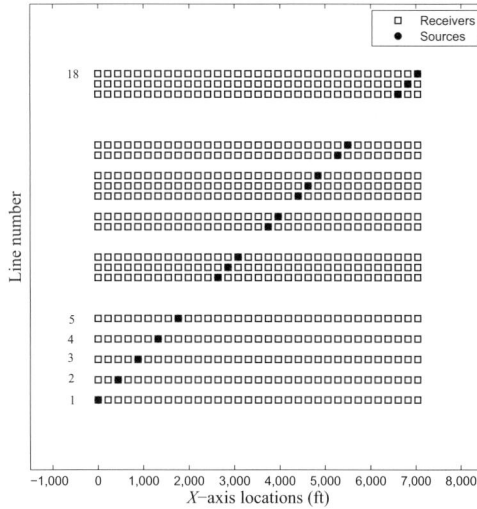

Figure 1.15 The stacking chart plot of the real data showing lines 1–18.

can be seen in Figure 1.15, where apparently the source is moving while the receiver locations were fixed.

Figure 1.16 displays using wiggle-variable area 5 out of 18 raw seismic shot gathers of the data set (Mousa and Al-Shuhail, 2011), where the amplitudes of the seismic data are concentrated near the source (at offset equal to 0) and is attenuated as the time increases (this will be discussed in Chapter 2).

2. Preprocessing is a stage that involves a series of steps to condition the seismic data and prepare it for further enhancement and processing. It includes demultiplexing, reformatting, trace editing, gain application, application of field statics, etc. Depending upon the data, some preprocessing steps are performed on the data. For the real data example, Figure 1.17 shows the data before and after applying gain compensation in order to compensate for various amplitude losses, as will be explained in Chapter 2. At this stage, one may start observing seismic energy and some accompanying noise, where filtering can be used to attenuate some of the noise in the data.

3. Filtering typically follows and is used to attenuate frequency/wavenumber components of the seismic signals based on some measurable property. It is an important step in order to proceed further with the other seismic data processing steps that will help geophysicists to better analyze and interpret the acquired data. For example, bandpass filtering every seismic trace, in a shot gather, can be used to attenuate low frequency components such as surface wave noise, while attenuating unwanted energy of high frequency components after certain seismic frequencies. See Figure 1.18 for an example of bandpass filtering of shot gather 7 of the given real data.

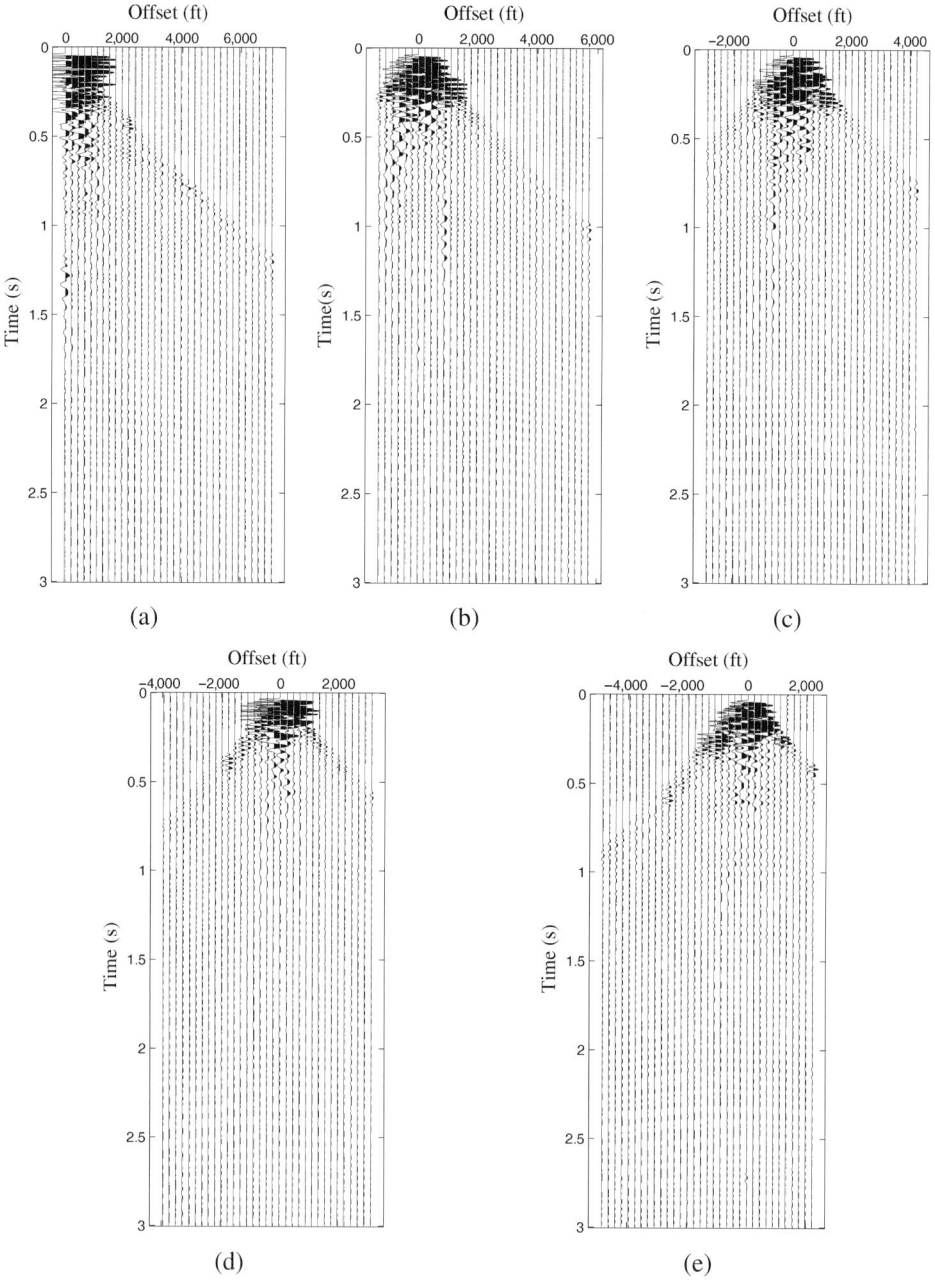

Figure 1.16 An example of five raw seismic shot gathers: (a) CSG 1, (b) CSG 4, (c) CSG 7, (d) CSG 10, and (e) CSG 13.

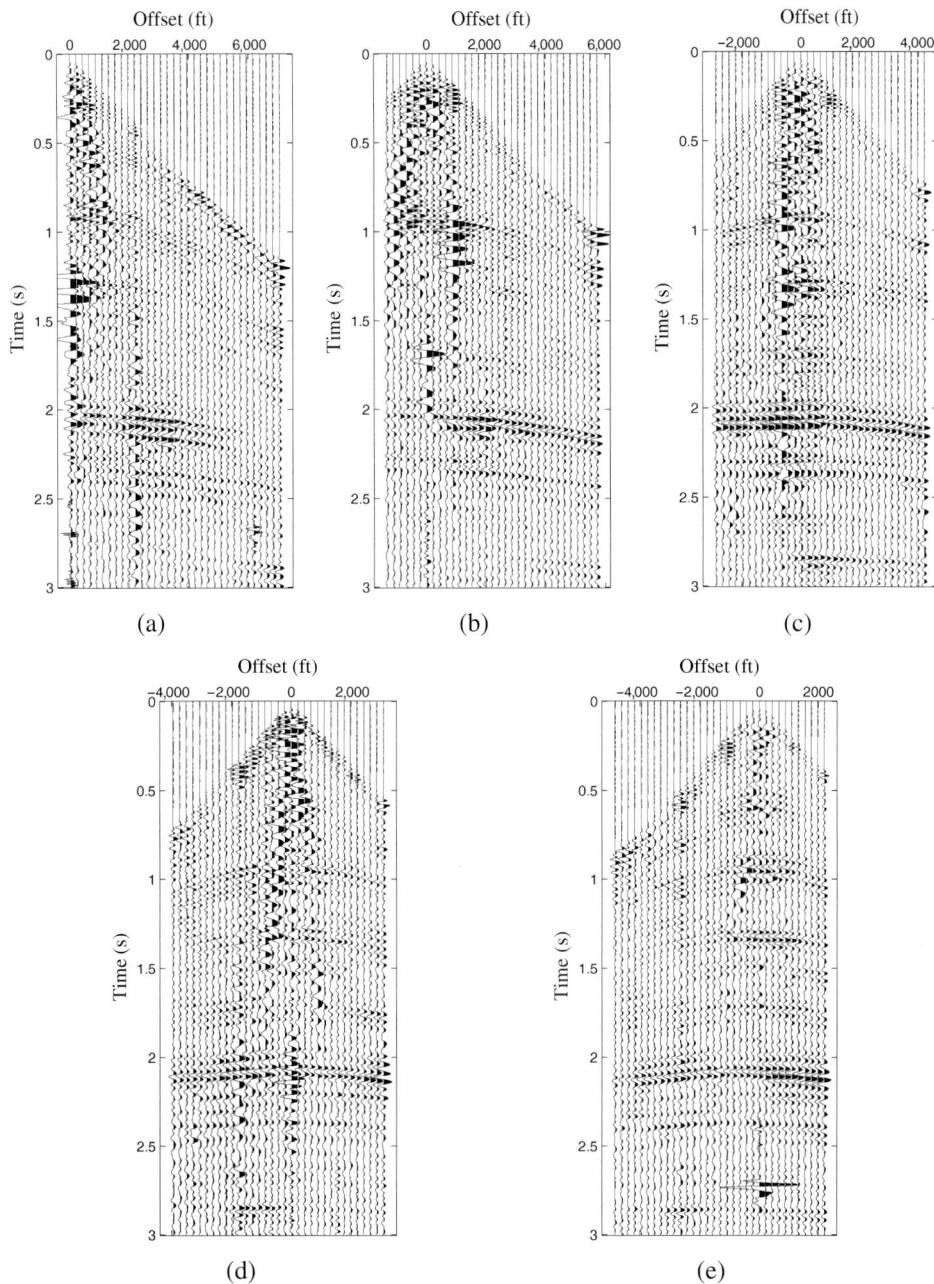

Figure 1.17 Seismic data shot gathers in Figure 1.16 after applying gain correction. (a) CSG 1, (b) CSG 4, (c) CSG 7, (d) CSG 10, and (e) CSG 13.

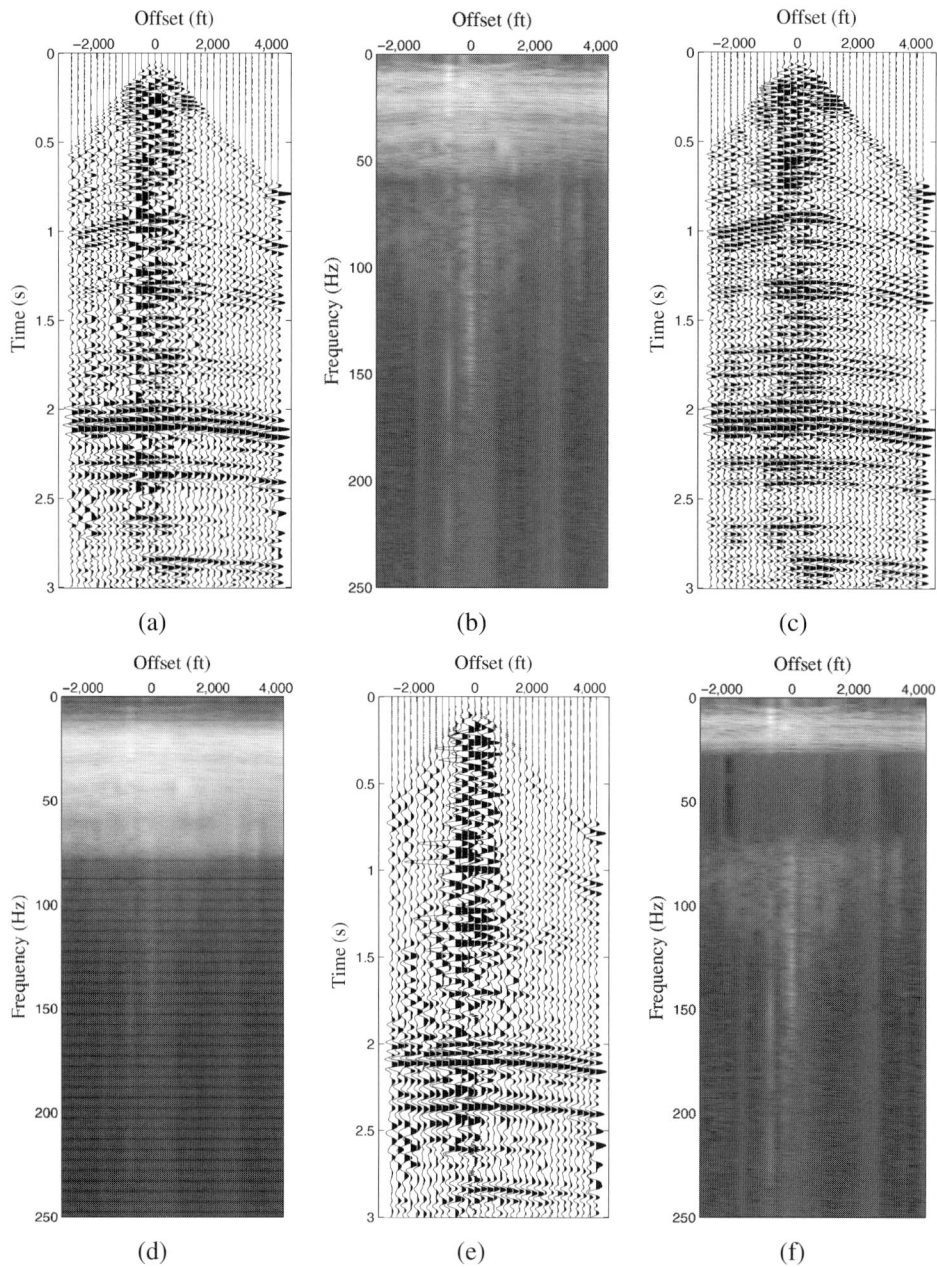

Figure 1.18 Seismic data shot gather 7 in Figure 1.17 after applying Bandpass filtering. (a) CSG 7, (b) its frequency-space spectrum, (c) CSG 7 after filtering, (d) its frequency-space spectrum, (e) difference between (a) and (c), and (f) its frequency-space spectrum.

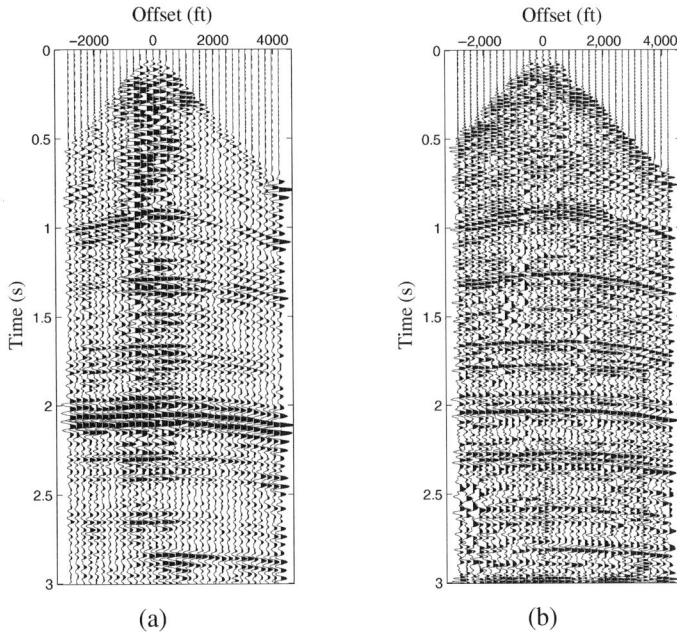

Figure 1.19 Seismic data shot 7 in Figure 1.18 (a) before and (b) after applying deconvolution.

4. Next comes seismic deconvolution. It is performed along the time axis to increase its vertical resolution by compressing the source wavelet to approximately a unit sample function, attenuating noise and unwanted coherent energy such as multi-path signals. Different types of deconvolution exist, as will be seen in Chapter 7. The so-called spiking deconvolution was used, in this case, to process the real data as seen in Figure 1.19 for shot gather 7 of the real data. Clearly, the vertical resolution of the short gather was improved compared to the short gather before applying spiking deconvolution.

5. CMP sorting takes place by which it transforms the data from shot gather configuration to CMP gather configuration using the field geometry information (refer to the stacking chart of the real data in Figure 1.15). From the real data header, one can see in Figure 1.20 the number traces per CMP called the fold versus the CMP gather number. Also, three CMP gathers of the real data are shown in Figure 1.21a–c.

6. In order to stack the seismic data, a CMP gather must have horizontal alignment for each reflection event. To do so, velocity analysis is performed on selected CMP gathers to estimate the stacking, root-mean squared (RMS), or NMO velocities to each reflector. Velocities are interpolated between the analyzed CMPs. They are then used to perform NMO correction.

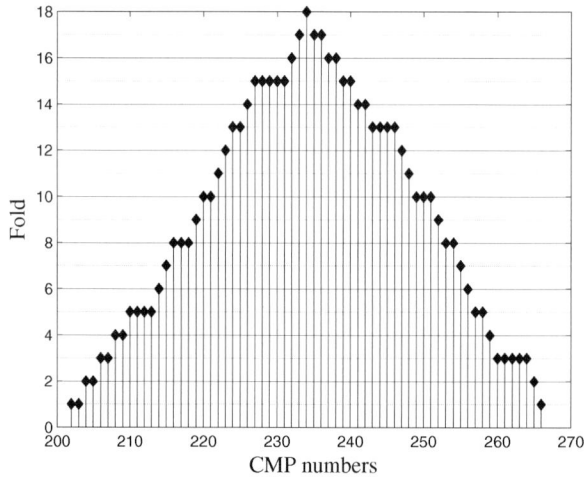

Figure 1.20 The fold versus the CMP numbers of the given real data example.

7. The stacking velocities are then used to flatten the reflections in each CMP gather. This is known as NMO correction. NMO also comes with muting, where it basically zeros out the parts of NMO-corrected traces that have been excessively stretched due to NMO correction. The same three real data CMP gathers in Figure 1.21a–c are shown in Figure 1.21d–f, respectively, but after NMO correction and muting where seismic reflection events are approximately flattened and ready for stacking.

8. The NMO-corrected and muted traces in each CMP gather are summed (stacked) over the offset to produce a single trace. This process is known as *Stacking*. Stacking M traces in a CMP increases the signal-to-noise ratio (SNR) of this CMP by $\sqrt{M}$ (Yilmaz, 2001). In fact, stacking will produce the first approximation of the geological image of the subsurface. The data after performing the above steps are presented in Figure 1.22a.

9. Poststack processing includes time-variant bandpass filtering, dip filtering, and other processes to enhance the stacked section, when needed.

10. Migration is an important step in improving the horizontal resolution of the stacked seismic section. In migration, dipping reflections are moved to their true subsurface positions and diffractions are collapsed by migrating the stacked section. One can choose from the many types of seismic migration: time/depth, or poststack/prestack migration, depending on the complexity of the geology of the subsurface (Biondi, 2006; Clearbout, 1985; Mousa et al., 2006; Yilmaz, 2001). The results of migration for the real data example are presented in Figure 1.22b.

In the remaining parts of this book, the focus will be on processing seismic data from a DSP point of view. Interested readers are recommended to look into, for

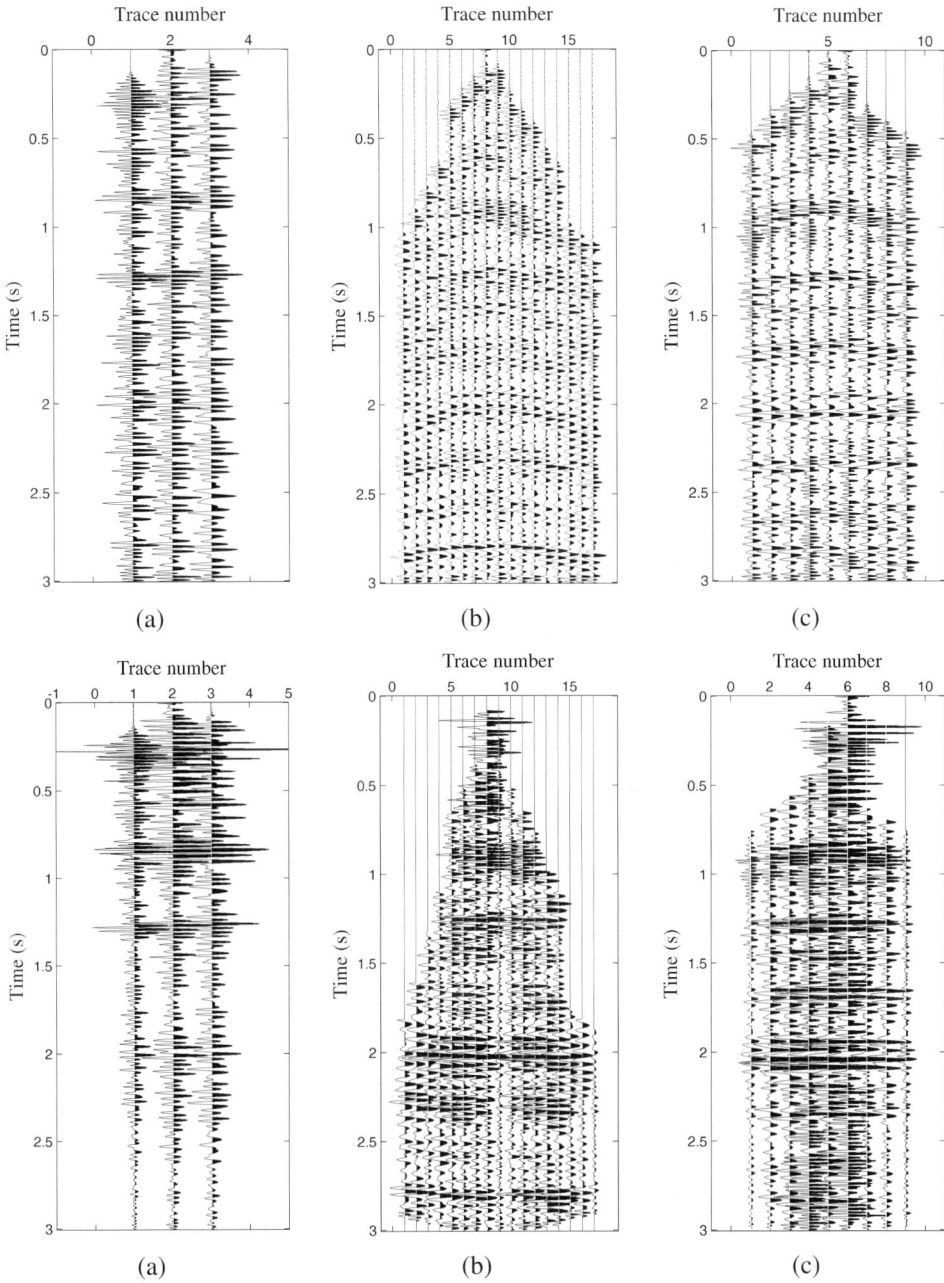

Figure 1.21 CMP gathers number (a) 207, (b) 235, and (c) 252 of the CMP sorted real seismic data example. The CMP gathers after NMO correction: (d) 207, (e) 235, and (f) 252.

38

CMP

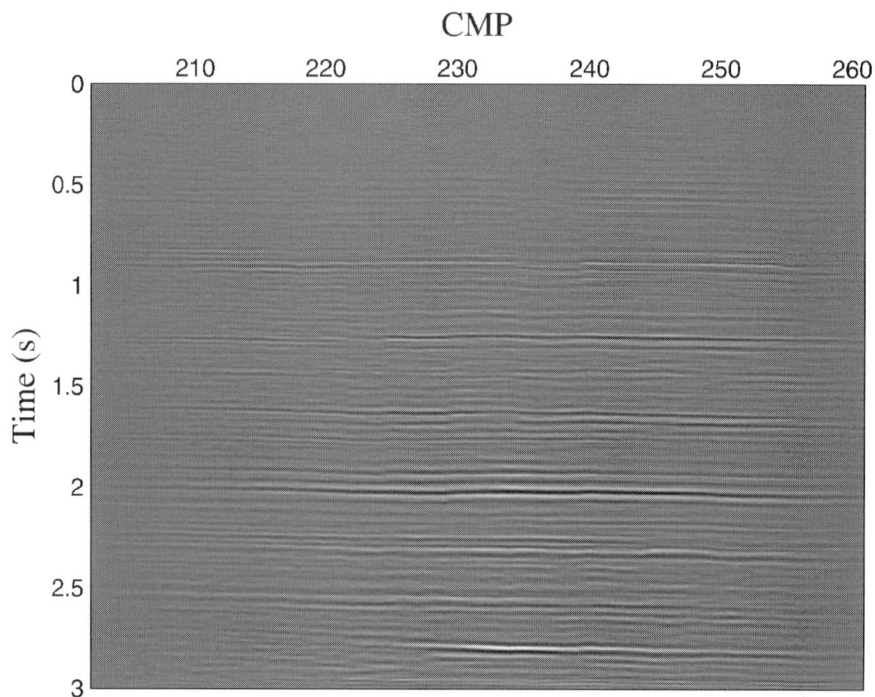

(a)

CMP

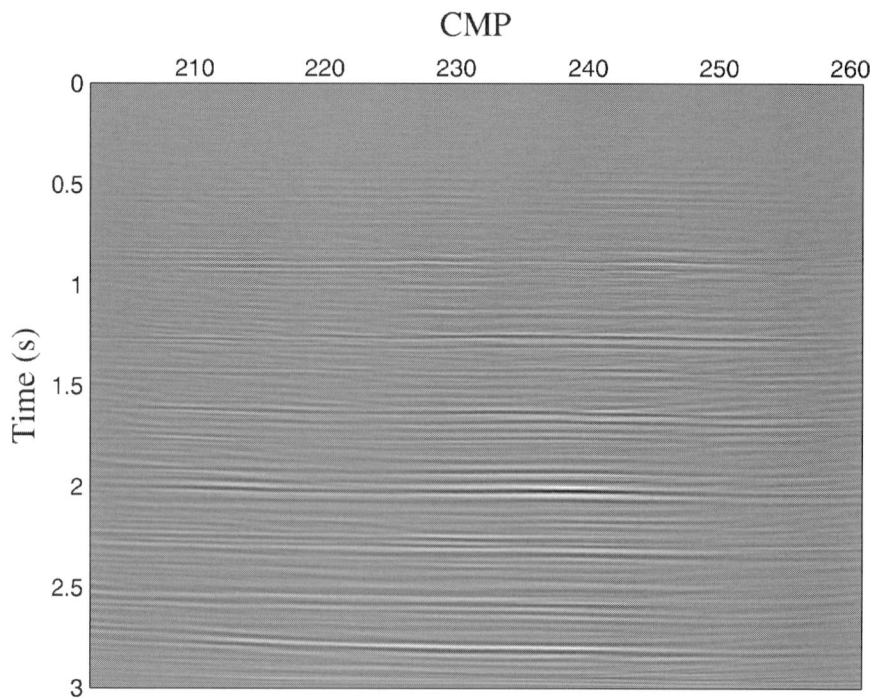

(b)

Figure 1.22 The real seismic data set: (a) its stacked section and (b) its post-stack migrated section.

example, Yilmaz (2001); Mousa and Al-Shuhail (2011) for more details on seismic data processing steps. Note that the previous steps are general and, depending on the data type, land or marine, the accompanying noise types, and/or the acquisition conditions, they may vary as well.

1.9 Seismic Data Interpretation

Seismic data interpretation aims to extract all available geological information from the processed and imaged seismic data such as structure, stratigraphy, and rock properties. The interpretation involves sophisticated image processing algorithms, thanks to the advances in computers and image processing techniques. The interpretation is performed based on the reservoir types: structural or stratigraphic. Considering the rock structures, it is very important to understand the regional tectonic settings and similar ones existing around the world. Additionally, the interpreter has to know the style of the geological structure such as a fault, a salt dome, an anticline and identify the complexity of the structures in the imaged seismic data. Figure 1.23 shows a seismic image that contains a salt diapir by which it is segmented from the seismic image using image processing techniques (Al-Shuhail et al., 2017).

When considering the rocks' stratigraphy, the interpreter has to understand the age relationships and depositional environment of important geological formations of oil/gas exploration interest. Also, he or she must know the time, space, and production relationship between the structure and stratigraphy in the area of interest. Finally, the interpretation process should consider the rock properties and lithology by knowing the rock types such as sandstone, carbonate, the existence of salt, anhydrite, limestone, or dolomite in the shallow section of the data, where they can distort the seismic signals. He or she must know the weathering layer lithology and how variable it is with respect to the survey area. For more information on seismic interpretation, readers are advised to look, for example, into the books by Yilmaz (2001); Bacon et al. (2003); Avseth et al. (2010); Herron (2012); Al-Shuhail et al. (2017).

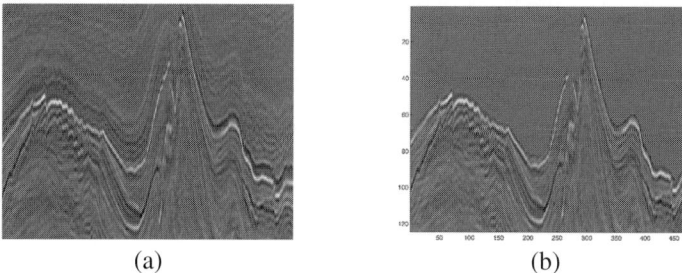

(a)   (b)

Figure 1.23 (a) A seismic image showing a salt diapir, and (b) the segmented salt diapir (courtesy of Al-Shuhail et al., 2017).

1.10 Summary

This chapter provided a basic introduction on oil/gas and their reservoirs. Whether the reservoirs are structural or stratigraphic, they represent probable locations for oil/gas. Before drilling, maps of the subsurface must be generated via geophysical exploration methods. Seismic exploration is the commonly used exploration method and was introduced in this chapter. With many examples, the basics of seismic acquisition, processing, and interpretation were presented. The convolution model by which seismic data are acquired and recorded presents the backbone for understanding how to process seismic data using digital signal processing techniques and algorithms.

Exercises

1.1 Identify the type of expected oil/gas reservoir seen in Figure 1.24.

1.2 Consider the seismic geophone LTI system represented by Equations (1.1) and (1.2). Assume that with $\Lambda = 6 \times 10^6$ Volts/m/s, $\omega_0 = 20\sqrt{3}$ rad/s and $h_0 = 20/\omega_0$ s/(kg $\times$ rad). Then, find and plot the response of the system due to a sinusoidal input $x_1(t) = \sin(100\pi t)$. Now, assume that $h_0 = 1$ s/(kg $\times$ rad). What will be the overall impulse and frequency responses of the geophone system? Explain.

1.3 Construct a stacking chart for a 2-D seismic acquisition setup. Note that the intended seismic line length is 100 m. The number of seismic channels is 5 with 25 m for receiver group spacing and 25 m source spacing. What is the CMP spacing and fold of coverage?

1.4 Using MATLAB, load the seismic data Yilmaz_data_25_g.mat. Then:

- Explore the seismic data and obtain the following information:

Information	Values
The number of rows (time samples or x-lines samples)	
The number of columns (traces or in-line samples)	
The maximum value of the data	
The minimum value of the data	
The mean value of the data	

- Display the seismic section in:

   - Wiggle variable area.
   - Gray-scaled variable density.
   - Colored variable density.

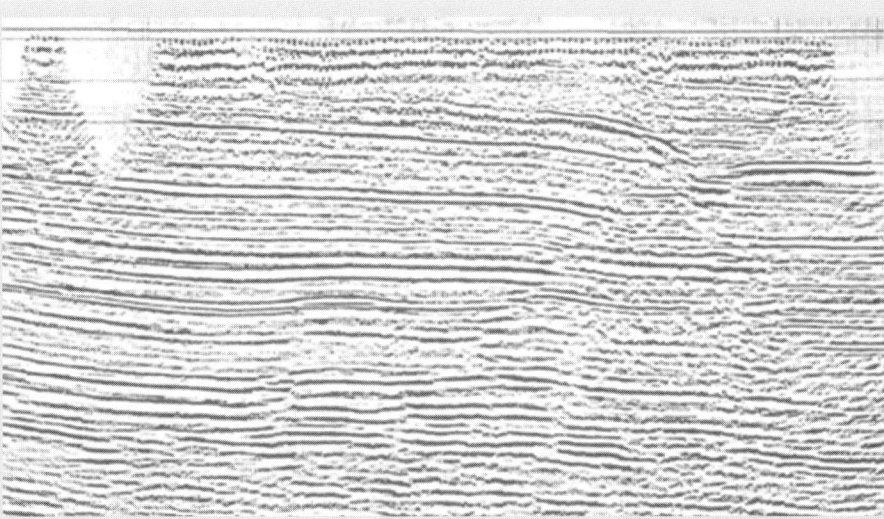

(a)

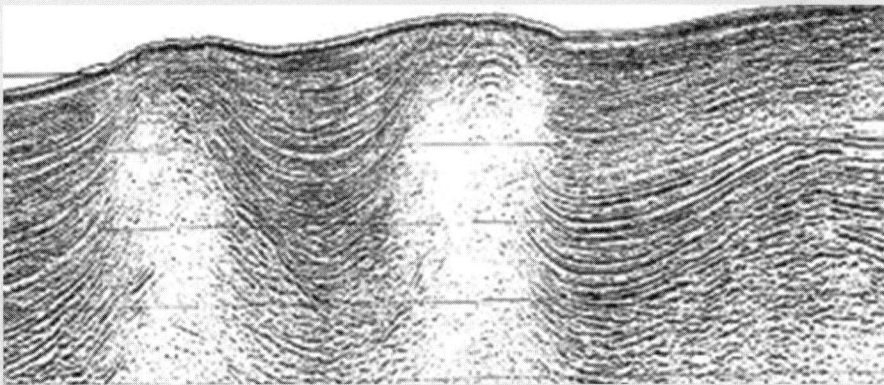

(b)

Figure 1.24 Problem 1.1.

- Display the overlay for the seismic wiggle variable area display of the data over the gray variable density display. Display also the overlay only for seismic traces 10–25.
- Display the overlay for the seismic wiggle variable area display of the data over the colored variable density display. Display also the overlay only for seismic traces 10–25.
- Repeat steps 1–6 only but with the time slice data Yilmaz_data_30.mat.

2

Seismic Theory and Reflection Surveying
A Necessary Background

2.1 Introduction

The physical properties of Earth are generally very complicated. Usually, a mathematical model is used to approximate wave propagation and accompanying effects in order to make useful seismic modeling and processing. Earth has various mathematical models: acoustic, elastic, or even more complicated ones such as visco-elastic (Sheriff and Geldart, 1995; Telford et al., 1990). The assumptions of elastic models are more realistic than those of acoustic models, where the elastic effects can be seen in real seismic data such as shearing and mode-converted waves. Also, they are less complicated than visco-elastic models. Hence, to gain the most comprehensive knowledge of the details of wave propagation in the Earth, it is necessary to study the theory of elasticity in its basic forms for a DSP audience. Additionally, the fundamental physical principles on which seismic methods are based will be reviewed. This includes (a) the nature of seismic waves and (b) the mode of propagation through the Earth subsurface layers, with particular reference to reflection and refraction at interfaces between different rock types. This requires not only the understanding of elasticity but also the understanding of different types of seismic waves that propagate through the subsurface layers away from a seismic source. This, in addition, calls for some elementary concepts of stress and strain to be considered prior to that. However, the theory of elasticity is mathematically complicated, and simplifications can be done if one considers the geometric theory of wave propagation that deals with seismic wavefronts and seismic raypaths and only uses the velocity property of rocks.

Seismic wave amplitudes and velocities are affected as they propagate deep in the Earth layers. Of course, whenever considering Earth layers, we have interfaces between the rock layers and, therefore, change of velocities due to the differences in the physical properties between the layers. These waves propagating within those layers will be reflected, transmitted, and may be also refracted. Once reflections and/or refractions occur, the seismic sensors will record them along with other types of waves (energy) and noise, where each type of those additional waves possesses its own characteristics. It is hoped by the end of this chapter that readers will be able to recognize and distinguish seismic signals of primary reflections from other signals representing undesirable signals, all of which are usually seen on seismic records.

2.2 Theory of Elasticity

The theory of elasticity is a vast subject that deals with the elastic properties of physical media, where, in addition, it attempts to explain the manner in which solids are deformed under applied stresses. When a solid is acted on by external agencies, this theory seeks to explain the possible types and modes of reactions that result. Furthermore, it describes in details how motion, momentum, and energy, initiated by a local disturbance, are propagated to other parts of an external solid. In particular, it provides a detailed exposition of the complex processes of elastic wave propagation in solids.

Elasticity of a body can simply be defined as the ability of that given body to resist changes in size or shape and return to the undeformed condition, when external forces are removed. A perfectly elastic body is one that recovers completely after being deformed. Many rocks can be considered perfectly elastic, when seismic waves are propagating within, except near the seismic source. The theory of elasticity relates the applied forces to the resulting changes in size and shape of body under those forces. The relationships between the applied forces and the deformations caused can be expressed in terms of *stress* and *strain* concepts. Both do not exist independently, where any quantitative description of seismic wave propagation requires the ability to characterize the internal forces (stress) and the deformations (strain) in solid material, such as the case of rocks. Note that both are linked together through constitute relationships, which describe the nature of elastic solids. For more mathematical details, the readers may refer to the work of Malvern (1969) and Sheriff and Geldart (1995).

2.2.1 Stress

Stress denoted by σ can be considered simply as the intensity of a deforming force acting on a body in terms of force per unit area with units of pressure such as Pascal (N/m^2). So when a force is applied to a body, the stress is the ratio of the force to the area on which the force is applied. Conceptually, balanced internal forces are setup within a body when external forces are applied to it and, therefore, stress is a measure of the intensity of these balanced internal forces. A stress acting on an area of any surface within a body may be resolved into (a) a component of normal stress

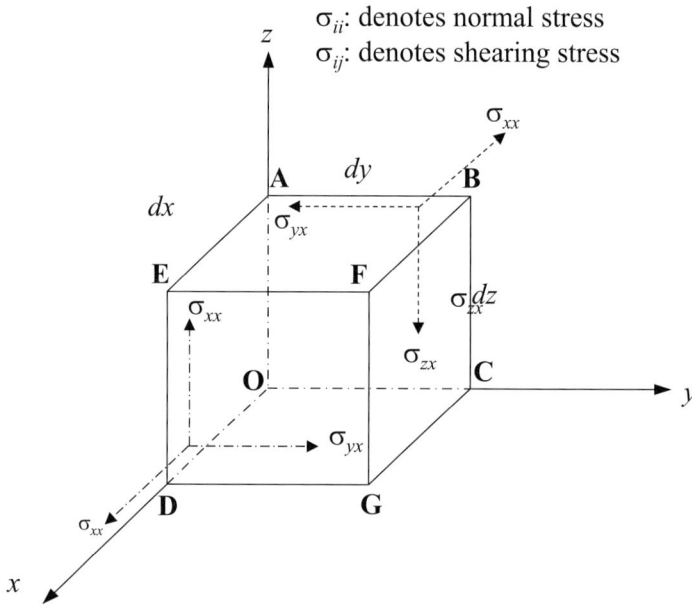

Figure 2.1 Stress for a body (modified after Sheriff and Geldart, 1995).

perpendicular to the surface and (b) a component of shearing stress in the plane of the surface (Figure 2.1). Stress can be classified into:

- Compressive if the forces are directed toward each other, causing the object under stress to become compressed (change in size).
- Tensile if the forces are directed away from each other, causing the object under stress to become tensed (change in size).
- Shearing if the forces are directed away from (or toward) each other, causing the object under stress to change its shape.

Note that a fluid body cannot sustain shearing stresses.

2.2.2 Strain

A body subjected to stress undergoes a change of shape and/or size known as strain. That is, strain ε can be defined as the relative (fractional) change in dimension or shape of a body due to the application of stress. Note that strain is dimensionless. In general, three cases for strain exist:

1. Elastic strain: Up to a certain limiting value of stress, the strain is directly proportional to the applied stress. This follows Hook's law,[1] which states that for

---

[1] Robert Hook is an English physicist who lived between 1635 and 1703 and discovered the law of elasticity.

small strains (or deformations), usually $<10^{-6}$, the stress is linearly proportional to the strain :

$$\sigma = c\varepsilon, \tag{2.1}$$

where c is the elastic constant, modulus, or stiffness tensor. The elastic constant can be a combination of two or more elastic constants such as Lame's constant λ_r or the rigidity of rocks μ_r, depending on the rock type. For this type of strain, the strain is reversible so that removal of stress leads to a removal of strain. The maximum limiting point for an elastic strain to hold is called the *Yield Strength*.

2. Plastic or Ductile strain: If the Yield strength is exceeded then Equation (2.1) no longer holds, where the relationship between stress and strain becomes nonlinear. In this case, permanent strain results may be obtained and the strain becomes partly irreversible.

3. Fracture (Breaking) Point: If the stress is increased still further then, in this case, the stressed body fails by a fracture.

Figure 2.2 shows a typical stress–strain curve for a solid body (Kearey et al., 2002). Now, in the following discussions, the elastic field is considered. The linear relationship between stress and strain is specified for any material by its various elastic constants known as the *Elastic Moduli*, each of which expresses the ratio of a particular type of stress to the resultant strain, as seen in Equation (2.1).

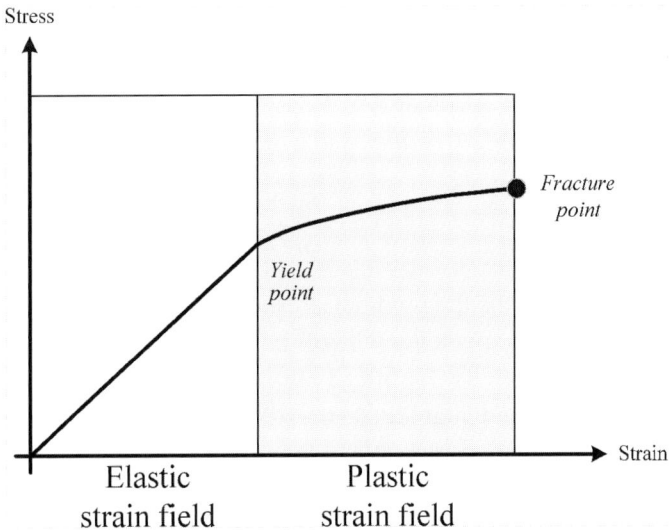

Figure 2.2 Stress and strain relationship for a solid body (courtesy of Kearey et al., 2002).

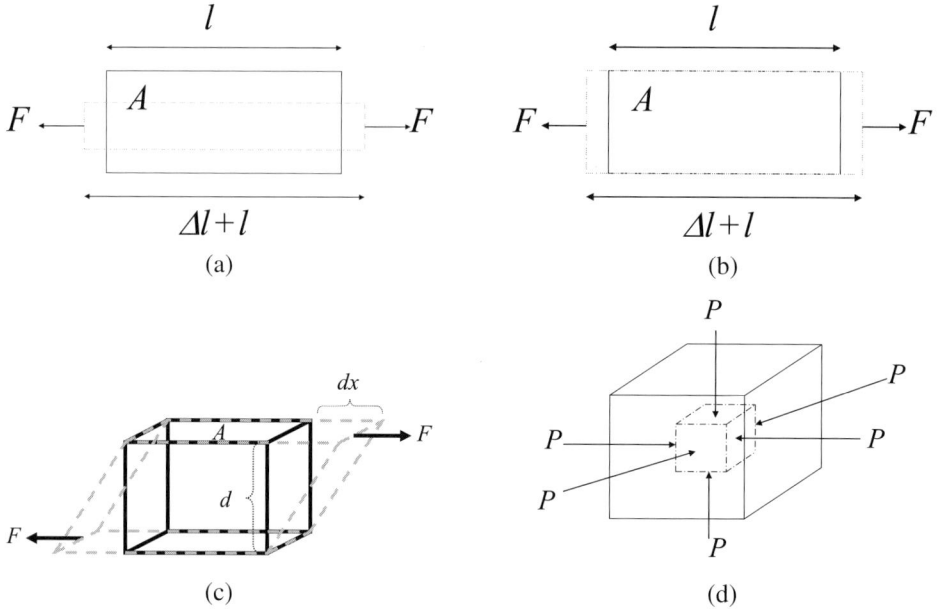

Figure 2.3 Types of Moduli: (a) Young's Modulus, (b) Axial Modulus, (c) Shear Modulus, and (d) Bulk Modulus.

Hence, many types of moduli exist but a few of them are going to be considered for illustration purposes as follows:

1. Young's Modulus (elasticity of length): Measures the resistance of a solid (rocks in this case) to a change in its length. Consider a rod of length *l* and a cross section area *A*, which is extended by an increment Δl through an application of a stretching force *F* to its end faces as seen in Figure 2.3a. The resultant modulus is called *Young's Modulus* and is denoted *Y* where:

$$Y = \frac{F/A}{\Delta l/l},\tag{2.2}$$

where F/A represents the longitudinal stress and $\Delta l/l$ is the longitudinal strain. Note that the extension of such a rod will be accompanied by a reduction in its diameter. That is, the rod will suffer both lateral and longitudinal strain. The ratio of the lateral to the longitudinal strain is called *Poisson's ratio*.

2. Axial Modulus: This defines the ratio of longitudinal stress to longitudinal strain in the case of no lateral strain (which is different from Young's modulus) as seen in Figure 2.3b.

3. Shear Modulus (elasticity of shape): Measures the resistance to motion of planes of a solid sliding past each other. A shearing stress changes the object shape, not

its volume or length. Consider Figure 2.3c, where it shows a rectangular block acted upon by shear forces F. The shearing stress is F/A and the shearing strain is equal to the angle of shear, say ϕ_s in radians. Under the assumption that ϕ_s is always small, its value can, therefore, be approximated by the ration of the displacement dx of the block's faces and the distance d between these faces. Hence, the shear modulus, denoted μ_r, will be:

$$\mu_r = \frac{F/A}{dx/d}. \tag{2.3}$$

4. Bulk Modulus (elasticity of volume): This modulus, denoted by K, expresses the stress strain ratio in the case of a simple hydrostatic pressure P, which is equal to the compressive force per unit area F/A, to a cubic element such as that seen in Figure 2.3d. So when compressive forces act over the entire surface of a body (in this case the cube), the body volume decreases. In this case, the resultant volume strain is the change of volume $\Delta\xi$ divided by the original volume ξ and, hence,

$$K = -\frac{P}{\Delta\xi/\xi}. \tag{2.4}$$

The minus sign is inserted here so that K becomes positive, since an increase in the volume stress will lead to a decrease in the volume of the object under the compressive forces.

Example 2.1 A load of 100 kg is supported by a wire of length 1 m and cross-sectional area of 0.1×10^{-4} m^2. The wire is stretched by 0.2×10^{-2} m. Find the tensile stress, strain, and Young's modulus for the wire. Note that the gravity acceleration g is equal to 9.8 m/s^2.

Solution: The tensile stress is:

$$\sigma = \frac{F}{A} = \frac{Mg}{A}$$
$$= \frac{100 \times 9.8}{0.1 \times 10^{-4}}$$
$$= 9.8 \times 10^7 \text{ N/m}^2,$$

where M stands for the mass in kg. Now, the tensile strain is:

$$\varepsilon = \frac{\Delta l}{l} = \frac{0.2 \times 10^{-2}}{1} = 0.2 \times 10^{-2}.$$

Hence, the Young's modulus will be:

$$Y = \frac{\sigma}{\varepsilon} = \frac{9.8 \times 10^7}{0.2 \times 10^{-2}} = 490 \times 10^9 \text{ N/m}^2.$$

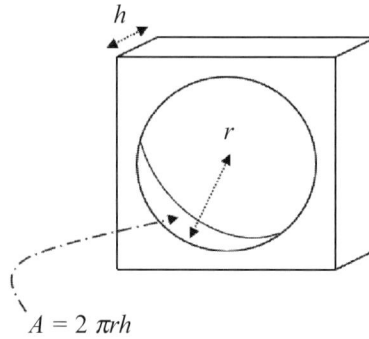

Figure 2.4 A steel sheet that has a hole of radius *r* in Example 2.2.

Example 2.2 Consider Figure 2.4. How much force is required to punch a hole of radius *r* equal to 0.2 m in a steel sheet of thickness *h* equal to 0.5 m, whose shearing stress σ is 5×10^4 N/m²?

Solution: From Figure 2.4, it is noticed that the shearing stress is exerted over a cylindrical surface, which represents the boundary of the hole. The area of this surface is:

$$A = 2\pi r h = 2\pi \times 0.2 \times 0.5 = 0.628 \text{ m}^2.$$

Now, since the minimum shear stress σ_{min} needed to rupture the steel is 5×10^4 N/m², then:

$$F = \sigma_{min} \times A = 5 \times 10^4 \times 0.628 = 61,400 \text{ N}.$$

Example 2.3 A solid lead sphere of volume 0.8 m³ is lowered to a depth in the ocean where the water pressure is equal to 2×10^7 N/m². The bulk modulus of the lead is 7.7×10^9 N/m². Obtain the change in volume of the sphere.

Solution: Solving for the change in volume $\Delta\xi$ of sphere using Equation (2.4), one obtains:

$$\Delta\xi = -\frac{P}{K/\xi} = -\frac{P\xi}{K} = \frac{-(2 \times 10^7)(0.8)}{7.7 \times 10^9} = -2.078 \times 10^{-2} \text{ m}^3,$$

where the negative sign indicates a decrease in the sphere's volume.

For more details on the elastic moduli it is recommended to see the book by Sheriff and Geldart (1995).

2.3 The Wave Equation and d'Alembert's Solution

Seismic waves are parcels of elastic strain energy that propagates outward from a seismic source. Such an elastic strain energy causes displacements of Earth particles in space and time as a seismic wave passes where these displacements are governed (related) by the so-called *wave equation*. The wave equation was first found and solved by the French mathematician d'Alembert during the eighteenth century, preceding Fourier by over a half century (O'Neil, 1995). It is based on two fundamental laws, namely, Hook's law, which states that stress is proportional to strain as seen in Equation (2.1), and Newton's law, which says that force is equal to mass times acceleration (Sheriff and Geldart, 1995; Telford et al., 1990).

The wave equation can be used to predict the existence of seismic waves like compressional or shear waves as well as their properties that may include Snell's law of reflection and refraction, the partition of energy at an interface into compressional and shear components, surface waves, diffractions, etc. (to be discussed in the section 2.4). In fact, seismic migration, which seeks to correct for wave propagation geometrical effects, invokes solving of the wave equation. The general form of the wave equation can be given by (Sheriff and Geldart, 1995):

$$\nabla^2 u(x, y, z, t) = \frac{1}{v^2} \frac{\partial^2 u(x, y, z, t)}{\partial t^2}, \tag{2.5}$$

where v is a constant that denotes the velocity of the seismic waves, $u(x, y, z, t)$ is the particle displacement, wavefield, sometimes called the disturbance at distances x, y and z, t stands for time, and ∇^2 is the Laplacian operator. A seismic wave that propagates only along the x-axis is governed by the following one-dimensional (1-D) wave equation:

$$\frac{\partial^2 u(x,t)}{\partial x^2} = \frac{1}{v^2} \frac{\partial^2 u(x,t)}{\partial t^2}. \tag{2.6}$$

Example 2.4 Show that the wave described by the displacement function:

$$u(x,t) = \sin\left(\frac{n\pi x}{L}\right) \cos\left(\frac{n\pi vt}{L}\right),$$

with n being an integer and L being a positive constant, satisfies the 1-D wave equation given by Equation (2.6).

Solution: The left-hand side of Equation (2.6) can be obtained as follows:

$$\frac{\partial^2 u(x,t)}{\partial x^2} = \frac{n\pi}{L} \cos\left(\frac{n\pi vt}{L}\right) \frac{\partial}{\partial x}\left[\cos\left(\frac{n\pi x}{L}\right)\right],$$

$$= -\frac{n^2\pi^2}{L^2} \cos\left(\frac{n\pi vt}{L}\right) \sin\left(\frac{n\pi x}{L}\right),$$

$$= -\frac{n^2\pi^2}{L^2} u(x,t).$$

On the other hand, consider the right-hand side of Equation (2.6), where:

$$\frac{1}{v^2}\frac{\partial^2 u(x,t)}{\partial t^2} = -\frac{n\pi}{vL}\sin\left(\frac{n\pi x}{L}\right)\frac{\partial}{\partial x}\left[\sin\left(\frac{n\pi vt}{L}\right)\right],$$

$$= -\frac{n^2\pi^2}{L^2}\sin\left(\frac{n\pi x}{L}\right)\cos\left(\frac{n\pi vt}{L}\right),$$

$$= -\frac{n^2\pi^2}{L^2}u(x,t).$$

Therefore, $u(x,t) = \sin(\frac{n\pi x}{L})\cos(\frac{n\pi vt}{L})$ is a valid solution of the 1-D wave equation given by Equation (2.6).

Example 2.5 Verify that the following is a solution of the wave equation:

$$u(x,t) = A\,\exp\left[j2\pi f\left(t - \frac{x}{v}\right)\right].$$

Solution: The left-hand side of Equation (2.6) can be calculated as follows:

$$\frac{\partial u(x,t)}{\partial x} = -\frac{j2\pi f}{v}A\,\exp\left[j2\pi f\left(t - \frac{x}{v}\right)\right].$$

$$\implies \frac{\partial^2 u(x,t)}{\partial x^2} = -\frac{4\pi^2 f^2}{v^2}A\,\exp\left[j2\pi f\left(t - \frac{x}{v}\right)\right].$$

Also, the right-hand side of Equation (2.6) is given by:

$$\frac{\partial u(x,t)}{\partial t} = j2\pi f A\,\exp\left[j2\pi f\left(t - \frac{x}{v}\right)\right].$$

$$\implies \frac{1}{v^2}\frac{\partial^2 u(x,t)}{\partial t^2} = -\frac{4\pi^2 f^2}{v^2}A\,\exp\left[j2\pi f\left(t - \frac{x}{v}\right)\right].$$

Therefore, $u(x,t)$ is a solution of the wave equation.

d'Alembert was able to obtain a general implicit solution of Equation (2.6) that is given by Sheriff and Geldart (1995) and Robinson and Clark (1987):

$$u(x,t) = B(x + vt) + F(x - vt), \tag{2.7}$$

where $B(.)$ and $F(.)$ can be arbitrary functions. In the case of a seismic wave propagating along the x-axis, $B(.)$ and $F(.)$ represent two waves traveling along the x-axis in opposite directions (refer to Figure 2.5) with velocity v but not necessarily of the same profile. Note that the given solution in Equation (2.7) is known as d'Alembert's formula.

Moving toward +ve x-axis   Moving toward −ve x-axis

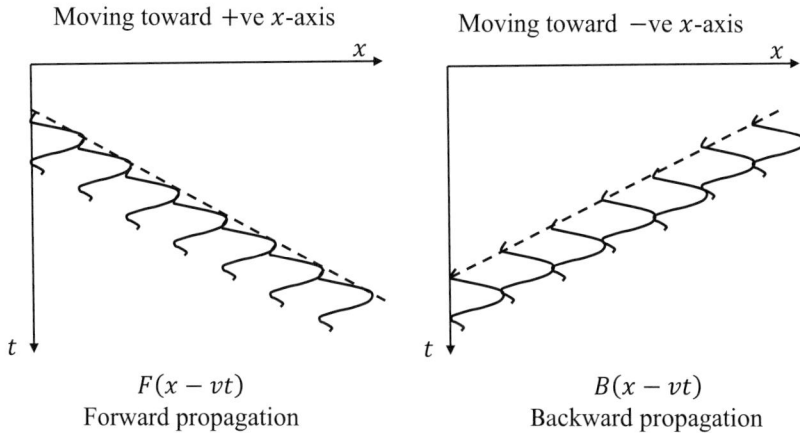

$F(x - vt)$

Forward propagation

$B(x - vt)$

Backward propagation

Figure 2.5 Plane wave solution of the one-dimensional scalar wave equation (modified after Sheriff and Geldart, 1995).

Equation (2.7) is now investigated for the case of 1-D waves. Suppose that $u(x,t)$ is a wave that travels in the positive x-direction with a constant positive velocity v, i.e.,

$$u(x,t) = F(x - vt). \tag{2.8}$$

The shape of $u(x,t)$ at any time instant can be found by holding t to be a constant, e.g., $t = 0$, then:

$$u(x,0) = F(x), \tag{2.9}$$

which represents the 1-D wave profile at time instant $t = 0$. Assuming that $u(x,t)$ does not change its profile while propagating through the space. That is, no attenuation occurs, and a coordinate frame is introduced that travels along with the disturbance at speed v. The displacement, in this case, is no longer a function of time but instead a time-invariant constant profile with the same functional form as the waveform at $t = 0$. So the displacement in this case becomes:

$$u(x,t) = F(x - vt), \tag{2.10}$$

where it represents a general form of the 1-D wave equation solution. In this case, a profile $F(x)$ needs to be chosen and then $x - vt$ substituted for x in $F(x)$, and the resulting expression for the displacement $u(x,t)$ describes a wave traveling in the positive x-direction. One may think of $F(x - vt)$ as a forward wave, translating $F(x - vt)$ by vt units to the positive x-axis (to the right) with velocity v. Likewise, similar remarks can be drawn if the solution of Equation (2.6) is traveling in the negative x-direction. That is, $B(x + vt)$ as a backward wave, translating $B(x + vt)$

by vt units to the left with velocity v (see Figure 2.5). And regardless of the wave profile, the variables x and t must appear in the general function as a unit $x \pm vt$.

It is also worth looking into the displacement $u(x,t)$ explicitly. The simplest waveform is that of a harmonic wave, that is, a wave that involves sine or cosine expressions (Telford et al., 1990). Examples are:

$$u(x,t) = A \sin[k(x - vt)], \tag{2.11}$$

$$u(x,t) = A \cos[k(x - vt)], \tag{2.12}$$

where k is a positive constant known as the wavenumber in cycles/unit distance (such as rad/m) and A is the amplitude of the sinusoidal waves. One can show that Equations (2.11) and (2.12) are solutions of Equation (2.6) and either is equal to $F(x - vt)$. Holding either x or t fixed will result in a sinusoidal disturbance and, hence, either of these sinusoidal waves is periodic in both space and time. Thus, consider the spatial and temporal periods.

> **Definition 2.6** The spatial period is known as the wavelength and is denoted by λ, where the period units are distance units such as meters (m). Figure 2.6 demonstrates a sinusoidal with period λ.

Therefore, having defined λ while fixing t implies that:

$$u(x,t) = u(x \pm \lambda, t), \tag{2.13}$$

and if one considers, for example, Equation (2.11), then:

$$\sin[k(x - vt)] = \sin[k(x \pm \lambda - vt)],$$
$$= \sin[k(x - vt) \pm k\lambda]. \tag{2.14}$$

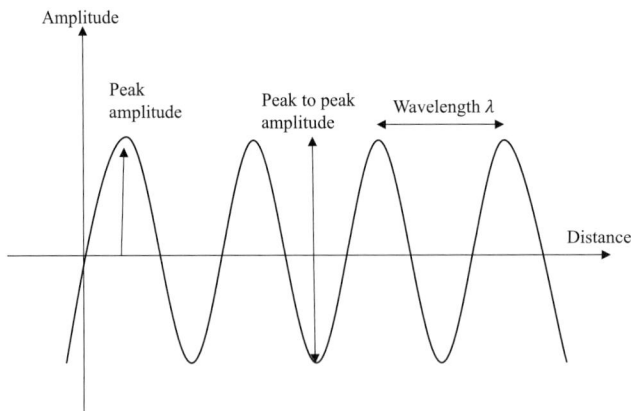

Figure 2.6 A spatial sinusoidal wave. The wave is a function of distance x and can be distinguished by its wavelength λ and peak (or its peak-to-peak) amplitude.

It is also known that an increase or decrease in the argument by an amount 2π leaves the sine or cosine representations unaltered as shown here:

$$\sin[k(x - vt)] = \sin[k(x - vt) \pm 2\pi]. \tag{2.15}$$

Therefore, it can be seen from Equations (2.14) and (2.15) that:

$$|k\lambda| = 2\pi. \tag{2.16}$$

However, both parameters k and λ are positive quantities, then the following fundamental equation is obtained:

$$k = \frac{2\pi}{\lambda}. \tag{2.17}$$

On the other hand, consider the temporal period by fixing x in this case.

Definition 2.7 The temporal period T (in seconds) is the amount of time one complete sinusoidal oscillation takes to pass a stationary observer, i.e., the number of units time per cycle.

Figure 2.7 shows a temporal sinusoidal wave with period T. Hence,

$$u(x,t) = u(x, t \pm T), \tag{2.18}$$

and for the harmonic function in Equation (2.11):

$$\sin[k(x - vt)] = \sin[k(x - v\{t \pm T\})], \tag{2.19}$$
$$= \sin[k(x - vt) \mp kvT], \tag{2.20}$$
$$= \sin[k(x - vt) \pm 2\pi]. \tag{2.21}$$

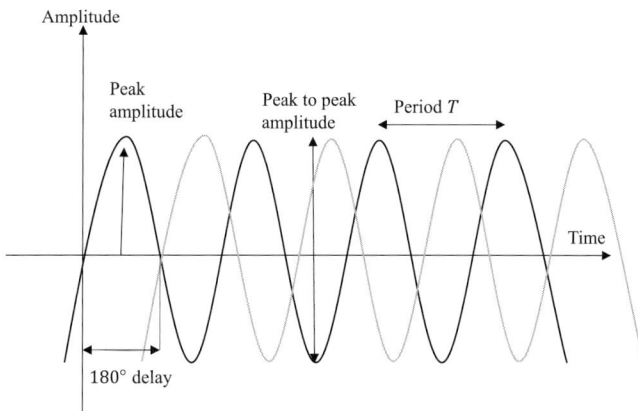

Figure 2.7 A temporal sinusoidal wave. The wave is a function of time t and can be distinguished by its period T and peak (or its peak-to-peak) amplitude. Also, the phase shift is another parameter that distinguishes a sinusoidal wave with a period T from another with the same period.

Taking into account that the quantities k, v, and T are positive, it can be deduced that:

$$kvT = 2\pi, \tag{2.22}$$

and considering Equation (2.17) one obtains:

$$v = \frac{\lambda}{T}. \tag{2.23}$$

Definition 2.8 The reciprocal of the period T is called the frequency and is denoted by f, that is, f is the number of cycles per unit time and it is in Hertz (Hz).

Hence,

$$f = \frac{1}{T}, \tag{2.24}$$

and

$$v = \lambda f. \tag{2.25}$$

If a medium with constant velocity is considered, this equation says that a wave with a short wavelength has a high frequency. Likewise, a wave with a long wavelength has a low frequency. Another frequently used quantity known as the angular frequency ω is defined as the repetition rate measured in rad/s. Thus:

$$\omega = 2\pi f = \frac{2\pi}{T}. \tag{2.26}$$

Example 2.9 A seismic wave is propagating, causing a sinusoidal disturbance $u(x,t)$ of the particles, where the wiggles recorded by a seismic geophone are 4 ms apart. What is the wavelength and wavenumber of the recorded seismic data if the velocity of the propagating wave was 2,500 m/s?

Solution: From Equation (2.23), it can be followed that the wavelength of the propagating wave is $\lambda = 2,500 \times 0.004 = 10$ m. Hence, the wavenumber can be obtained using Equation (2.17), where $k = \pi/5$ rad/m.

Example 2.10 The velocity of seismic waves propagating in a sandstone rock layer is 3,000 m/s. Obtain the wavelength in the sandstone rock layer of the seismic wave whose frequency is 45 Hz.

Solution: The wavelength can be obtained via Equation (2.25), where $\lambda = v/f = 3,000/45 = 66.67$ m.

Example 2.11 A seismic wave of frequency 15 Hz and wavelength of 134 m goes from a medium of sandstone rocks into a dolomite medium, in which its velocity is twice the velocity of the sandstone rocks medium. Find the frequency and wavelength of the seismic wave propagating in the dolomite medium.

Solution: Consider Equation (2.25) and solve for the wavelength of the second (dolomite rocks) medium, $\lambda_2 = v_2/f_2$. However, since it is given that the seismic wave velocity of the second medium is twice that for the first (sandstone) medium, $\lambda_2 = 2v_1/f_2 = 2\lambda_1 f_1/f_2$. Hence, $\lambda_2 f_2 = 2\lambda_1 f_1$. In this case, $\lambda_2 = 2\lambda_1 = 268$ m and $f_2 = f_1 = 15$ Hz.

In summary, the quantities λ, k, T, and f all describe aspects of the repetitive nature of waves that are periodic functions of space and time. This includes other periodic waves than sinusoids. This is obvious since one can represent any arbitrary periodic waveform by the sum of harmonically related sinusoidal terms using the Fourier series (O'Neil, 1995; Ziemer et al., 1998a). Also, based on these defined quantities, Equation (2.11) can, for example, be written in the following equivalent forms:

$$u(x,t) = A \sin[k(x - vt)], \tag{2.27}$$

$$= A \sin\left[\frac{2\pi}{\lambda}(x - vt)\right], \tag{2.28}$$

$$= A \sin[kx - \omega t], \tag{2.29}$$

$$= A \sin\left[\omega\left(\frac{x}{v} - t\right)\right], \tag{2.30}$$

$$= A \sin\left[2\pi\left(\frac{x}{\lambda} - ft\right)\right], \tag{2.31}$$

$$= A \sin[kx - 2\pi ft]. \tag{2.32}$$

Example 2.12 Find the following and state the units of each when necessary for $u(x,t) = 10 \sin(0.1\pi x - 20\pi t)$:

(a) The amplitude.
(b) The wavenumber.
(c) The angular frequency.
(d) The frequency.
(e) The velocity.
(f) The period.
(g) The wavelength.

Solution: The given harmonic function follows the form seen in Equation (2.29). Hence:

(a) The amplitude $A = 10$.
(b) The wavenumber $k = 0.1\pi$ rad/m.
(c) The angular frequency $\omega = 20\pi$ rad/s.
(d) The frequency $f = \omega/(2\pi) = 10$ Hz.
(e) The velocity $v = \omega/k = 200$ m/s.
(f) The period $T = 1/f = 0.1$ s.
(g) The wavelength $\lambda = v/f = 20$ m.

An important note to mention particularly about the harmonic waves is that Equations' (2.27)–(2.32) waveforms are said to be *monochromatic* since each has a constant frequency. Other examples exist as well that are solutions of the wave equation (Equation [2.6] such as the monochromatic plane wave):

$$u(x,t) = A \exp\left[j2\pi f\left(t - \frac{x}{v}\right)\right],\tag{2.33}$$

and it is characterized by its amplitude A, frequency f, and velocity v. Also, in seismic data, one deals with a *Wavelet* (a wave pulse), which contains many frequencies and, hence, a dominant frequency (or period) exists. So for a given wavelet, if f_1 and f_2 are, respectively, the minimum and maximum signal frequencies, then the dominant frequency will be:

$$f_{dom} = \frac{f_1 + f_2}{2}.\tag{2.34}$$

An example of continuous-time wavelets is:

$$w(t) = (1 - 2\pi^2 f_{dom}^2 t^2)e^{-\pi^2 f_{dom}^2 t^2},\tag{2.35}$$

which is one mathematical version of the so-called *Ricker* wavelet (Buttkus, 2000). Another example is the Morlet wavelet (Yilmaz, 2001):

$$w(t) = \frac{1}{\sqrt{2\pi}}e^{-j2\pi f_{dom}t}e^{-t^2/2}.\tag{2.36}$$

The bandwidth (BW) of a wavelet ($f_2 - f_1$) is usually described in Hz but also is given in octaves:

$$BW_{octave} = \frac{\ln(f_2/f_1)}{\ln(2)},\tag{2.37}$$

which is a measure of how dominant the central peak of a given wavelet is relative to its side-lobes (Liner, 1999b). Figure 2.8 shows an example of wavelet in both the time and frequency domains. More details on wavelets will be discussed in Chapter 8.

> **Example 2.13** One requires a dominant wavelength of 50 m, and if a seismic wave is propagating with a seismic velocity of 2,300 m/s, find the dominant frequency and the bandwidth in Hz of the seismic wavelet, where the bandwidth in octaves is 2.4.
>
> **Solution:** Using Equation (2.25), the dominant frequency will be $f_{dom} = v/\lambda_{dom} = 2,300/5 = 46$ m. Now, using Equation (2.37), $f_2/f_1 = 1.66$ or $f_2 = 1.66 f_1$. Also, from Equation (2.34), one can see that $f_2 = 2 f_{dom} - f_1 = 92 - f_1$. Hence, solving both equations will yield $f_1 = 90.34$ Hz and $f_2 = 149.96$ Hz. Therefore, $BW = 59.62$ Hz.

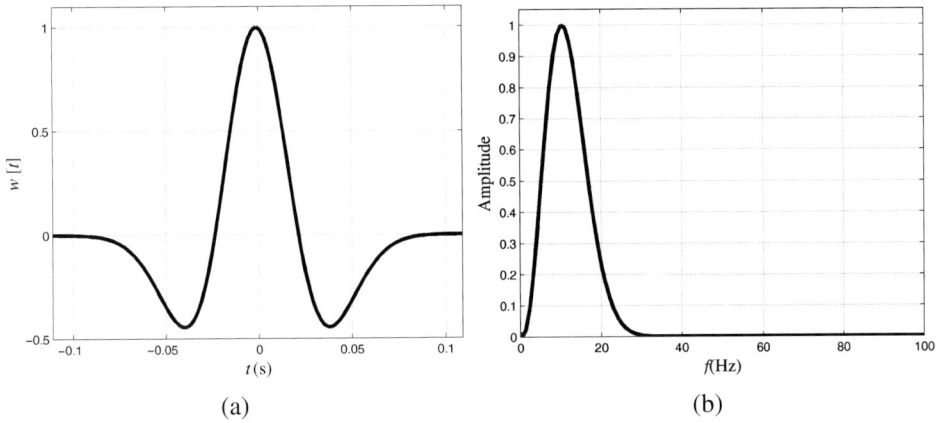

(a)   (b)

Figure 2.8 The (a) time and (b) frequency domains of a Ricker wavelet with a dominant frequency of 10 Hz.

Definition 2.14 The argument of the waveform in Equation (2.29) is known as the phase of the wave and is denoted with the symbol ϕ. That is:

$$\phi(x,t) = kx - \omega t. \tag{2.38}$$

Note that ϕ is a function of x and t where kx is called the spatial phase and ωt is the temporal phase. Consider a more general form of Equation (2.29) which is:

$$u(x,t) = A \sin[kx - \omega t + \gamma], \tag{2.39}$$

where γ is the initial phase. Hence, Equation (2.38) becomes:

$$\phi(x,t) = kx - \omega t + \gamma. \tag{2.40}$$

To obtain the rate of change of ϕ with respect to t, the partial derivative of Equation (2.40) is taken with respect to t while holding x, in this case, as a constant. That is:

$$\frac{\partial \phi}{\partial t} = -\omega. \tag{2.41}$$

Similarly, to obtain the rate of change of ϕ with respect to x, the partial derivative of Equation (2.40) is taken with respect to x (in this case, t is assumed to be a constant) where:

$$\frac{\partial \phi}{\partial x} = k. \tag{2.42}$$

If $\phi(x,t)$ in Equation (2.40) is a constant, then:

$$\partial \phi(x,t) = k\partial x - \omega \partial t + 0 = 0,$$

$$\Leftrightarrow \frac{\partial x}{\partial t} = \frac{\omega}{k}. \tag{2.43}$$

However, if Equation (2.41) is divided over Equation (2.42) and considering Equation (2.43), the following relation is deduced:

$$\frac{\partial x}{\partial t} = \frac{-\partial \phi / \partial t}{\partial \phi / \partial x} = \frac{\omega}{k} = v, \tag{2.44}$$

which represents the velocity of propagation, in this case, of ϕ being constant. Because the only variable in the sinusoidal wave function in Equation (2.29) is the phase, it is fixed at a value producing the constant displacement u at the chosen point. Then, the point moves along with the profile at velocity v. Hence, assume that ϕ being a constant is valid. The velocity in Equation (2.44) is called the phase velocity, which is the velocity of any given phase or a wave of a single frequency. It basically carries a positive sign when the wave moves in the direction of increasing x, and it carries a negative sign when the wave moves in the direction of decreasing x.

Example 2.15 Compute the phase velocity for $u(x,t) = 5\sin(0.04\pi x - 25\pi t)$.

Solution: Based on the sinusoidal parameters of the given displacement function, the phase velocity is equal to $v = \omega/k = 25\pi/(0.04\pi) = 625$ m/s.

2.4 Seismic Waves

Two general ways by which mechanical energy can be transferred from one place to another exist: the passage of matter from one place to another, and the passage of energy through a material medium in such a way that the medium is left essentially unchanged after the transfer of energy. The latter way can be achieved by mechanical traveling waves like the case of seismic. So for seismic waves to exist, there must be:

1. A source of disturbance such as an explosion or a vibrating seismic truck.
2. A medium that can be disturbed, where its adjacent particles possess some physical connection by which they influence one another.

To see the picture more clearly, consider the following description for the passage of seismic waves:

- A source of energy makes an initial rock particle oscillate around its equilibrium position.
- This oscillation begins to push and pull on the second particle so that it oscillates about its equilibrium position.

- The seismic energy is transferred from the first particle to the second particle. Then, the oscillation of the second rock particle causes the third rock particle to oscillate around its equilibrium position. Again, the seismic energy transfer occurs.

- This process continues constructively where each individual particle acts to displace the adjacent particle so that it begins to oscillate as well.

In this sequential manner, a disturbance travels through the medium and thereby transports kinetic energy. Now, assuming that strains associated with the passage of seismic waves (energy) are elastic except in the immediate vicinity of the seismic source, the seismic wave propagation velocities will be determined by: (a) the elastic moduli, and (b) the densities of the material through which seismic waves pass. So, in summary, seismic energy propagates through the Earth in the form of elastic waves, where they disturb particles in a locality in a periodic manner. The displacement of these particles is oscillatory about their undisturbed locations. The boundary conditions that limit the wave propagation and the nature of the displacement classify the type of the passing seismic waves. It is worth mentioning here that most of the reflected seismic energy used for petroleum exploration is contained within a frequency range of $2-120$ Hz, where the dominant frequency range f_{dom}, in this case, is $5-15$ Hz and the dominant wavelength λ_{dom} range is $30-400$ m (Sheriff and Geldart, 1995). Two groups of seismic waves exist, namely, body waves and surface waves, which will be discussed in the following subsections. Figure 2.9 summarizes the breakdown of the seismic waves.

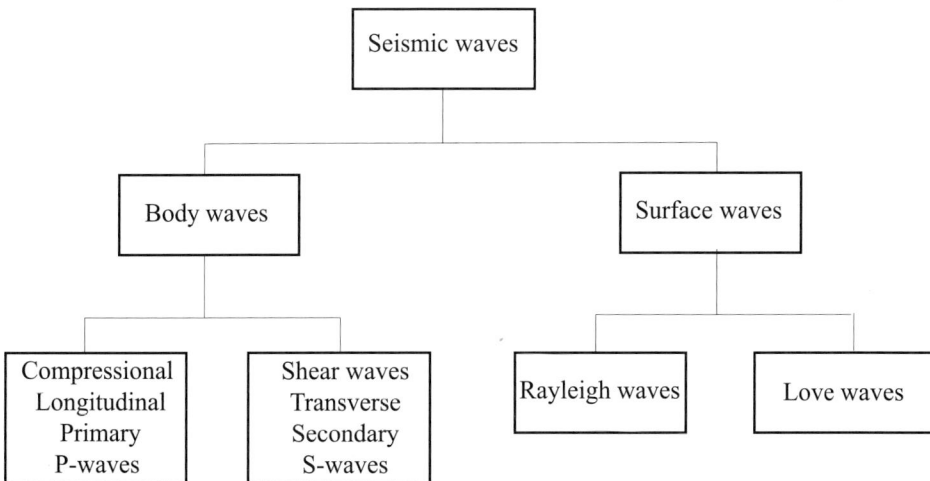

Figure 2.9 Types of seismic waves.

2.4.1 Body Waves

Body waves can travel through the body of the medium, i.e., they propagate through the internal volume of an elastic solid. Consider the medium in which seismic body waves are propagating (which are rocks in this case); they can be *homogeneous*. That is, the medium is a material with properties (such as velocity) that are independent of position. Otherwise, the material is said to be *inhomogeneous*. Also, rocks can be *isotropic*, where their properties are independent of direction (angle) of wave travel. On the other hand, rocks can be *anisotropic*, where one physical property or more depends on the direction of wave travel (the wave direction or angle). Since velocity is the only rock physical property used in seismic theory, another type of inhomogeneity of rocks exists. They can be isotropic everywhere; however, the absolute velocity can vary from one place to another. For example, the velocity of sedimentary rocks, where most of the oil is found, increases with depth. In addition, it can change laterally and, hence, lateral velocity variations exist.

In general, the velocity of wave propagation of any body wave in any homogeneous and isotropic material (in a material with velocity that is independent of both position and direction) can be obtained using:

$$v = \frac{c}{\rho}, \tag{2.45}$$

where c represents the elastic modulus (see Section 2.2) of the material and ρ stands for the density of the material. Body waves are *non-dispersive*, where their velocities are independent of frequency. An important remark to address here is that when the propagation velocity of a given seismic wave is discussed, this velocity is meant to be the one with which the seismic energy travels through a given medium. This is, on the other hand, different from the velocity of particles within the medium that are disturbed by the passage of the propagating seismic waves. Body waves are of two types based on the particle motion:

- Compressional Waves: These are also known as longitudinal, primary, or P-waves. They propagate by compressional and dilatational uni-axial strains in the direction of wave travel. In this case, the particle motion associated with the passage of a compressional wave involves oscillation about a fixed point in the direction of (or parallel to) wave propagation , i.e., the direction of energy transmission (see Figure 2.10).
- Shear Waves: Also called transverse, secondary, or S-waves. They propagate by a pure shear strain in the direction perpendicular to the wave travel. That is, the individual particle motions involve oscillation about a fixed point in a plane at right angles to the direction of wave propagation as seen in Figure 2.11. Shear waves are also classified into two types, depending on particle motion. That is,

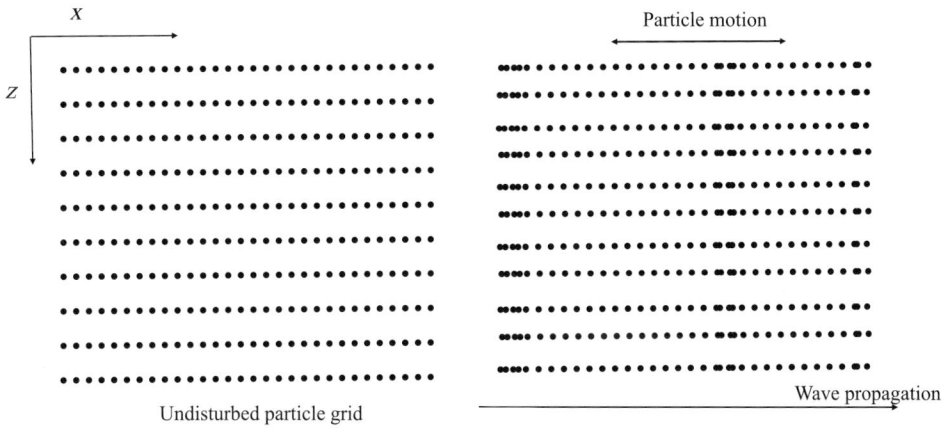

Figure 2.10 A set of material particles at rest (left) and the same set of particles oscillating about a fixed point in the direction of wave propagation (right). Such a wave is called compressional.

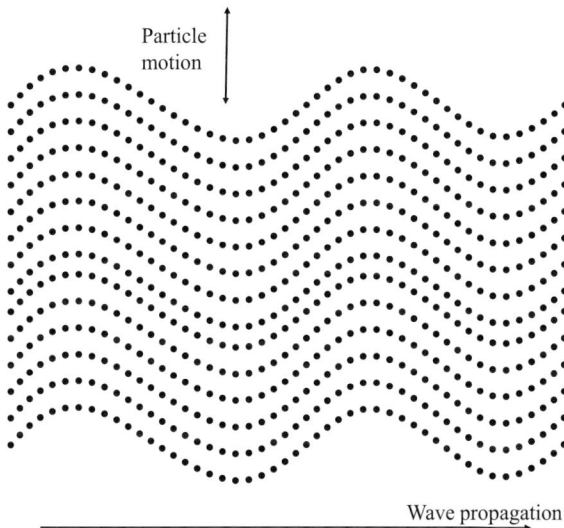

Figure 2.11 A set of material particles oscillating (horizontally or vertically) about a fixed point that is perpendicular to the direction of wave propagation. Such a wave is called shear.

one may have vertical shear (SV) waves, where vertical particle motion exists. On the other hand, one may also have horizontal shear (SH) waves, where horizontal particle motion propagates.

The compressional wave velocity (v_p) is a function of the so-called Lame's constant λ_r, and the rigidity (or sometimes called the shear modulus) μ_r, which is

Table 2.1 *Typical compressional and shear wave velocity, and density values for different material (Sheriff and Geldart, 1995)*

Material Type	v_p (m/s)	v_s (m/s)	ρ (kg/m^3)
Air	331	0	3.8
Limestone	3,500–6,100	2,000–3,300	2,400–2,700
Dolomite	3,500–6,500	1,900–3,600	2,500–2,900
Sandstone	2,000–4,300	700–2,800	2,100–2,400
Soil	300–900	120–360	1,700–2,400
Stiff mud	1,600	0	1,500
Water	1,500	0	1,000

the ability of the material to resist shear deformation, and the material density ρ. It is, therefore, given by:

$$v_p = \sqrt{\frac{\lambda_r + 2\mu_r}{\rho}}. \tag{2.46}$$

Note that λ_r and μ_r are known as the elastic constants for isotropic media and they are defined to be positive numbers (Sheriff and Geldart, 1995). On the other hand, the shear velocity v_s is given by:

$$v_s = \sqrt{\frac{\mu_r}{\rho}}. \tag{2.47}$$

Compressional waves always travel faster than shear waves since their velocities are higher when compared with shear wave velocities. Typically, the shear wave velocities are about one-half those of compressional wave velocities. Table 2.1 shows typical values for different material types including sedimentary rocks.

Example 2.16 Compute Lame's constant and the rigidity parameters for the following materials:

(a) Air.
(b) Limestone. Assume that $v_p = 3,500$ m/s, $v_s = 1,950$ m/s, and $\rho = 2,450$ kg/m^3.
(c) Water.

Solution: From Equation (2.46)

$$\lambda_r = v_p^2 \rho - 2\mu_r, \tag{2.48}$$

and from Equation (2.47)

$$\mu_r = v_s^2 \rho = 1,950^2 \times 2,450 = 9.316 \times 10^9 \text{ kg/ms}^2. \tag{2.49}$$

Hence, $\lambda_r = 3,500^2 \times 2,450 - 2 \times 9.316 \times 10^9 = 11,380.25 \times 10^6 \text{ kg/ms}^2$.

Practically, most of the seismic surveys have used only compressional waves, mainly for two reasons that are related to easier acquisition. The first is related to geophones, where they only record the vertical ground motion and, hence, can be used. So there is no need to triple the recording effort. The second reason is that compressional waves are easier to recognize since $v_p > v_s$ and, hence, reach the geophones faster. In addition, and not a main reason for not acquiring shear waves, some researchers believe that the processing of compressional waves are much easier when compared to the shear ones. However, with advances in the seismic acquisition technologies and an increasing demand for fossil fuel energy, multi-component seismic surveys are becoming part of the standard in seismic land data acquisition. For example, shear waves are important for obtaining other measurements that are related to near-surface analysis. In such analysis, Poisson's ratio can be calculated and may also be used to estimate the elastic moduli (Sheriff and Geldart, 1995). Another example where shear waves are of great interest is that their velocities v_s can be used in detecting gas-filled sediments (Telford et al., 1990).

2.4.2 Surface Waves

Surface waves, sometimes called boundary waves, are waves with energy that travels along or near the surface or Earth's boundary in a bounded elastic solid. In general, the motion of such waves falls off rapidly with distance from the surface, and their particle motion is complex. Many types of surface waves exist, such as Rayleigh and love waves. Rayleigh waves are considered here, since they are commonly seen in seismic land exploration records. Rayleigh waves exist only in the vicinity of an elastic free-surface – a surface where the contact is between an elastic solid and vacuum. Although the interface between Earth and air does not represent an exact free-surface, the interface sufficiently allows the existence of Rayleigh waves. The particle motion of Rayleigh waves is elliptical in a plane perpendicular to the surface and containing the direction of propagation (see Figure 2.12). Generally, their velocities are less than 1,000 m/s, are of frequencies below 15 Hz, and are of high amplitudes when compared to other recorded seismic energy. The amplitude of Rayleigh waves decreases exponentially with distance below the surface (with depth), where at a depth approximately equal to one wavelength, the Rayleigh wave particle motion is negligible. They have propagation velocities v_r that are lower than shear wave velocities, where they are typically equal to $v_r = 0.92v_s$, under the condition that $v_s/v_p \approx 0.577$ (Kearey et al., 2002; Sheriff and Geldart, 1995). A more accurate approximation to determine v_r is given by Liner (1999a):

$$v_r = v_s \frac{20 - \left[256\left(\frac{v_s}{v_p}\right)^4 - 336\left(\frac{v_s}{v_p}\right)^2 + 130\right]^{1/2}}{16\left(\frac{v_s}{v_p}\right)^2 + 9}. \tag{2.50}$$

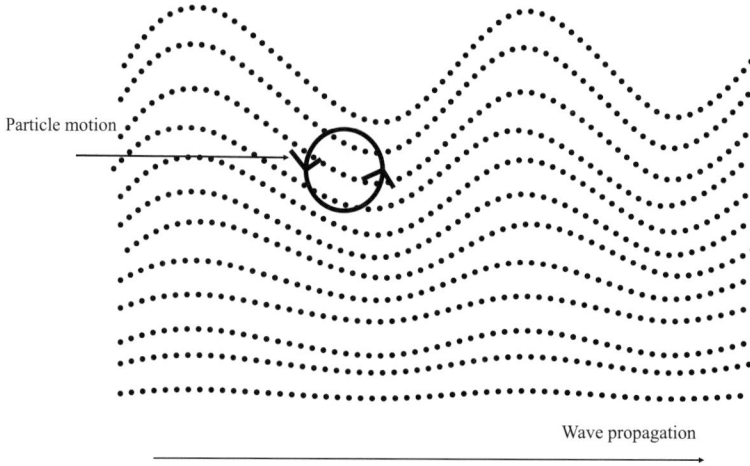

Figure 2.12 A set of material particles which are oscillating about a fixed point in an elliptical manner as the wave propagates. Such a wave is called a surface wave.

Rayleigh waves can also be dispersive (more mathematical details regarding dispersion are given in Section 2.7), which is due to the fact that their waveform undergoes progressive changes during propagation as a result of the different frequency components traveling at different velocities. Such a dispersion indicates the existence of velocity variations with depth in the Earth's subsurface. At the surface, they are grouped together as *Ground roll* and they appear as noise overlying compressional waves, as will be seen in Section 2.10.

2.5 Seismic Wavefronts and Raypaths

When a disturbance occurs, energy radiated outward in the medium via disturbances of the particles and the geometry of the propagating energy is described equally usefully either as *wavefronts* or *raypaths*.

A seismic source generates seismic waves that propagate outward at a velocity controlled by the physical properties of the surrounding rocks. If the waves travel at the same velocity in all directions away from the source, i.e., the medium is isotropic, then at any subsequent time instant, the locus of all points that seismic waves have reached at any time defines a sphere and the locus itself, which is of equal time, is called the *wavefront*. In 3-D a wavefront is a surface and in 2-D it is a curve, over which the phase of traveling wave disturbance is the same. As seen in Figure 2.13a, it is the surface over which the phase of a traveling wave disturbance is the same and it moves perpendicular to itself as the disturbance travels in an isotropic medium. Wavefronts are a very useful way of understanding the propagation of seismic waves through rocks, particularly, when discussing seismic migration or imaging.

Wavefronts   Raypaths

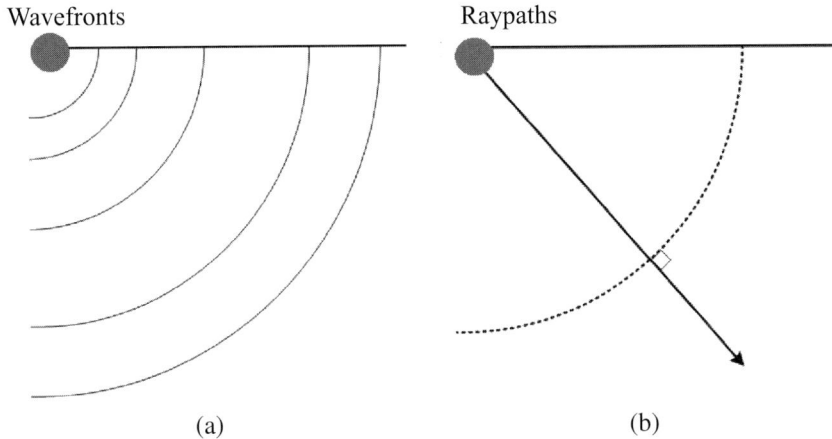

(a)   (b)

Figure 2.13 Seismic wavefronts and raypaths. (a) A seismic wave that propagates outward at a constant velocity, and (b) a raypath is perpendicular to the wave in an isotropic media.

Raypaths, sometimes called seismic rays or simply rays, are the equal of wavefronts. They are paths over which the energy of the disturbance travels from the seismic source to any wavefront or from any wavefront to a later one. In an isotropic medium, raypaths are everywhere perpendicular to wavefronts. They generally have no physical significance, but they are very useful when we describe travel paths of seismic energy through the Earth layers, i.e., they provide understanding of various wave phenomena such as reflections (see Figure 2.13b). They, finally, allow for a much easier computation of seismic wave propagation times of specific seismic phases. This is due to the ability of explicitly constructing paths along which seismic waves have traveled before being recorded by seismic receiver.

2.6 Seismic Wave Velocity of Rocks

Rocks have various compositions, texture, porosities, etc., which distinguish a rock type from another and, hence, they are different in terms of their elastic modulus c's as well as their densities ρ's and, in consequence, their seismic wave velocities v (refer to Equation 2.45). The seismic wave velocity is the rate at which a seismic wave travels through a medium, such as the case of rocks. If one is interested in (a) converting seismic wave travel times into depths, which are very important in reflection seismology, and (b) indicating the lithology of rocks, then it would be required to know v_p (and v_s). Recall from Section 2.2 that Poisson's ratio (denoted here α) is given by the ratio of the lateral contraction of an elastic body to its longitudinal extension when it is stretched. This quantity is useful in relating v_p with v_s through:

$$v_s^2 = v_p^2 \frac{1 - 2\alpha}{2(1 - \alpha)}, \qquad (2.51)$$

where for an ideal solid $\alpha = 0.25$. Therefore,

$$v_s = \frac{\sqrt{3}}{3} v_p. \qquad (2.52)$$

Typical range of α for rocks is $0.2{-}0.4$ (Kearey et al., 2002). As a special case, when $\alpha = 0.5$ (which constitutes an upper bound in this case), v_s is zero and no shear waves exist. Hence, $\alpha = 0.5$ for fluids and gases. The problem now is how to determine, in general, seismic wave velocities.

Example 2.17 If a seismic wave is propagating in a dolomite rock with a compressional wave velocity v_p of 3,600 m/s and a shear wave velocity v_s of 2,100 m/s, then determine the dolomite rock Poisson's ratio α.

Solution: From Equation (2.51), one can solve for the Poisson's ratio α, which is:

$$\alpha = \frac{1 - 2\left(\frac{v_s}{v_p}\right)^2}{2\left[1 - \left(\frac{v_s}{v_p}\right)^2\right]}$$

$$= \frac{1 - 2\left(\frac{2,100}{3,600}\right)^2}{2\left[1 - \left(\frac{2,100}{3,600}\right)^2\right]}$$

$$= 0.242.$$

Rocks are complicated objects and no accurate simple equations exist that relate rock properties to seismic velocities. However, the rock's seismic wave velocities can be estimated. Various methods exist for deriving velocity profiles of the subsurface layer structures from the seismic reflection, or even refraction data (Sheriff and Geldart, 1995; Yilmaz, 2001). Additionally, velocities can be measured in the lab through rock core samples, which are cylindrical rock units taken from within a borehole near the surveyed area, and then correlating the results with those that were estimated. A more direct method in the field is to use a logging tool in the borehole-like acoustic logs. Even vertical seismic profiles (VSP)s, which are measurements of the response of seismic geophone sensors at various depth in a borehole to sources on the surface, can instead be acquired to infer seismic velocities. A less expensive way to obtain a simple time-to-depth relationship at a well location is by using a source on the surface and a geophone sensor in the borehole (Telford et al., 1990). Note that many types of seismic velocities exist that are used in reflection seismology, including the following:

- Apparent velocity.
- Average velocity.

- Root-mean-square (RMS) velocity.
- Normal move-out (NMO) velocity.
- Stacking velocity.

Some of those velocities will be encountered in the coming few sections of this chapter.

2.7 Propagation Effects on Seismic Waves Amplitudes

When considering the propagation of seismic waves, it was assumed in previous sections that Earth rocks are perfectly elastic. However, to some extent, real rocks are imperfectly elastic and this directly affects seismic wave amplitudes as seismic waves propagate within the Earth layers. The seismic wave amplitudes simply become attenuated. Also, in reflection seismology, situations may be encountered where the seismic wave velocities are dispersive (velocities are functions of frequency) and, therefore, affect seismic wave amplitudes or its wavetrain shape. In this section, a few of these related, but rather important, effects due to wave propagation within the Earth rocks will be discussed.

2.7.1 Amplitude Attenuation

Attenuation is considered the reduction in amplitude (or energy) caused by the transmitting medium. It usually includes geometric divergence effects, as waves spread out from a source, conversion of energy into heat, and other factors that affect the seismic wave amplitude like mode conversion (compressional waves to shear waves or vice versa). In general, attenuation has three main causes:

1. Transmission Losses: occur at interfaces because of reflection, diffraction, mode conversion, and scattering. Here, there is no loss in terms of motion energy, since the lost energy merely travels somewhere else. When considering reflection, for example, some of the incident waves or motion energy are reflected and part of are transmitted through the interface. Also, compressional wave energy converting into shear wave energy or vice-versa, which is known as *mode conversion*, occurs when a wave arrives at an interface at an obliquely incident angle to that interface (as will be discussed in Section 2.8) (Sheriff and Geldart, 1995).
2. Geometric Divergence: affects wave spread out from a source, as seen in Figure 2.14. Assuming that the medium is homogeneous, the energy of a wave propagating within that medium is proportional to the square of its amplitude (Telford et al., 1990). The seismic energy transmitted outward from the source becomes distributed over a spherical shell of expanding radius, say r. Thus, the energy E contained within a unit area of the shell is:

$$\frac{E}{(4\pi r^2)}.$$

(2.53)

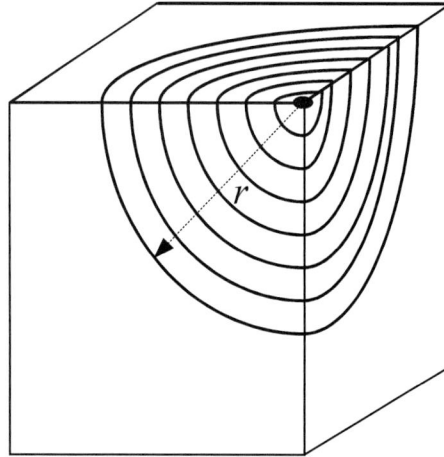

Figure 2.14 Geometric divergence occurs as seismic waves propagate away from the source, where their amplitudes will be inversely proportional to the sphere radius r.

With the increasing distance along a raypath (see Figure 2.14), the energy contained in the ray falls off as $1/r^2$ due to the geometrical (sometimes called spherical) spreading-loss of amplitude because a wave spreads out. Hence, the seismic wave amplitude falls-off by $1/r$ (Sheriff and Geldart, 1995; Telford et al., 1990; Yilmaz, 2001).

3. Absorption: absorption of elastic waves in rocks represents an important problem in seismic exploration. It occurs where kinetic energy, which is the energy of motion, is converted into heat by frictional dissipation. The loss from absorption turns out to be approximately exponential with distance (Sheriff and Geldart, 1995; Yilmaz, 2001).

Both the geometrical divergence and absorption mechanisms of amplitude attenuation can be combined in the following equation:

$$A = A_o \frac{1}{r} e^{-\beta r}, \tag{2.54}$$

where A is the amplitude at radius r from the source, A_o is the amplitude at the source, and β is the absorption coefficient. Note that the quantity $1/r$ indicates geometric divergence. Also, β indicates absorption, where it defines the portion of energy lost during transmission through a distance equivalent to a complete wavelength λ. It depends on the rock material, and common values for Earth material ranges between 0.25 to 0.75 dB/λ. Also, β is related as well to the so-called *Quality Factor Q* by:

$$\beta = \frac{\pi f}{Q v}. \tag{2.55}$$

The quality factor Q can be used to describe the attenuation characteristics of rocks. Note that, from Equation (2.55), waves with higher frequencies tend to be absorbed more than waves with lower frequencies are absorbed.

Example 2.18 Given the following propagating seismic plane wave:

$$u(x,t) = e^{-\beta r} \sin\left[\frac{2\pi}{156}(x - 3,048t)\right],$$ (2.56)

and assume no spherical spreading loss exists. Then compute the amplitude loss of this plane wave when the seismic wave travels a distance of 1,524 m and $Q = 100$. Also, by how much should one multiply the recorded seismic signal in order to restore its original amplitude?

Solution: The given $u(x,t)$ follows from Equation (2.28), which implies that $\lambda = 156$ m and $v = 3,048$ m/s. However, one can see from both Equations (2.55) and (2.25) that:

$$\beta = \frac{\pi}{Q\lambda}.$$ (2.57)

Hence, $\beta = \pi/15,600$ and the loss in the amplitude of the plane wave will, therefore, be:

$$e^{\left[-\frac{\pi}{15,600} \times 1,524\right]} \approx 0.7357.$$

In order now to compensate for the loss, one should multiply the recorded seismic signal by $1/e^{\left[-\frac{\pi}{15,600} \times 1,524\right]} = 1/0.7357 = 1.3593$.

Example 2.19 Consider two seismic waves with frequencies of 10 Hz and 100 Hz to propagate through a rock in which $v_p = 2,000$ m/s and $\beta = 0.5$ dB/λ. By how many dB will each wave be absorbed?

Solution: From Equation (2.25), the wavelength associated with the 10 Hz seismic wave is $\lambda_{10} = 2,000/10 = 200$ m and this wave will be attenuated due to absorption by 0.5 dB. In the case of the 100 Hz seismic wave, $\lambda_{100} = 2,000/100 = 20$ m and this wave will be, in this case, attenuated by 5 dB over a distance equal to $10 \times \lambda_{10} = 200$ m, which is over the same distance as the 10 Hz seismic wave.

From the last example, it can be seen that the shape of a seismic wave amplitude with a broad frequency content changes continuously during propagation due to the progressive loss of higher frequencies. Hence, the effect of absorption is to produce, in general, a progressive lengthening of the seismic waves. To correct the amplitude of the seismic data, data-independent and dependent amplitude correction schemes exist (Mousa and Al-Shuhail, 2011; Yilmaz, 2001), which are important when processing seismic data as seen in Chapter 1. Figure 2.15 shows an example of a real seismic shot gather before and after applying gain correction techniques. Clearly, seismic events became visible for further analysis and processing considerations.

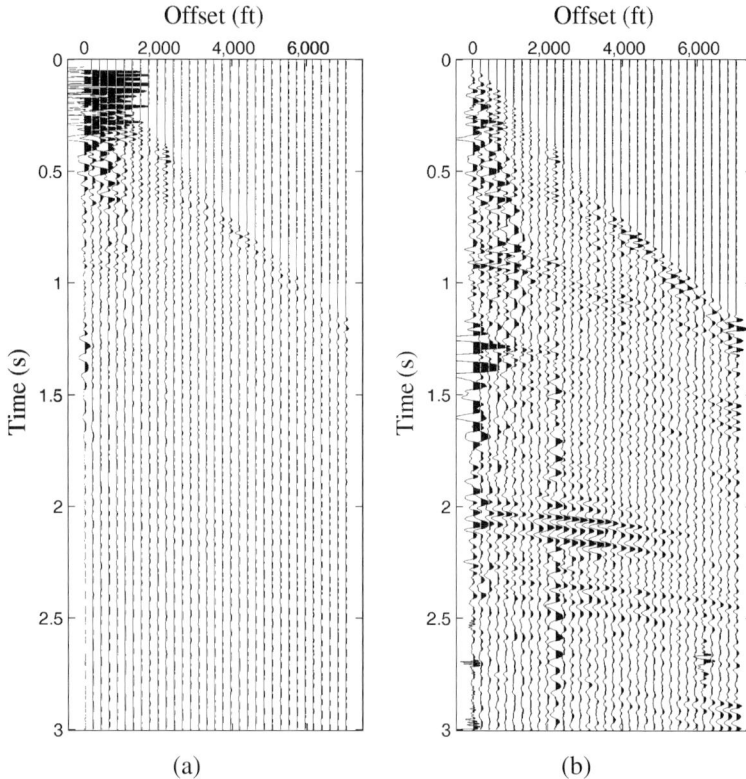

Figure 2.15 A seismic data shot gather (Mousa and Al-Shuhail, 2011) (a) before and (b) after applying gain correction.

2.7.2 Dispersion

Dispersion can be considered the variation of velocity with frequency. It distorts the shape of seismic wave amplitudes where peaks or troughs advance toward the beginning of the wave, as the seismic wave travels in the media. Recall that v is called the phase velocity because it is the distance traveled per unit time by a point of constant phase, such as a peak or trough. The group velocity v_u, on the other hand, is the speed with which a pulse of energy travels. In dispersive media, v and v_u are different. Dispersion is not a dominant feature of exploration seismology because most rocks exhibit little variation of velocity with frequency in the seismic frequency range. However, dispersion is important in connection with surface waves like ground-roll, where they tend to travel with travel velocities (Telford et al., 1990). Figure 2.16 shows an example of a real seismic shot gather that contains dispersive ground-roll waves.

When different frequency components travel at different phase velocities, seismic pulse shape will not remain the same as they travel but they become dispersed

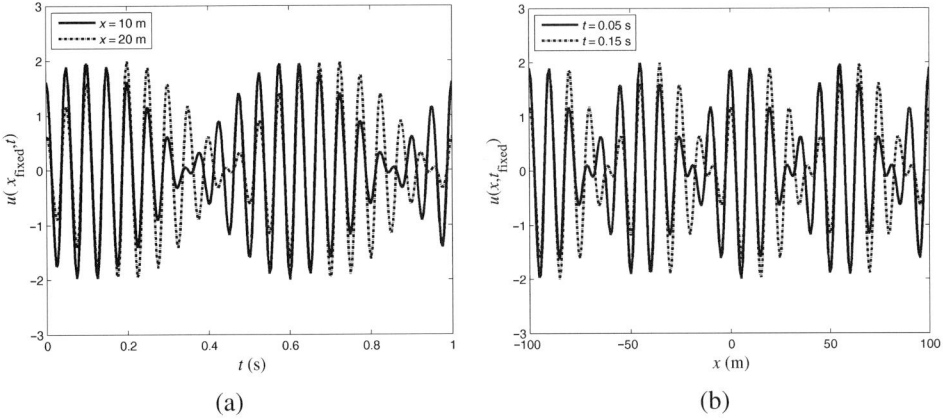

Figure 2.16 An example of a dispersed (modulated) sinusoid: (a) when t is varying, and (b) when x is varying.

as the frequencies separate. Interference will then take part, where it will cancel the wave energy except at times defined by the group velocity of the seismic wave. This might be illustrated by considering the following example. Consider the following disturbance function or wavefield:

$$u(x,t) = \cos(\omega_1 t - k_1 x) + \cos(\omega_2 t - k_2 x), \qquad (2.58)$$

where $u(x,t)$ is the sum of two sinusoids of slightly different frequency and wavenumber values. Relative to an average frequency ω and wavenumber k, one can write:

$$\omega_1 = \omega - \Delta\omega \text{ and } \omega_2 = \omega + \Delta\omega, \qquad (2.59)$$

$$k_1 = k - \Delta k \text{ and } k_2 = k + \Delta k. \qquad (2.60)$$

Substituting Equations (2.59) and (2.60) into Equation (2.58) will result in:

$$
\begin{aligned}
u(x,t) &= \cos[(\omega - \Delta\omega)t - (k - \Delta k)x] + \cos[(\omega + \Delta\omega)t - (k + \Delta k)x] \\
&= \cos[(\omega t - kx) - (\Delta\omega t - \Delta k x)] + \cos[(\omega t - kx) + (\Delta\omega t - \Delta k x)] \\
&= 2\cos(\omega t - kx)\cos(\Delta\omega t - \Delta k x). \qquad (2.61)
\end{aligned}
$$

That is, Equation (2.61), can be seen as a waveform which consists of a signal with the frequency ω whose amplitude is modulated by a carrier signal with frequency $\Delta\omega$ (its amplitude is modulated by a longer period sinusoid carrier). An example is shown in Figure 2.16.

Note that the short period wave $\cos(\omega t - kx)$ travels at velocity ω/k, which is known as the phase velocity v. On the other hand, the longer period envelope of the wave $\cos(\Delta\omega t - \Delta k x)$ travels at velocity $\Delta\omega/\Delta k$ that is defined to be the group velocity v_u. As both $\Delta\omega$ and Δk approach zero in the limit, the group velocity v_u will be:

$$v_u = \frac{d\omega}{dk}.$$ (2.62)

Based on the Equation (2.62) and using the various relationships between harmonic wave parameters that was described in Section 2.3, v_u can be given as:

$$v_u = v + k\frac{dv}{dk},$$ (2.63)

or

$$v_u = v - \lambda\left(\frac{dv}{d\lambda}\right) = v + f\frac{dv}{df}.$$ (2.64)

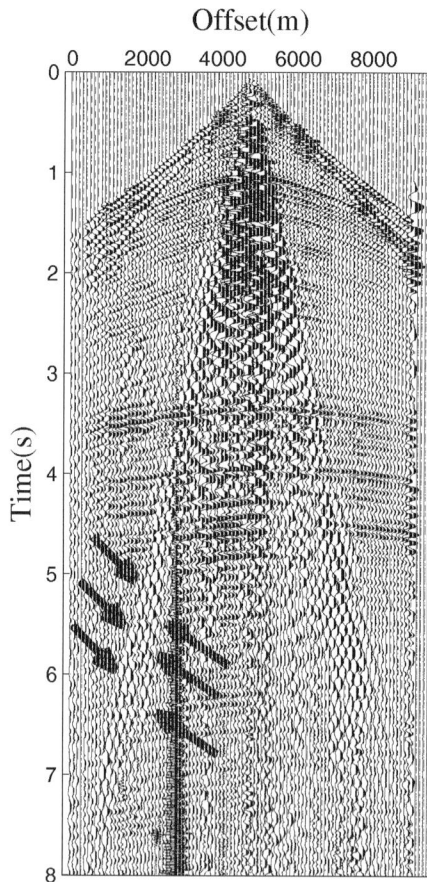

Figure 2.17 A real seismic shot gather example indicating with arrows dispersive surface waves (modified from Yilmaz, 2001).

A real seismic shot gather example indicating dispersive surface waves is shown in Figure 2.17.

2.8 Raypaths in Layered Media

There is generally a change of propagation velocity resulting from the difference in physical properties of two rock layers at the interface between them as illustrated in Figure 2.18. One can define an interface as the boundary where the properties of the Earth media change in terms of particle structure, density, and/or seismic wave velocity, i.e., an interface is a common surface separating two different media in contact. The energy within an incident seismic wave can be partitioned into transmitted and reflected waves at such an interface, where their amplitudes depend on (a) the velocities and densities of the two layers in between the interface, and (b) the angle of incidence on that interface. Hence, several cases for which the seismic wave energy generally interacts with rock interfaces or sometimes media discontinuities will be considered in this section. These include reflection and transmission of normally incident seismic wave rays, reflection and refraction of obliquely incident seismic wave rays, critical refraction, and diffractions.

2.8.1 Reflection and Transmission of Normally Incident Seismic Wave Rays

Consider Figure 2.19, where it assumed that a compressional (wave) ray of amplitude A_0 is normally incident on an interface between two media of differing

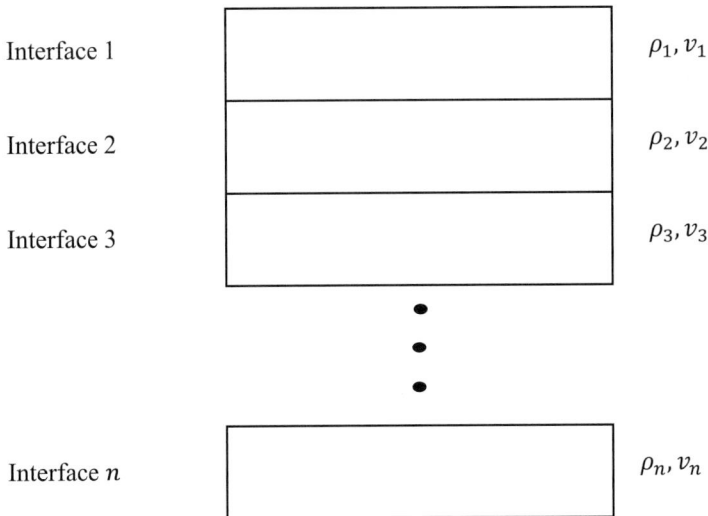

Interface 1 ρ_1, v_1

Interface 2 ρ_2, v_2

Interface 3 ρ_3, v_3

Interface n ρ_n, v_n

Figure 2.18 Raypaths in layered media.

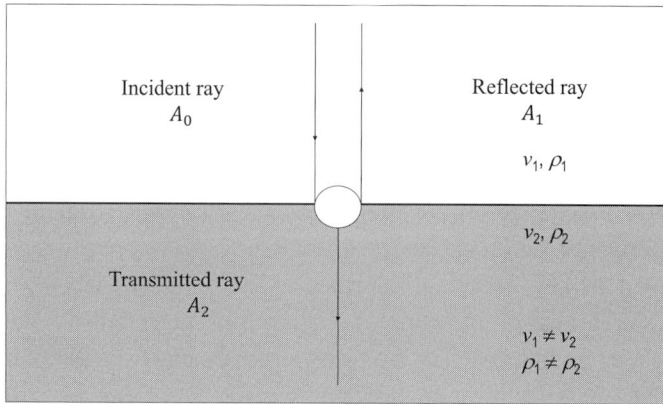

Figure 2.19 Reflection and transmission of normally incident seismic rays for two layers.

velocities and densities, that is, $v_1 \neq v_2$ and $\rho_1 \neq \rho_2$. A transmitted ray of amplitude A_2 travels through the interface in the same direction as the incident seismic ray. At the same time, a reflected seismic ray of amplitude A_1 will return back along the path of the incident ray. The total energy of the transmitted and reflected seismic rays must therefore be equal to the energy of the normally incident seismic ray. The contrast of the so-called *Acoustic Impedance Z* across the interface determines the relative proportion of seismic energy transmitted and reflected. Z is the physical property whose change determines the reflection coefficient (as will be defined later) at normal incidence. For a given rock, it can be given by the product of the rock density ρ and its compressional seismic wave velocity v_p:

$$Z = \rho v_p. \tag{2.65}$$

In the cases of nonnormal incidence, the term elastic impedance may be used, although an equation for it is hardly ever used (Telford et al., 1990). In general, although it is difficult to relate Z to a tangible rock property, the harder a rock is, the higher is its acoustic impedance. Here is a list of points that relates the acoustic impedance Z with the propagating seismic energy:

1. The smaller the contrast in Z across a rock interface, the greater is the proportion of energy transmitted through that interface.
2. All the energy will be transmitted if the given rock material is the same on both sides of the interface.
3. The more reflected seismic energy, the greater the contrast.
4. If one requires maximum transmission of seismic energy then one will definitely require a matching of acoustic impedance between the interfaces.

The acoustic impedance determines the change of the reflection coefficient or reflectivity at normal incidences, where the reflection coefficient can be defined as follows:

Definition 2.20 The reflection coefficient (or reflectivity) denoted by R is a numerical measure of the effect of an interface on wave propagation and is calculated as the ratio of the amplitude of the reflected seismic ray to that of the incident seismic ray.

Considering Figure 2.19, R is given by:

$$R = \frac{A_1}{A_0}. \tag{2.66}$$

This relationship is obtained by solving boundary condition equations that express the continuity of displacement and stress at the boundary (Sheriff and Geldart, 1995). To relate R to physical properties of the material at the interface is a complex problem. The propagation of compressional waves, for example, depends on the elastic moduli and the material density. At the boundary, the stress and strain in the two material must be considered, where the relations between stress and strain in different materials become different and the orientation of stress and strain to the interface is also important. However, the relationship of physical properties of rocks to the reflection coefficient R can be described using Zoepprits equation (Sheriff and Geldart, 1995) for a normally incident seismic ray:

$$R = \frac{\rho_2 v_2 - \rho_1 v_1}{\rho_2 v_2 + \rho_1 v_1} = \frac{Z_2 - Z_1}{Z_2 + Z_1}, \tag{2.67}$$

where Z_i, ρ_i, and v_i are, respectively, the acoustic impedance, density, and compressional wave velocity of layer i. From Equation (2.67), it follows that:

$$-1 \le R \le 1. \tag{2.68}$$

A negative value of R implies phase inversion (a phase shift of $180°$ in the reflected seismic ray) and a compressional ray is reflected as a refraction. If $R = 0$, then all the incident wave energy is transmitted where this occurs and there is no contrast of Z across an interface even when $v_1 \ne v_2$ and $\rho_1 \ne \rho_2$ of two layers and $Z_1 = Z_2$. There exist situations where $R = 1$ or -1 and, in this case, all the energy is reflected.

Example 2.21 A seismic marine acquisition air gun source triggers a seismic pulse within the water layer. Determine the reflection coefficient R between the water and air medium and comment on the resultant value of R.

Solution: If one calculates R based on the shown velocity and density parameters based on Table 2.1, then the following R will be obtained:

$$R = \frac{3.8 \times 331 - 1{,}000 \times 1{,}500}{3.8 \times 331 + 1{,}000 \times 1{,}500} = \frac{-1{,}498{,}742.2}{1{,}501{,}257.8} \approx -0.9983. \tag{2.69}$$

Hence, seismic rays traveling upward from an explosion in a water layer are almost totally reflected back from the water surface with a phase change of $180°$ and $R \approx -1$.

Typical values of R for interfaces between rocks are less than ±0.2 and rarely exceed ±0.5 (Sheriff and Geldart, 1995). This implies that most of the seismic energy (wave) incident on a rock interface is transmitted and only a small portion is reflected. That may be one good reason for the recorded seismic reflected waves to be weak. Note that for less laterally heterogeneous media, it is possible to estimate the reflection coefficient R from the velocity information by (Kearey et al., 2002):

$$R = 0.625 \ln\left(\frac{v_1}{v_2}\right). \tag{2.70}$$

Example 2.22 If $R = 0.2$ and v_1 was $1{,}500$ m/s, then determine the value of v_2, assuming that the encountered two layers are homogenous?

Solution: Following to Equation (2.70), one can express v_2 in terms of v_1 and R, where:

$$v_2 = v_1 e^{-R/0.625} = 1{,}500 e^{-0.2/0.625} = 4{,}687.5 \text{ m/s.}$$

On the other hand, the transmission coefficient T can be defined as the ratio of the amplitude of the transmitted seismic ray to that of the incident ray. Considering Figure 2.19, T can be found using:

$$T = \frac{A_2}{A_0}, \tag{2.71}$$

and, for normally incident seismic rays, it can be calculated using the following Zoepprits equation (Telford et al., 1990):

$$T = \frac{2Z_1}{Z_1 + Z_2}. \tag{2.72}$$

Assuming no internal losses, the downgoing (source) and upgoing (receiver) incident waves for N interfaces t_N are presented in Figure 2.20. The downgoing and upgoing transmission factor can be given by (Robinson and Treitel, 2000):

$$T_{N,d} = (1 + R_1)(1 + R_2)\cdots(1 + R_N), \tag{2.73}$$

$$T_{N,u} = (1 - R_1)(1 - R_2)\cdots(1 - R_N). \tag{2.74}$$

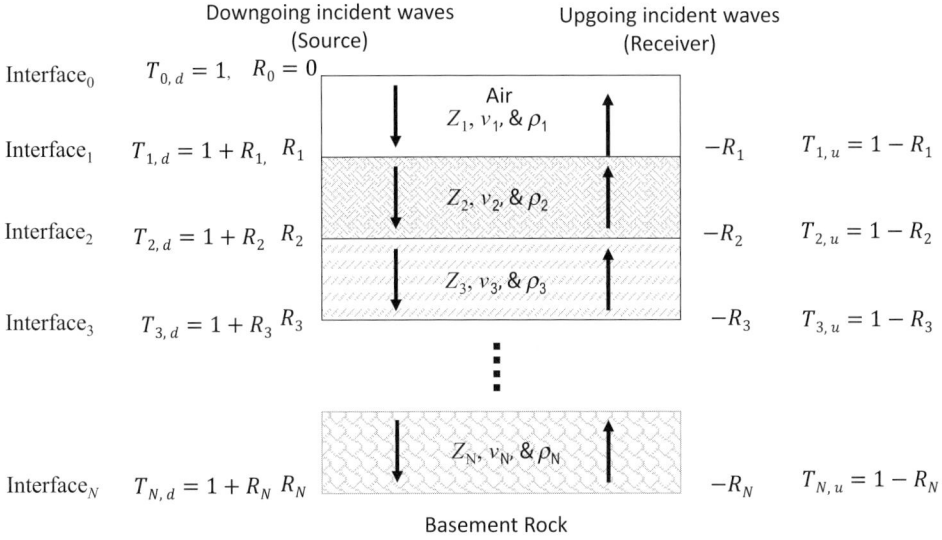

Figure 2.20 A system of N layers with no internal losses (modified after Robinson and Treitel, 2000).

Hence, the two-way transmission factor:

$$T_N = (1 - R_1^2)(1 - R_2^2) \cdots (1 - R_N^2). \tag{2.75}$$

Therefore, the amplitude of the seismic wave reflected from interface i is given by (assuming a unity amplitude at the source):

$$A_i = (1 - R_1^2)(1 - R_2^2) \cdots (1 - R_{i-1}^2) R_i,$$

The amplitude of the seismic wave reflected from interface i is given by (assuming an amplitude A_0 at the source):

$$A_i = A_0(1 - R_1^2)(1 - R_2^2) \cdots (1 - R_{i-1}^2) R_i.$$

Example 2.23 Compute the reflection coefficients for the model shown in Figure 2.20 with $N = 3$, where $\rho_1 = 1.2 \text{ kg/m}^3$, $\rho_2 = 1,000 \text{ kg/m}^3$, and $\rho_3 = 2,400 \text{ kg/m}^3$. Also, $v_1 = 360$ m/s, $v_2 = 1,500$ m/s, and $v_3 = 4,300$ m/s. Assuming a normally incident seismic wave with amplitude 10 at layer 1. Calculate the amplitude for each interface for the model shown.

Solution: First, one need to compute R_1 and R_2 as follows:

$$R_1 = \frac{\rho_2 v_2 - \rho_1 v_1}{\rho_2 v_2 + \rho_1 v_1} = \frac{1,000 \times 1,500 - 1.2 \times 360}{1,000 \times 1,500 + 1.2 \times 360} = 0.994,$$

$$R_2 = \frac{\rho_3 v_3 - \rho_2 v_2}{\rho_3 v_3 + \rho_2 v_2} = \frac{2,400 \times 4,300 - 1,000 \times 1,500}{2,400 \times 4,300 + 1,000 \times 1,500} = 0.7462.$$

Hence, the amplitude at interface 1 and 2 is, respectively, given as follows:

$$A_1 = A_0 R_1 = 10 \times 0.994 = 9.994,$$

$$A_2 = A_0(1 - R_1^2)R_2 = 10(1 - 0.994^2)0.7462 = 8.951 \times 10^{-3}.$$

2.8.2 *Reflection and Refraction of Obliquely Incident Rays*

Consider a more general case, where instead of having a normally incident seismic ray, one will have an obliquely incident seismic ray as shown in Figure 2.21. When a compressional (or even shear) ray is obliquely incident on an interface of acoustic impedance contrast, reflected and transmitted compressional (shear) seismic rays are generated similar to the normally incident seismic rays case. In this case, the transmitted ray is considered as a refracted compressional (shear) wave, since the direction of seismic wave propagation is changed. This situation is directly analogous to the behavior of a light ray that is obliquely incident on the boundary between, for example, air and water. Additionally, the so-called mode conversion will occur where reflected and, in this case, refracted shear (compressional) rays will be generated. Mode conversion is normally significant for large angles of incidence (with respect to the normal axis – see Figure 2.21) beyond those that are obtained in reflection seismology.

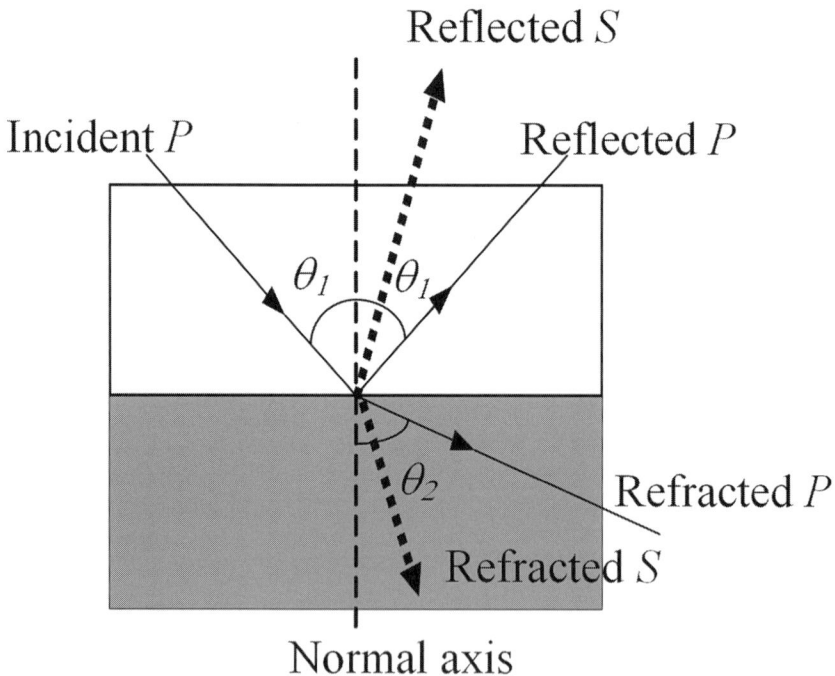

Figure 2.21 Reflection and refraction of obliquely incident rays.

Here, Snell's law of refraction can be applied to the case of seismic. Snell defines the ray parameter p as the angle of inclination of the ray θ_i in a layer in which it is traveling with a velocity v (v_p or v_s). That is:

$$p = \frac{\sin \theta_i}{v}. \tag{2.76}$$

The ray parameter remains constant along any one ray according to Snell's law. Hence, for an obliquely incident compressional wave ray shown in Figure 2.21:

$$\frac{\sin \theta_1}{v_1} = \frac{\sin \theta_2}{v_2}. \tag{2.77}$$

From Equation (2.77) and considering Figure 2.21, if the compressional wave velocity of the second layer v_2 is greater than that of the first layer v_1, then the seismic ray is refracted away from the normal axis with respect to the interface. In other words, the angle of the refracted ray θ_2 with respect to the normal axis will be greater than the incident angle θ_1 with respect to the normal axis. Also, for the reflected ray, it follows from Equation (2.77) that its angle is equal to the angle of incidence. In this case, the reflection coefficient can be found using the following Zoepprits equation:

$$R = \frac{\cos \theta_2 \rho_1 v_1 - \cos \theta_1 \rho_2 v_2}{\cos \theta_1 \rho_2 v_2 + \cos \theta_2 \rho_1 v_1}, \tag{2.78}$$

or,

$$R = \frac{\cos \theta_2 Z_1 - \cos \theta_1 Z_2}{\cos \theta_1 Z_2 + \cos \theta_2 Z_1}. \tag{2.79}$$

Example 2.24 If an incident compressional wave travels with a velocity $v_{p1} = 2350$ m/s at $\theta_{p1} = 35°$ with respect to the normal axis. Assume mode conversion occurred. In this case, what are the angles of the transmitted shear waves, i.e., θ_{s2}, when $v_{s2} = 590$ m/s?

Solution: Based on Snell's law of refraction given by Equation (2.77), the angle of the transmitted shear wave θ_{s2} will be:

$$\theta_{s2} = v_{s2} \arcsin \left(\frac{\sin \theta_{p1}}{v_{p1}} \right) = 590 \arcsin \left(\frac{\sin 35°}{2350} \right) \approx 8.25°.$$

Example 2.25 Assuming an obliquely incident seismic compressional wave of $\theta_{p1} = 25°$ from the normal, compute the reflection coefficient R for the model shown in Figure 2.22?

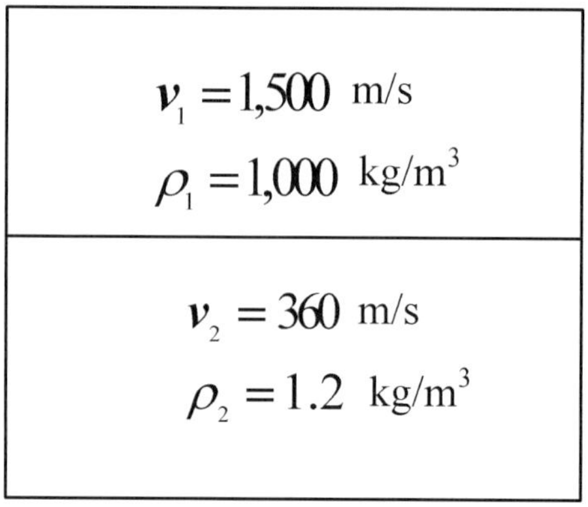

$$v_1 = 1{,}500 \text{ m/s}$$
$$\rho_1 = 1{,}000 \text{ kg/m}^3$$

$$v_2 = 360 \text{ m/s}$$
$$\rho_2 = 1.2 \text{ kg/m}^3$$

Figure 2.22 Model for Example 2.25.

Solution: One needs to first compute θ_{p2} as follows:

$$\theta_{p2} = v_{p2} \arcsin\left(\frac{\sin\theta_{p1}}{v_{p1}}\right) = 360\arcsin\left(\frac{\sin 60°}{1500}\right) \approx 0.0036°.$$

Then, R can be obtained using Equation (2.78):

$$R = \frac{\cos\theta_2\rho_1 v_1 - \cos\theta_1\rho_2 v_2}{\cos\theta_1\rho_2 v_2 + \cos\theta_2\rho_1 v_1},$$
$$= \frac{\cos 0.0036°(1{,}000)(1{,}500) - \cos 60°(1.2)(360)}{\cos 60°(1.2)(360) + \cos\theta_2(1{,}000)(1{,}500)},$$
$$\approx 0.9997,$$

where most of the seismic wave energy will be reflected.

2.8.3 Critical Refraction

In some cases when the seismic wave velocity is higher in the underlaying layer, a particular angle of incidence exists for which the angle of refraction with respect to the normal axis of the interface boundary becomes equal to 90°. That is known as the *critical* angle and is denoted by θ_c. In this case, a critically refracted seismic ray is obtained, which travels along the interface boundary at the higher velocity of the underlaying layer as shown in Figure 2.23. Based on Snell's law (Equation 2.77), θ_c can be found using:

$$\frac{\sin\theta_c}{v_1} = \frac{\sin 90°}{v_2} = \frac{1}{v_2}, \tag{2.80}$$

or

$$\theta_c = \arcsin\left(\frac{v_1}{v_2}\right). \tag{2.81}$$

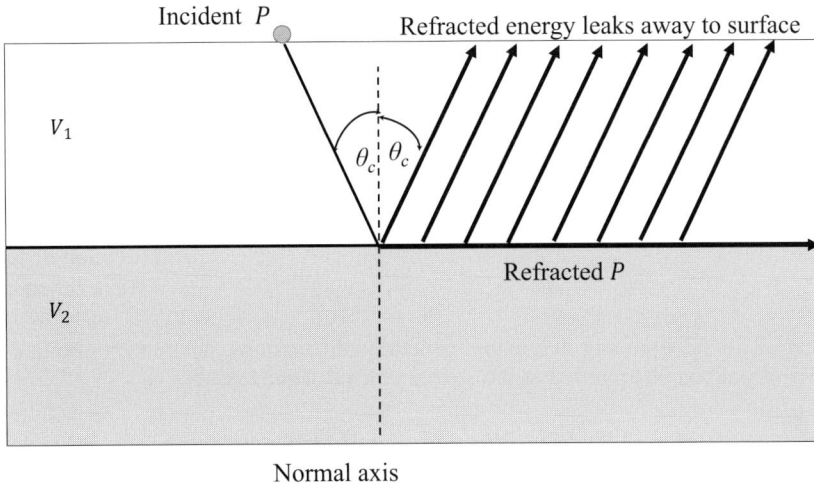

Figure 2.23 Critical refraction occurs when a particular angle of incidence for which the angle of refraction with respect to the normal axis of the interface boundary becomes equal to 90°.

The refracted wave is known as the *interface boundary wave*, the *head wave*, or, simply, *refraction*.

Refractions or head waves are important in near-surface seismic analysis, where the first break velocities can be obtained. They leak a way to the surface when considering exploration seismology and are recorded on the seismic profiles. Also, they can be either compressional or shear waves and tend to decay in their amplitudes rapidly as their energy leaks into the upper or overlaying layer. So as the angle of incidence increases, it will reach the critical angle, and the refracted seismic waves will travel along the interface boundary. However, a total internal reflection of the seismic wave energy (ray) will be obtained and, hence, no incident energy will be transmitted. More description of refractions will be seen in Section 2.10.

Example 2.26 At which angle is an incident seismic wave going to be totally reflected, where $v_1 = 2,000$ m/s and $v_2 = 4,000$ m/s?

Solution: The angle at which an incident seismic wave is going to be totally reflected will be any angle that is greater than the critical angle θ_c. Hence, using Equation (2.81), $\theta_c = \arcsin(0.5) = 30°$. So the total reflection occurs for incident seismic waves with an incident angle $\theta > 30°$.

Example 2.27 The critical angle for air waves that are impinging on water is equal to 12.25°. Determine the compressional velocity of air.

Solution: From Equation (2.81), $v_{air} = v_{water} \sin \theta_c = 1,500 \sin 12.25° \approx 318.27$ m/s.

Envelope defining a wavefront

Envelope
defining new
wavefronts

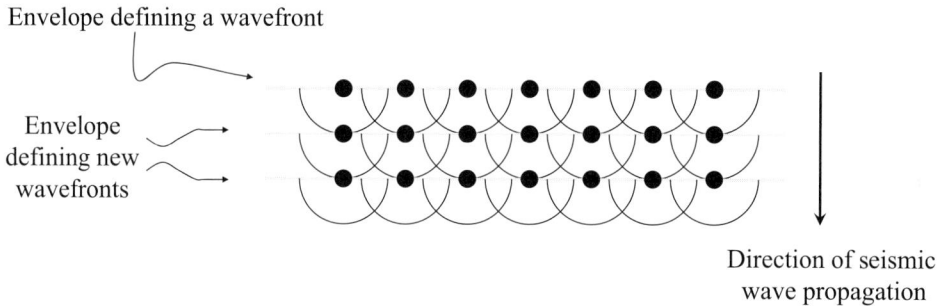

Direction of seismic
wave propagation

Figure 2.24 Huygen's principle and generated diffractions, where every point on a wavefront can be regarded as the source of a subsequent wave.

2.8.4 The Phenomena of Diffraction

In preceding discussions, it was assumed that the interfaces between geological layers are continuous and approximately planar. Hence, reflections and transmissions of seismic energy at such interfaces of acoustic impedance contrasts follow Snell's law. In general, seismic energy travels along other paths besides those given by Snell's law. The ordinary laws of reflection and refraction do not apply anymore whenever a seismic wave encounters a geological discontinuity or a feature whose radius of curvature is shorter than the wavelength of the incident seismic waves. This phenomenon gives rise to a radial scattering of incident seismic energy known as *diffraction*.

Although their laws are complex, diffraction as a process is important because seismic wavelengths are often greater than 100 m, compared with many geological dimensions. At discontinuities greater than several wavelengths from the diffracting source, the diffracted wavefront is essentially described by Huygen's principle[2] (Trorey, 1970). His principle states that every point on a wavefront can be regarded as the source of a subsequent wave. The principle itself is very useful for treating such problems – i.e., when irregular discontinuities exist in a uniform medium, they give rise to diffractions (Sheriff and Geldart, 1995). See Figure 2.24 for illustrating the principle of Huygens. Common sources of diffractions within the subsurface structures may include edges of faulted layers as well was small isolated objects such as boulders within a homogeneous given layer.

2.9 Seismic Events Geometry in Layered Media

In seismic reflection surveys, seismic energy rays can arrive directly to seismic detectors and can be reflected from subsurface interfaces and are recorded by the

[2] Christiaan Huygens was a Dutch mathematician, astronomer, physicist, and horologist who lived between 1629 and 1695 and is remembered for his wave theory of light.

detectors' near-normal incidence at the surface. The travel times are measured and can be converted into estimates of depths to the interfaces. Recall that such seismic surveys are mostly carried out in areas of shallowly dipping sedimentary sequences (layers). In such cases, velocity varies as a function of depth due to the dipping physical properties of the individual layers. Velocity may also vary horizontally due to lateral lithological changes with the individual layers. As a first approximation, the horizontal variations of velocity may be ignored.

2.9.1 Geometry of Direct Raypaths

Assuming that a seismic source triggers seismic energy in an isotropic elastic media with constant velocities v_p and v_s as seen in Figure 2.25a, four wave types are going to be generated, in such a medium, which share a common type of raypaths. The raypaths that they follow are direct, in the form of a sound wave in the air with a velocity, say v_a, a direct compressional wave with a velocity v_p, a direct shear wave with a velocity v_s, and a Rayleigh wave with a velocity v_r. Note that, in the case of marine acquisition, there are neither direct shear waves nor Rayleigh waves, and air waves are rarely recorded on hydrophones due to the types of sources used. Assuming that the spacing Δx between any two consecutive seismic receivers (geophones or hydrophones) is constant, then all the direct rays behave linearly in seismic shot gathers such that the direct wave arrival time or the time–distance curve (for any of the aforementioned four types) is:

$$t(x) = \frac{x}{v},\tag{2.82}$$

where x denotes the source–receiver offset value and v is the corresponding velocity of any of these direct arrival types. If one considers the time slope of any direct arrival, it can be given by:

$$\frac{\Delta x}{\Delta t} = \frac{1}{v},\tag{2.83}$$

which basically infers that slow direct arrivals correspond to steep time–distance curves on a given seismic shot gather. Note that Δt represents the corresponding time difference for which Δx was computed (Figure 2.25b). Figure 2.25c shows four direct arrival curve envelopes of compressional, shear, Rayleigh, and air waves. Note that the wave with the highest velocity should arrive faster and should be recorded first by seismic detectors.

2.9.2 Single Layer Reflector

In Figure 2.26, a simple case of a single layer, known henceforth as a single reflector, lying at a depth z beneath a homogeneous top layer of velocity v (it can be

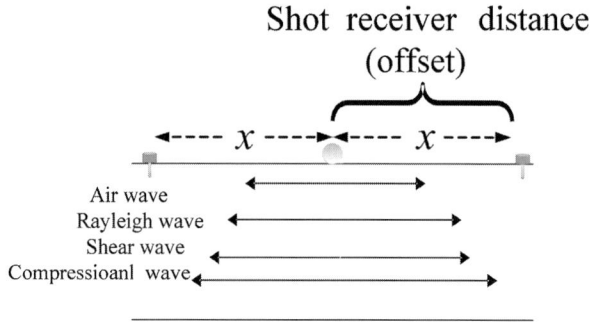

(a)

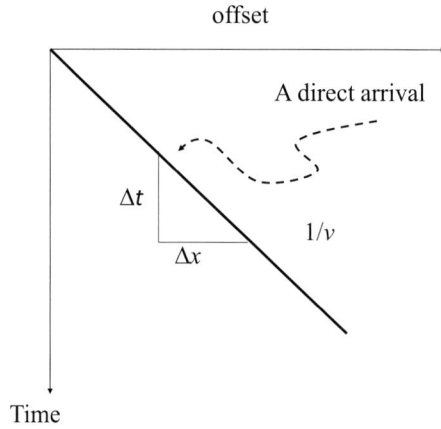

(b)

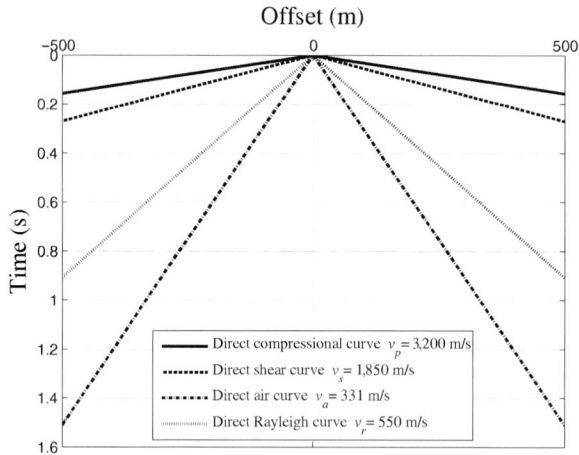

(c)

Figure 2.25 (a) Types and geometry of direct arrivals, (b) their time–distance curve with a slope equal to $1/v$, and (c) seismic direct arrivals time–distance curve envelopes.

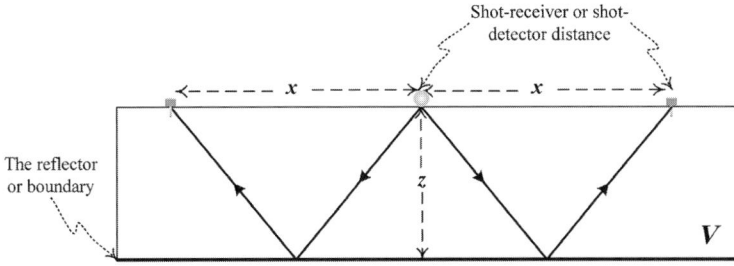

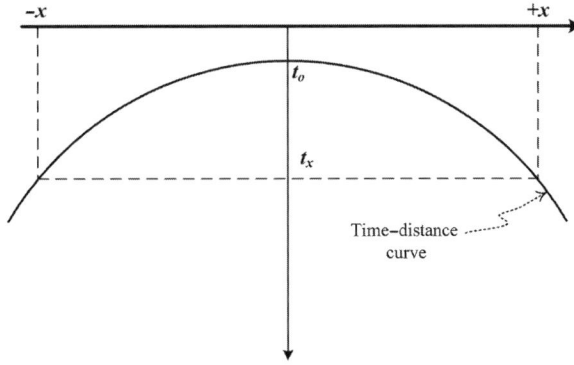

(a)

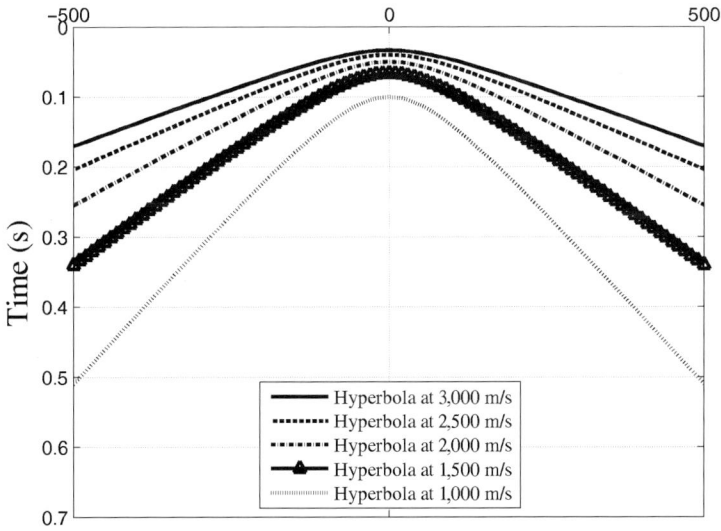

(b)

Figure 2.26 (a) A single layer reflector model along with its time–distance curve. The curve represents the reflection as a hyperbola whose axis of symmetry is the time axis. (b) Various reflections with different velocities for a depth of 50 m.

v_p or v_s depending on the recorded seismic wave type) is presented. The equation for the travel time t of the reflected ray from a seismic source to a detector (e.g., a geophone) at horizontal offset x is given by the ratio of the travel path length to the velocity:

$$t(x) = \frac{\sqrt{x^2 + 4z^2}}{v}. \tag{2.84}$$

The reflection time is measured at an offset distance x. However, it is still required to obtain two other unknowns that are related to the subsurface structure, namely, z and v. If many reflection times t are measured at different offsets x, there will be enough information to solve Equation (2.84) for both of these unknown values. Also, Figure 2.26 shows a graph of the time–distance curve of a reflector, where, on a given shot gather, the reflection will appear as a hyperbola whose axis of symmetry is the time axis. At zero-offset (when substituting $x = 0$ in Equation 2.84), the travel time t_0 of a vertically reflected ray will be:

$$t_0 = t(0) = \frac{2z}{v}, \tag{2.85}$$

which is the intercept on the time-axis of the time–distance curve. Note that Equation (2.84) can be rewritten as:

$$t^2(x) = \frac{4z^2}{v^2} + \frac{x^2}{v^2}. \tag{2.86}$$

This form of the travel time equation suggests the simplest way of determining the velocity:

- If t^2 is plotted versus x^2, the graph will produce a straight line of slope $(1/v^2)$.
- The intercept on the time axis will also give the vertical two-way time, t_0, from which the depth to the reflector can be found.

However, in practice, this method is unsatisfactory since the range of values of x is restricted and the slope of the best-fit straight line has a large uncertainty. A much better method of determining velocity is by considering the increase of reflected travel time with offset distance. This is known as Moveout, which is going to be defined after the following mathematical treatment of Equation (2.84). Hence, rearranging Equation (2.84) yields the following:

$$t(x) = \frac{1}{v}\sqrt{4z^2 + x^2},$$

$$= \frac{1}{v}\sqrt{4z^2\left(1 + \frac{x^2}{4z^2}\right)},$$

$$= \frac{2z}{v}\sqrt{1 + \frac{x^2}{4z^2}}, \tag{2.87}$$

and, therefore, from Equation (2.85):

$$t(x) = t_0 \sqrt{1 + \left(\frac{x}{t_0 v}\right)^2}.$$

(2.88)

This form of the equation is useful. It indicates clearly that the travel time at any offset will be the vertical travel-time plus an additional amount that increases as x increases provided that t_0 and v are constants. Now, by using the standard binomial expression of Equation (2.88), the time–distance relationship can be given by:

$$t(x) = t_0 \left[1 + \frac{1}{2} \left(\frac{x}{v t_0}\right)^2 - \frac{1}{8} \left(\frac{x}{v t_0}\right)^4 + \cdots \right].$$

(2.89)

Note that when considering Equation (2.85), the term $x/(v t_0)$ can be written as $x/(2z)$, in Equation (2.89). Additionally, in the case of small offset/depth ratios, i.e., when $x/z \ll 1$, Equation (2.89) may be truncated after the first term to obtain:

$$t(x) \approx t_0 \left[1 + \frac{1}{2} \left(\frac{x}{v t_0}\right)^2 \right] = t_0 + \frac{1}{2} \frac{x^2}{v^2 t_0}.$$

(2.90)

This is the most convenient form of the time–distance equation for reflected rays, and it is used extensively in the processing and interpretation of reflected data (Yilmaz, 2001).

Definition 2.28 Moveout is defined as the difference between travel times t_1 and t_2 of reflected ray arrivals recorded at two offset distances x_1 and x_2.

Using Equation (2.90) for t_1, x_1 and t_2, x_2:

$$t_1 \approx t_0 + \frac{1}{2} \frac{x_1^2}{v^2 t_0} \text{ and } t_2 \approx t_0 + \frac{1}{2} \frac{x_2^2}{v^2 t_0}.$$

(2.91)

Subtracting these equations, one obtains the time moveout between two different times recorded at two offset distances as:

$$t_2 - t_1 \approx \frac{x_2^2 - x_1^2}{2 v^2 t_0}.$$

(2.92)

Definition 2.29 Normal moveout (NMO) at an offset distance x is the difference in travel time ΔT between reflected arrivals at an offset x and at zero offset. It is calculated using:

$$\Delta T = t_x - t_0 = \frac{x^2}{2 v^2 t_0}.$$

(2.93)

Note that the NMO is a function of offset, velocity, and reflected depth z since $z = v t_0/2$, where its concept is fundamental to the recognition, correlation, and enhancement of reflection events as well as to the calculation of velocities using reflection data. It is used explicitly or implicitly at many stages in the processing and interpretation of reflection data. However, v still needs to be computed. Rearranging the terms of Equation (2.93) yields:

$$v^2 = \frac{x^2}{2\Delta T t_0},$$ (2.94)

which implies that:

$$v = \frac{x}{\sqrt{2\Delta T t_0}}.$$ (2.95)

Using this relationship, the velocity v above the reflector can be computed from the knowledge of the zero-offset reflection time t_0 and the NMO ΔT at a particular offset x. In practice, velocity values are obtained by computer analysis, which produces a statistical estimate based upon many such calculations using large numbers of reflected raypaths. So once the velocity has been derived, it can be used in conjunction with t_0 to compute the depth z to the reflector using $z = vt_0/2$.

Moreover, when Equation (2.84) is now rearranged to be in the following form:

$$t(x) = \frac{1}{v}\sqrt{4z^2 + x^2},$$

$$= \frac{1}{v}\sqrt{x^2\left(1 + \frac{4z^2}{x^2}\right)},$$

$$= \frac{x}{v}\sqrt{1 + 4\left(\frac{z}{x}\right)^2},$$ (2.96)

then when $x \gg z$, this implies that z/x approaches zero. In this case, the time–distance relationship becomes:

$$t(x) \approx \frac{x}{v}.$$ (2.97)

This equation states that at far offset distances, the direct arrivals (whether compressional or shear) and reflections (whether compressional or shear) are approximately equal to each other based on Equation (2.82).

Example 2.30 Assuming a single layer reflector, calculate the NMO for geophones 600, 1,200, and 3,600 m from the source for a reflection at $t_0 = 2.358$ s, given that the compressional velocity = 2,900 m/s. Also, what is the depth z of the layer?

Solution: From Equation (2.93),

$$\Delta T_{600} = \frac{600^2}{2(2,900)^2 \times 2.358} \simeq 0.0091 \text{ s}.$$

$$\Delta T_{1,200} = \frac{1,200^2}{2(2,900)^2 \times 2.358} \simeq 0.0363 \text{ s}.$$

$$\Delta T_{3,600} = \frac{3,600^2}{2(2,900)^2 \times 2.358} \simeq 0.3268 \text{ s}.$$

Similarly, from Equation (2.85)

$$z = \frac{t_0 v}{2} = \frac{2.358 \times 2,900}{2} = 3,419.1 \text{ m.}$$

2.9.3 Single Layer Refractions

Consider the single layer model in Figure 2.27. Now, when a seismic incident wave at a critical angle (θ_c) occurs, a refracted wave will be generated and recorded. One can show that (Sheriff and Geldart, 1995) the time–distance curve of a refracted wave can be given by the following straight line equation:

$$t(x) = \frac{x}{v_2} + \frac{2z \cos \theta_c}{v_1}, \tag{2.98}$$

where z is the layer thickness, and v_1 and v_2 are the velocities of the first and second layers, respectively.

2.9.4 Single Layer Dipping Reflectors

In case of a dipping reflector, the value of the dip θ enters the time–distance equation for reflections as an additional unknown, as seen in Figure 2.28. Similarly to the Equation derived to the horizontal layers, the following equation can be written to describe the time–distance curve for a dipping reflector:

$$t(x) = \frac{\sqrt{x^2 + 4z^2 + 4xz \sin \theta}}{v}. \tag{2.99}$$

This equation still has the form of a hyperbola, as for the case of the horizontal reflector, but the axis of symmetry of the hyperbola is no longer the time axis as seen in Figure 2.28. By the same methodology used for the approximation of Equation (2.84), the following equation is obtained for the dipping reflector case:

$$t(x) \approx t_0 + \frac{1}{2} \frac{x^2 + 4xz \sin \theta}{v^2 t_0}, \tag{2.100}$$

and the moveout related to such a dipping reflector model, called *Dip moveout*, and is defined as follows.

Definition 2.31 Dip moveout (DMO) ΔT_d is defined as the difference in travel times t_x and t_{-x} of rays reflected from the dipping interface to receivers at equal and opposite offsets x and $-x$.

The DMO can, therefore, be calculated using:

$$\Delta T_d = t_x - t_{-x} = \frac{2x \sin \theta}{v}. \tag{2.101}$$

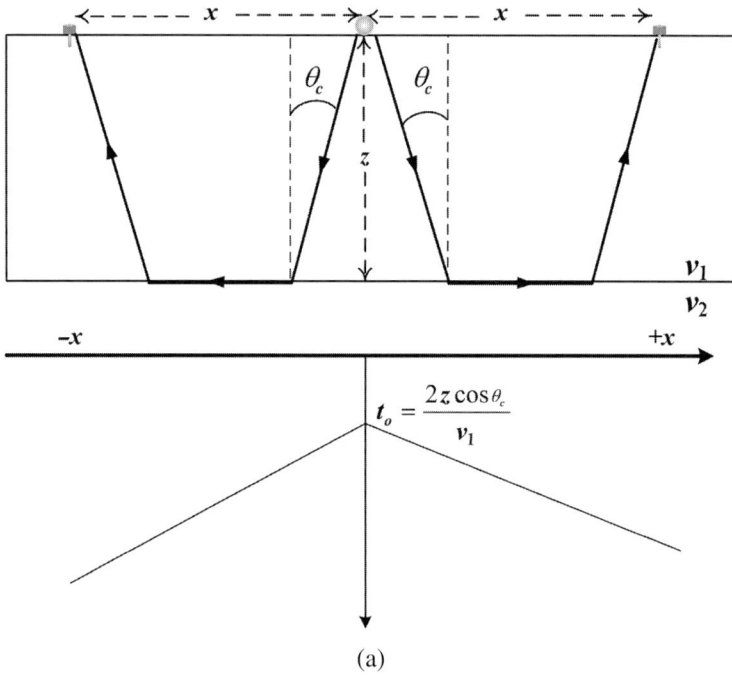

(a)

Offset (m)

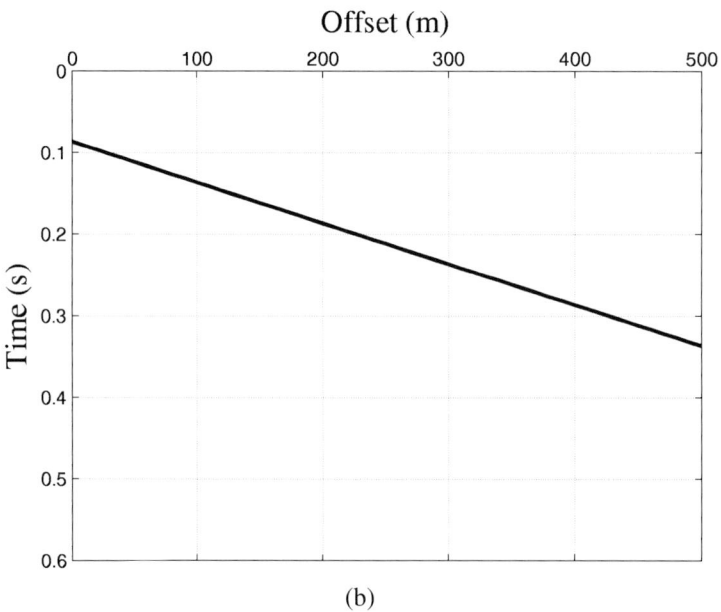

(b)

Figure 2.27 (a) A single layer with a seismic incident wave at a critical angle occurs that generates a refracted wave. (b) The time–distance curve with $\theta_c = 30°$, $v_1 = 1,500$ m/s, $v_2 = 2,000$ m/s, and depth $z = 50$ m.

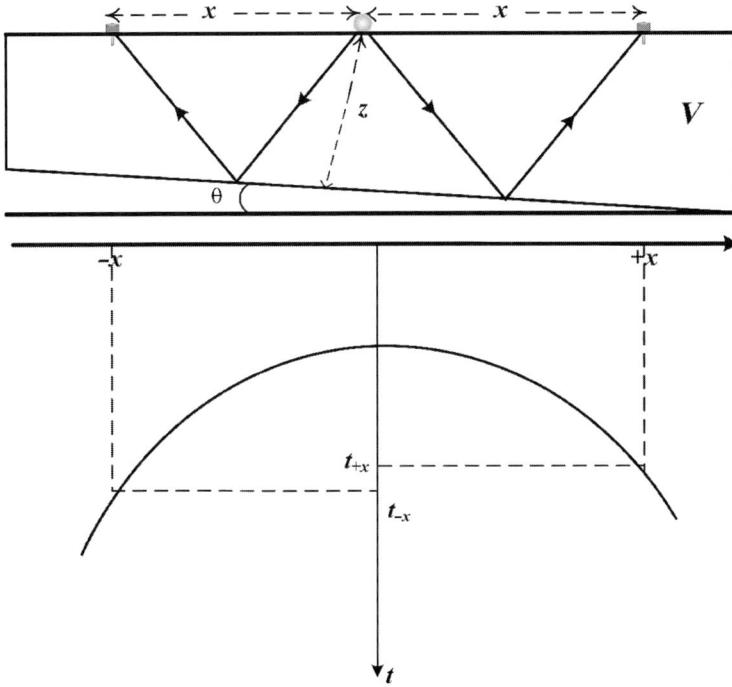

(a)

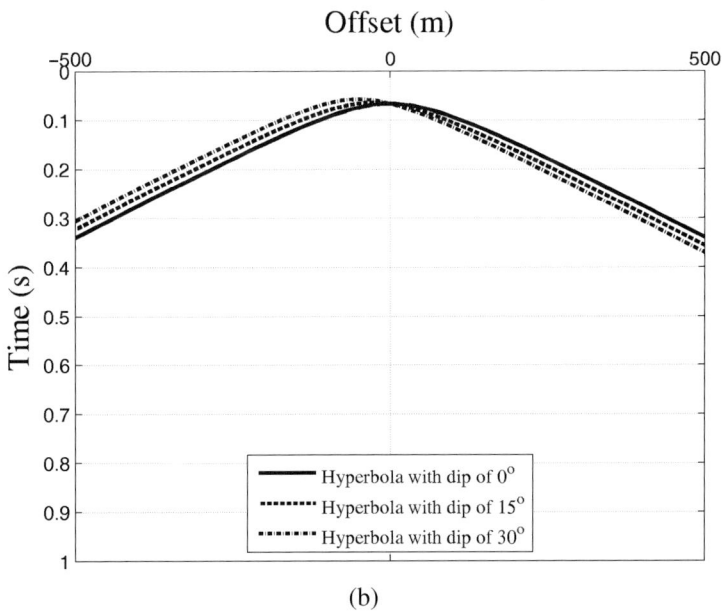

(b)

Figure 2.28 (a) A single dipping reflector with a dip angle of θ. (b) Reflections at $v_p = 1,500$ m/s with various dip angles.

Rearranging the terms and assuming that θ is very small, that is, $\sin\theta \approx \theta$, then:

$$\theta \approx \frac{v\Delta T_d}{2x}. \tag{2.102}$$

Hence, the DMO ΔT_d can be used to compute the reflector dip if v is known. Note that v can be derived via Equation (2.95) using the NMO. For small dips, the NMO may be obtained with sufficient accuracy by averaging the up-dip and down-dip moveouts as follows:

$$\Delta T \approx \frac{t_x - t_{-x} - 2t_0}{2}. \tag{2.103}$$

Note that the same equations can be used for the case of a sequence of dipping reflectors, which will be discussed later in this chapter.

2.9.5 Geometry of Diffractions

The same geometry that was used to describe a reflection from a horizontal reflector interface can be used in the case of wave scattering from a diffraction point. As explained in Section 2.8.4, the diffraction point takes an incoming seismic ray and generates scattered seismic rays in all directions. Based on Figure 2.29, the time–distance equation of a diffracted wave can be written in terms of half the offset value as follows:

$$t(x) = \frac{2}{v}\sqrt{\frac{x^2}{4} + z^2}, \tag{2.104}$$

where a diffraction is going to follow a hyperbola as the offset increases.

2.9.6 Sequences of Horizontal Reflectors

Consider a simple physical model of horizontally layered ground with vertical reflected raypaths from the various layer boundaries. The model here assumes that each layer is characterized by an interval velocity v_i, which may correspond to the uniform velocity within a homogeneous geological unit or the average velocity over a depth interval containing more than one unit. If z_i is the thickness of such an interval and t_i is the one-way travel time of a ray through it, then the interval velocity is given by:

$$v_i = \frac{z_i}{t_i}. \tag{2.105}$$

The interval velocity may be averaged over several depth intervals to yield a Time Average Velocity or simply, Average Velocity V_{avg}. Thus, V_{avg} of the n top layers in Figure 2.18 is given by:

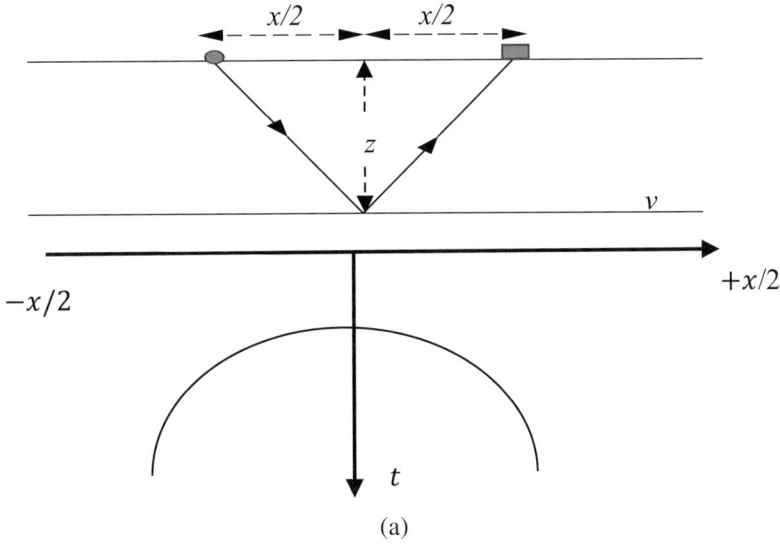

(a)

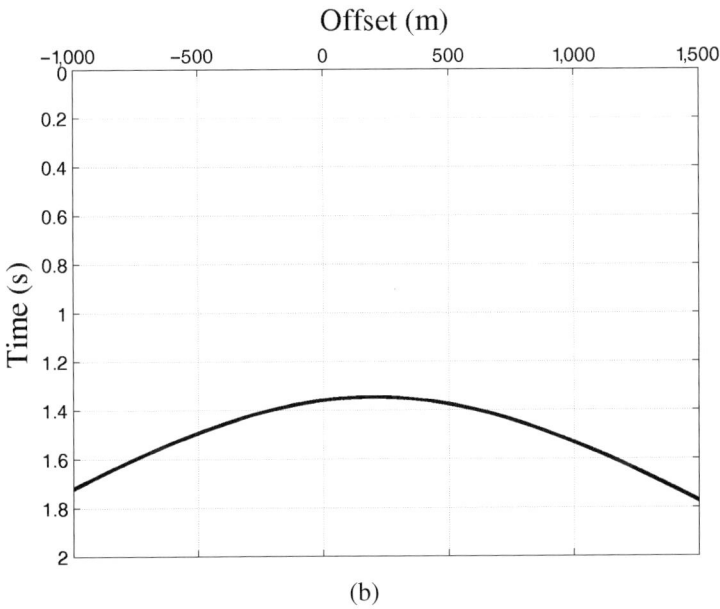

(b)

Figure 2.29 (a) A single layer model of a diffractor. (b) The time–distance curve of a diffraction wave with $v_p = 1,500$ m/s.

$$V_{avg} = \frac{\sum_{i=1}^{n} z_i}{\sum_{i=1}^{n} t_i} = \frac{\sum_{i=1}^{n} v_i t_i}{\sum_{i=1}^{n} t_i} = \sum_{i=1}^{n} v_i, \qquad (2.106)$$

which is basically the sum of the n layer velocities. In a multi-layered ground, inclined rays reflected from the nth interface undergo refraction at all higher interfaces to produce a complex travel path. At offset distances that are small compared to reflector depth ($x/z \ll 1$), the time–distance curve is still essentially hyperbolic but the homogeneous top layer velocity v in Equations (2.84) and (2.93) is replaced by the average velocity V_{avg} or to a considered closer approximation, namely, the Root-Mean-Square velocity V_{RMS} of the layers overlying the reflector. That is:

$$t(x) = \frac{\sqrt{x^2 + 4z^2}}{V_{RMS}}, \qquad (2.107)$$

and the NMO, in this case, is given by:

$$\Delta T = t_x - t_0 = \frac{x^2}{2V_{RMS}^2 t_0}. \qquad (2.108)$$

As the offset increases, the departure of the actual travel time curve from a hyperbolic becomes more marked, as seen in Figure 2.30, where Equation (2.86) becomes:

$$t^2(x) = t_0 + \frac{x^2}{V_{RMS}^2}. \qquad (2.109)$$

Note that the RMS velocity of the section of ground down to the nth interface can be given by:

$$V_{RMS,n} = \left[\frac{\sum_{i=1}^{n} v_i^2 t_i}{\sum_{i=1}^{n} t_i} \right]^{\frac{1}{2}}, \qquad (2.110)$$

where t_i is the one-way travel time of a ray through the ith layer and v_i is the interval velocity of the same ith layer. Thus, at small offsets ($x \ll z$), the total

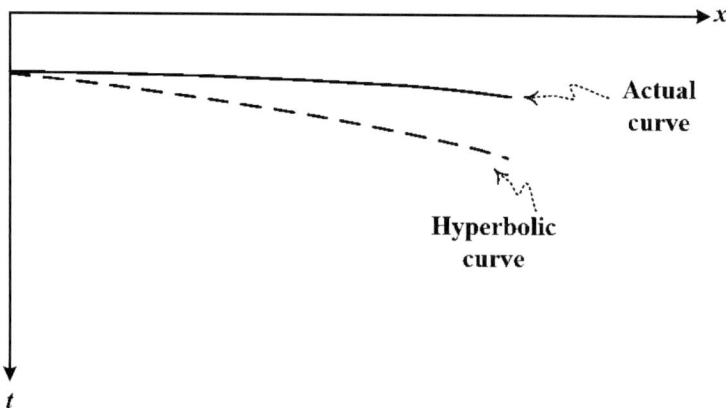

Figure 2.30 The time–distance curve of a sequence of horizontal reflectors.

travel time $t_n(x)$ of the ray reflected from the nth interface at depth z is given to a close approximation by:

$$t_n(x) = \frac{\sqrt{x^2 + 4z^2}}{V_{RMS,n}}. \tag{2.111}$$

On the other hand, at far offsets ($x \gg z$):

$$t_n(x) \approx \frac{x}{V_{RMS,n}}. \tag{2.112}$$

The NMO for the nth reflector is given by:

$$\Delta T_n \approx \frac{x^2}{2V_{RMS,n}^2 t_0}. \tag{2.113}$$

The individual NMO value associated with each reflection event may, therefore, be used to derive the $V_{RMS,n}$ for the layers above the nth reflector. Values of V_{RMS} down to different reflectors can also be used to compute interval velocities, which is the average velocity over some interval of travel path (Sheriff, 2006). Hence the interval velocity v_n for the nth interval is obtained using the so-called Dix Formula (Yilmaz, 2001):

$$v_n = \left[\frac{V_{RMS,n}^2 t_n - V_{RMS,n-1}^2 t_{n-1}}{t_n - t_{n-1}} \right]^{\frac{1}{2}}. \tag{2.114}$$

Example 2.32 Consider the given sequence of horizontal layers with constant interval velocities seen in Figure 2.31, where each layer is of thickness 300 m. Calculate for each of the layers shown its average and root mean-square velocities. Also, ray

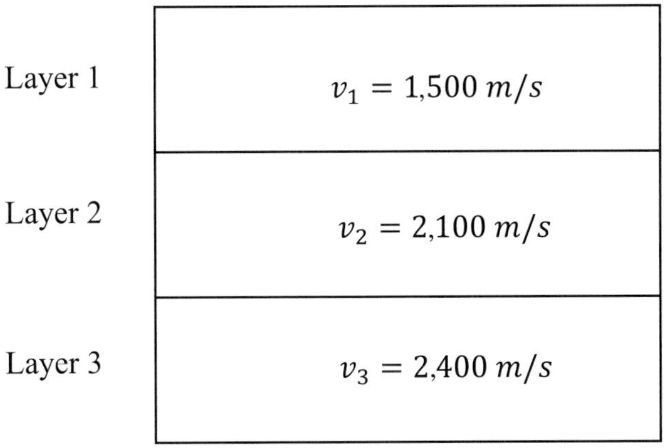

Layer 1	$v_1 = 1{,}500\ m/s$
Layer 2	$v_2 = 2{,}100\ m/s$
Layer 3	$v_3 = 2{,}400\ m/s$

Figure 2.31 Constant interval velocities of three layers for Example 2.32.

trace through the model to determine offset distances and arrival times for seismic rays that make an angle of incidence with the base of layer 2, 400 m/s of 0°.

Solution: Using Equation (2.106), the respective average velocity is calculated as follows:

$$V_{avg1} = 1,500 \text{ m/s, for layer 1}$$

$$V_{avg2} = \frac{1,500 + 2,100}{2} = 1,800 \text{ m/s for layer 2}$$

$$V_{avg3} = \frac{1,500 + 2,100 + 2,400}{3} = 2,000 \text{ m/s for layer 3.}$$

For computing the RMS velocity for each layer, first Equation (2.105) is needed to compute the one-way travel time for each layer. Then, using Equation (2.110), the RMS values for each layer are shown in the following table:

Layer	t_i (s)	V_{rms_i} (m/s)
1	$\frac{300}{1,500} = 0.2$	1,500
2	$\frac{300}{2,100} = 0.143$	$\sqrt{(\frac{1,500^2 \times 0.2 + 2,100^2 \times 0.143}{0.2+0.143})} = 1,775$
3	$\frac{300}{2,400} = 0.125$	$\sqrt{(\frac{1,500^2 \times 0.2 + 2,100^2 \times 0.143 + 2,400^2 \times 0.125}{0.2+0.143+0.125})} = 1,965.5$

For ray tracing, one has normal incidence of seismic rays and, hence, the following table offset distances and arrival times of respective layers:

Layer	Two-Way Travel Time (s)	Z (m)	x (m)
1	$2 \times 0.2 = 0.4$	300	0
2	$2 \times (0.2 + 0.143) = 0.686$	600	0
3	$2 \times (0.2 + 0.143 + 0.125) = 0.936$	900	0

Example 2.33 Assuming a normally incident seismic ray travels in medium with the specification shown in the following table:

Layer	Depth (m)	Two-Way Time (s)	Velocity (m/s)	Density (kg/m³)
1	1,000	1.0	2,000	2,100
2	2,500	2.0	3,000	2,500
3	4,800	3.1	4,200	2,900

If the normally incident travel in layer 1 is of amplitude equal to 15, then it can be shown that average and RMS velocities as well as the layer impedence are equal to the values given in the following table:

Layer	Average Velocity (m/s)	RMS Velocity (m/s)	Impedance kg/(m²s)
1	2,000	2,000	4.2×10^6
2	2,500	2,549.51	7.5×10^6
3	3,066.7	3,233.1	12.18×10^6

Also, the reflection and transmission coefficients as well as the reflected and transmitted ray amplitudes are calculated as in the following table:

Interface	Reflection (m) Coefficient	Transmission Coefficient	Reflected Ray Amplitude	Transmitted Ray Amplitude
1	0.282	0.718	4.23	10.77
2	0.238	0.762	2.56	8.21

At this stage, the reader hopefully is in a better position to identify seismic events and accompanying noise, as shall be described in the next section, based on the previously presented sections of this chapter.

2.10 Characteristics of Seismic Events and Accompanying Noise

The question, probably, that should come to the reader's mind at this stage is: Given a seismic exploration shot record, what are the main characteristics and information (signal and noise) that can be seen on such a record? By definition, seismic events are considered to be arrivals of new seismic waves that are usually ascertained by phase changes in addition to an increase in amplitudes noticed on a given seismic record such as a shot gather (Sheriff and Geldart, 1995). In the case of seismic reflection surveying, they can be in the form of useful energy (or signals), in the form of reflections, or in the form of, generally, unwanted energy (or maybe noise) like refractions, diffractions, surface waves, multiples, direct waves, random signals, and so on. The basic task of interpreting seismic reflection data sections is that of selecting those events on the seismic record that represent:

- Primary reflections.
- Translating their travel times into depths and dips.
- Mapping the reflectors.

The other types of seismic events are also important to consider since some of them may yield valuable information to the interpreter or, depending on their characteristics, one can attenuate them when necessary. According to Sheriff and Geldart (1995), seismic events can be recognized and identified based upon the following characteristics:

1. Coherence: this basically is a measurement of the similarity among more than two functions, i.e., the similarity in appearance from one trace to another. Seismic reflection events are coherent in a linear way with respect to a dip or linear seismic ray energy such as direct arrivals. They are coherent in a hyperbolic way with respect to normal moveout. Thus, it is considered to be a necessary condition to recognize any seismic event.
2. Amplitude standout: refers to an increase of amplitude such as results from the arrival of coherent energy, by which the amplitude of a given event exceeds the mean amplitude. If quality control is performed on seismic traces like gain correction, amplitude standout will not be noticed.
3. Character: refers to the recognizable aspect of a seismic event or the waveform, which distinguishes it from other events. It involves:

   - The shape of the envelope.
   - The number of cycles showing amplitude standout.
   - Irregularities in phase that result from signal interfaces among seismic event components.

4. Moveout: is defined as the difference in arrival times at different seismic sensor positions, where (as previously discussed in Section 2.9) one may have arrival time differences because of source-to-detector distance differences (NMO) and/or due to the existence of dipping reflectors (DMO). A final cause is due to the elevation and weathering of the near-surface layer, in general, and is called *statics*.

For more information on this, the reader is recommended to see Sheriff and Geldart (1995). Note that coherence and amplitude standout can both be used to determine whether or not a strong seismic event is present without knowing its type. On the other hand, the most distinctive criterion that can be used to identify and recognize the nature of seismic events is the moveout. In the following subsections, typical seismic events are presented and classified in to linear, hyperbolic, and noise. This classification, hopefully, will enable readers in recognizing, in a simple way, seismic events.

2.10.1 Linear Events

Direct Arrivals

Direct arrivals (sometimes called first arrivals) are considered to be the first arrivals seen on seismic records. They travel directly from the source to the receivers. They are usually seen in the time–distance curve as straight lines with an intercept at $t = 0$, except when near-surface geological anomalies exist; the time–distance direct arrivals curves tend to appear piece-wise linear. Figure 2.32 shows a real

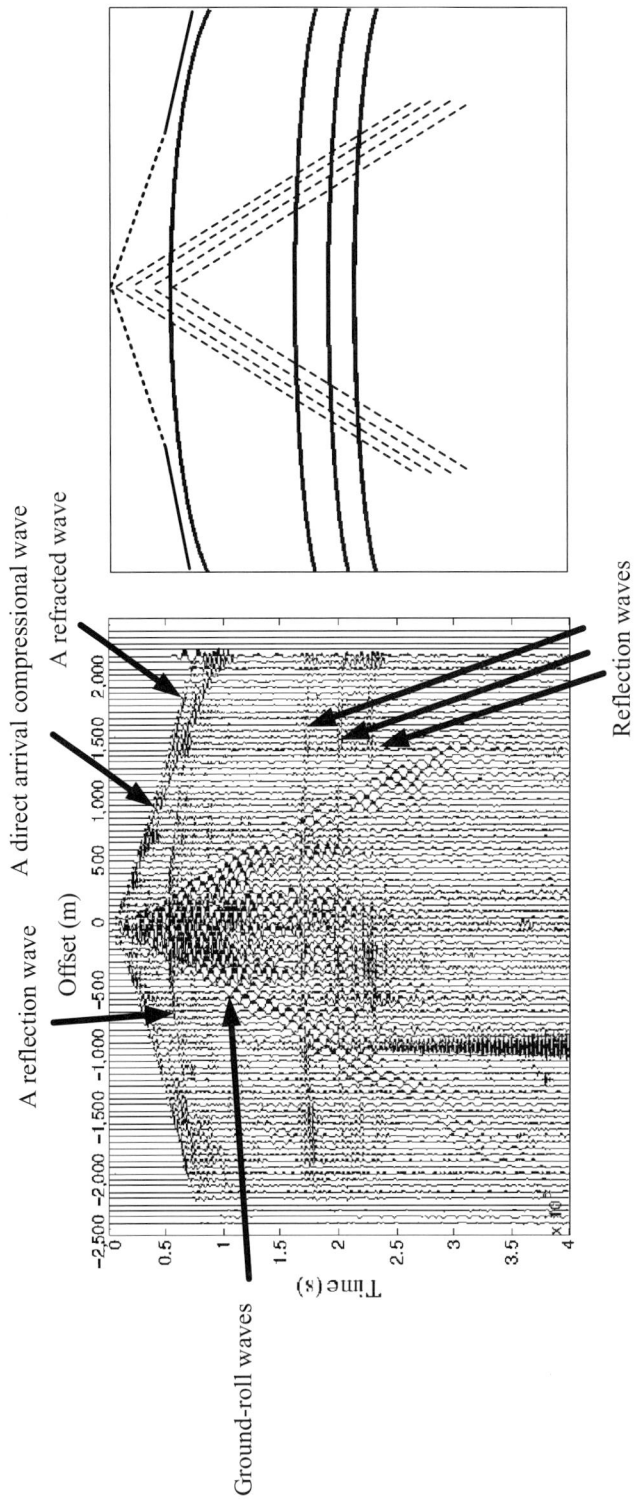

Figure 2.32 A land seismic shot record modified from Yilmaz (2001) containing linear (direct, refractions and surface waves), hyperbolic (reflections) events, as well as other unwanted energy.

example of seismic records containing direct arrivals. The picking of direct arrival travel times is among the important steps for near-surface analysis and modeling (Yilmaz, 2001). Many processing techniques exist for picking direct arrivals, where some are manual, semi-automatic, and automatic techniques, and others are trace by trace based or multi-trace based methods (Boschetti et al., 1996; Coppens, 1985; Criss et al., 2003; Hart, 1996; Keho and Zhu, 2009; Molyneux and Schmitt, 1999; Mousa and Al-Shuhail, 2012; Mousa et al., 2011; Murat and Rudman, 1992; Sabbione and Velis, 2010; Wong et al., 2009; Zhang et al., 2003).

Refractions

Refractions, sometimes called headwaves, are critically refracted waves (see Figure 2.32). They are events of relatively low frequencies. Also, they usually oscillate for more cycles and possess much smaller apparent velocities than reflections. Their time–distance curves are straight lines with intercept at $t \neq 0$ and are attenuated by stacking. Refractions involve long wavetrain of many cycles, where, as the offset distance increases, the number of cycles increases, resulting in shifting the peak energy to later in the wavetrain. This phenomenon is known as *Shingling*. This phenomena becomes more pronounced when the refractor is limited in its thickness and causes loss of visibility of early cycles with the increase of the range (Sheriff and Geldart, 1995). In general, refractions can be used for near-surface modeling that benefits engineering studies and determining static corrections for the processing of seismic reflection data sets (Telford et al., 1990).

Surface Waves and Air Waves

Surface waves travel along the ground surface (termed ground-roll) and are usually Rayleigh waves with (low) velocities ranging from 100–1000 m/s. Their frequencies are usually lower than reflections and refractions, often with energy concentrated below 10 Hz. Their time–distance $t - x$ curves are straight lines with low velocities and zero intercepts for an in-line source. Surface wave envelopes build up and decay very slowly and often include many cycles. Their energy is high enough even in the reflection band to override all except for the strongest reflections. There might be several modes of ground-rolls in the record because of their dispersive nature (i.e., different frequency components travel with different velocities). Figure 2.32 shows an example of ground-roll waves. They are attenuated using source and receiver arrays in the field (Telford et al., 1990) and various processing methods (e.g., frequency filtering, fan or $f-k$ filtering) (Ansari, 1987; Lu and Antoniou, 1992; Yilmaz, 2001). On the other hand, air waves travel along the ground surface and usually have very low velocities around 331 m/s, depending on the air temperature. They are linear with the steepest slope one can see on a seismic shot gather compared to other linear events. They can also be attenuated via various processing methods (e.g., fan or $f-k$ filtering) (Ansari, 1987; Lu and Antoniou, 1992; Yilmaz, 2001).

2.10.2 Hyperbolic Events

Reflections

Reflections, which are the most important part of seismic energy, usually result from the interface between reflection components from a closely spaced series of interfaces, and their signature depends on the spacing and magnitude of the individual acoustic impedance contrasts (Telford et al., 1990). They are generated by waves that are reflected once from an interface. Their time–distance curve are hyperbolic in nature, i.e., they have curved alignments. Additionally, they exhibit normal moveouts that must fall within certain limits set by the velocity distribution. Their apparent velocities, defined as the distance between two receivers divided by the difference in travel time, are very large ($\geq 5,000$ m/s), and they rarely involve more than two or three cycles. Reflections are often rich in frequency components in the range 15–50 (or 60) Hz except for shallow high frequency reflections and very deep reflections (Sheriff and Geldart, 1995). Usually, reflections are fairly short events with little ringing. Once reflections are identified and properly processed, they provide geophysicists with rich information about geological structures. Hence, after seismic imaging, it will be easier for the interpreters to identify possible locations for oil/gas traps such as faults or anticline structures. Figure 2.32 shows seismic records with identified reflections.

Diffractions

Diffraction events are caused by abrupt lateral changes in lithology. If one considers an irregularity, this feature acts as a point source and radiates waves in all directions, in accordance with Huygens' principle (Yilmaz, 2001). Similar to reflections, their time–distance curves are hyperbolic. They are indistinguishable from reflections on the basis of character. However, their amplitudes fall off rapidly away from their apex and they usually exhibit distinctive moveout as seen in Figure 2.33. Finally, unlike reflections, diffractions usually are not centered around the time axis. Some seismic data processing techniques exist to collapse diffractions to points such as seismic migration (Clearbout, 1985; Scales, 1997; Yilmaz, 2001), and data mapping to attenuate near-surface diffractions (Al-Lehyani, 2008).

Multiples

If seismic energy is reflected more than once in its path to the detector then a multiple reflection, or simply a multiple, is produced. Multiples arise when there are interfaces with large reflection coefficients, i.e., where there are large velocity and/or density changes. For example, sea bed and sea surface on marine, and a

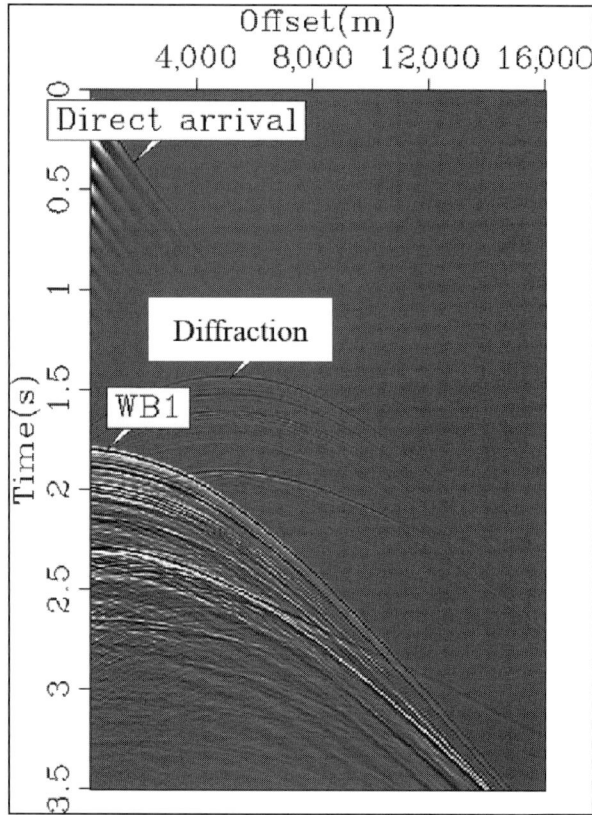

Figure 2.33 A shot record containing an example of a diffraction (from Yilmaz, 2001).

weathering layer on land. They possess curved alignment and usually exhibit more normal moveout than primary reflections with the same travel time, since their velocity generally increases with depth. However, the difference in normal moveout often is not large enough to identify multiples. Multiple reflections give rise to false seismic events or horizons that are recorded and displayed along with the primary (true) events. Note that the arrival time of the false event will be twice that of the true event. The event will appear on the shot record at a later time. However, because it traveled in the upper layer at low velocity, it's curvature will be greater. They can be classified into:

1. Long-path multiples: refers to those multiples whose travel path are long compared with primary reflections from the same deep interfaces, where they appear as separate events on seismic records (see Figure 2.34) and are easy to identify and attenuate. At zero offset, the multiples are periodic in time, as seen in Figure 2.35.

2. Short-path multiples: arrive so soon after the associated primary reflections that they interfere with and add tails to primary reflections. Hence, their effect is that of changing wave shape rather than producing separate events (see Figure 2.36). Short-path multiples come in many forms:

   • Peg-leg: they have reflected successfully from the top and the base of thin reflectors. They alternate the seismic wave shape by delaying part of the energy and, therefore, lengthening the seismic wavelet. They mostly tend to have the same polarity as primary reflections, which lowers the signal frequency as the time increases.

   • Ghosts: are seismic energy that travels upward from an energy release and then is reflected downward, such as those that occur at the base of the weathering layer in land seismic or at the water surface in marine seismic data. The interference between the ghost and primary reflection depends on the fraction of a wavelength represented by the difference in effective path length. This interference effect will vary for different frequency components since the seismic wavelet is made up of a range of frequencies.

   • Water reverberations (see Figure 2.35): are practical troublesome types of short-path multiples. They are frequently seen on marine seismic data but may also be seen on land seismic data. In the case of marine seismic data, a water reverberation occurs due to the multiple reflections in the water layer. The layer reflection coefficient at the top and bottom of this layer result in considerable energy being reflected back and forth repeatedly. This energy is thus reinforced periodically by reflected energy.

Various signal processing techniques exist to attenuate the effect of multiples (Yilmaz, 2001). A standard method is the use of deconvolution and linear prediction filtering based on Wiener's least square filtering (as will be explained in Chapters 6 and 7) (Robinson, 1967). Many variations of Wiener least square filtering followed after the work of Robinson (1967). Examples can be seen in various publications (Peacock and Treitel, 1969; Porsani and rn Ursin, 2007; Robinson, 1998; Sanchis and Hanssen, 2011; Yilmaz, 2001). Other techniques of multiple attenuation include processing them in the so-called $\tau - p$ transform domain, which will be presented in Chapter 4 (Landa et al., 1999; Yilmaz, 2001).

2.10.3 Noise

Noise in seismic records is variable in both time and space (Miao and Cheadle, 1998). Poor seismic records usually have SNR ratios less than one. One can define the signal of interest (coherent energy) as the energy that is coherent from trace to trace. Random noise, on the other hand, is the energy that is incoherent from trace to trace (Ulrych et al., 1999). Furthermore, data from seismic events are correlated

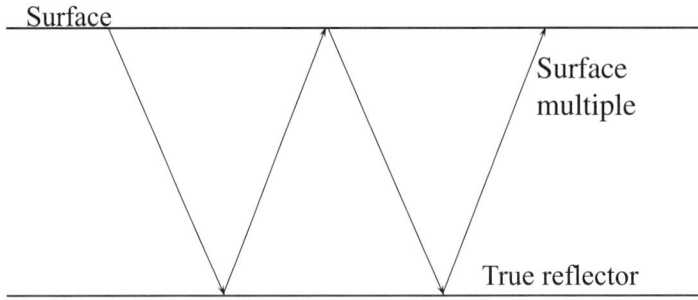

Figure 2.34 Illustration of long-path multiples.

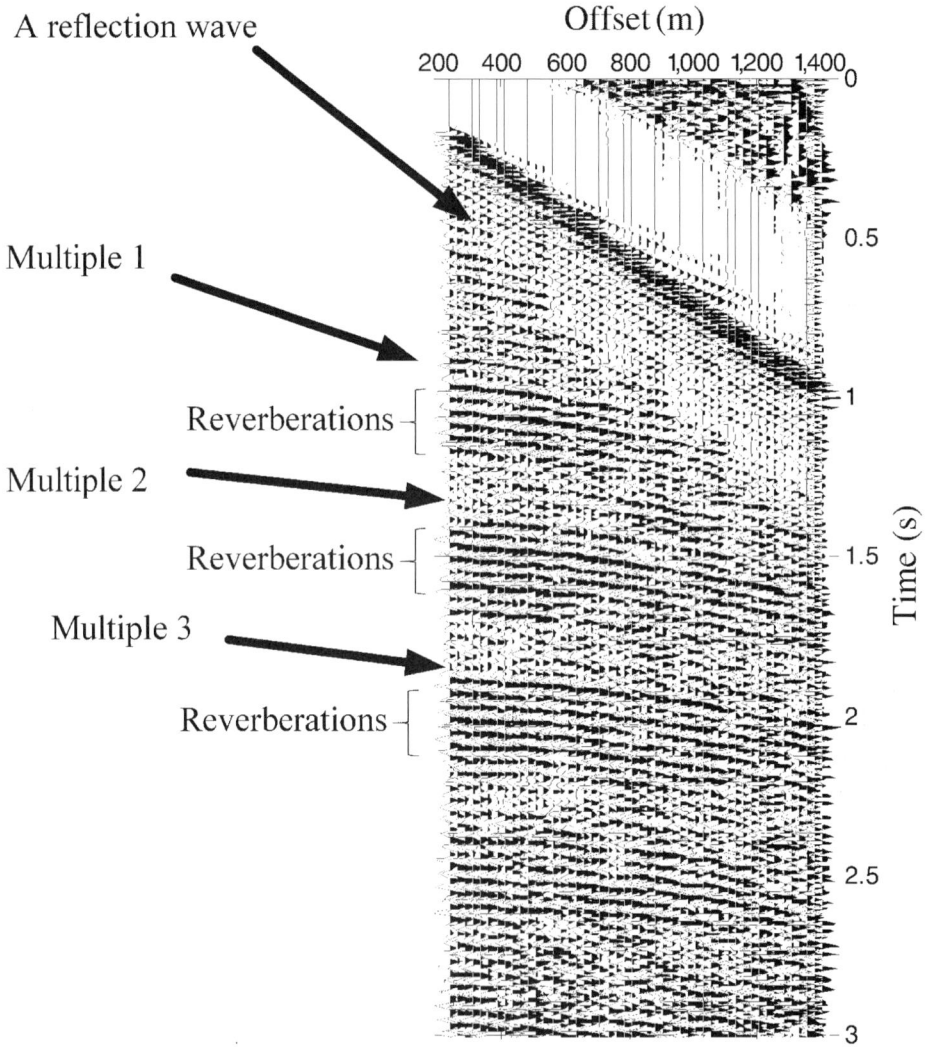

Figure 2.35 Real marine data from modified from Yilmaz (2001) with long-path multiples.

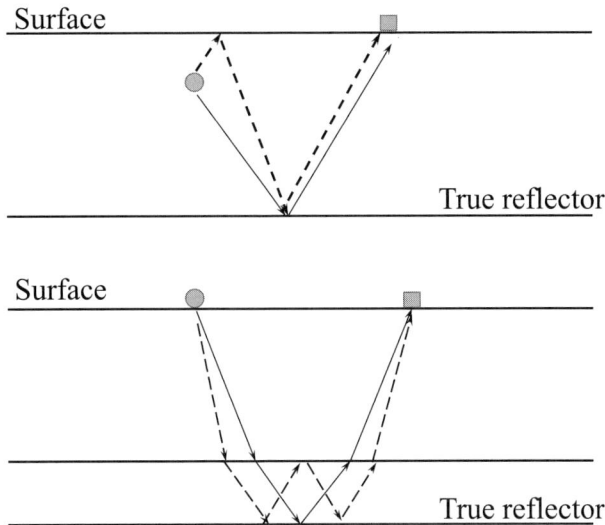

Figure 2.36 Illustration of short-path multiples.

and energy is generally concentrated in a fairly narrow band of frequencies, while noise is more uncorrelated and broadband (Spanias et al., 1991). However, this is only true for random noise. Spatially coherent noise is the most troublesome noise and can be highly correlated and sometimes aliased with the signal (Ulrych et al., 1999) and (Özbek, 2000a). In general, noise can be considered anything other than the desired signal. A more proper definition of noise contaminating seismic signals can be stated by defining the type of signals one is interested in.

Coherent Noise

Spatially coherent noise is the energy that is generated by the source. It is an undesirable energy that is added to the primary signals. Such energy shows consistent phases from trace to trace. Examples of such a type in land seismic records are (Yilmaz, 2001): multiple reflections or multiples, surface waves like ground roll and, air waves, coherent scattered waves, dynamite ghosts, etc. Improper removal of coherent noise can affect nearly all the processing techniques and complicates interpretation of geological structures (see for example (Kearey et al., 2002; Linville and Meek, 1995; Yilmaz, 2001)). Loads of techniques that deal with the problem of attenuating coherent noise that contaminates seismic data exist (Ansari, 1987; Bamberger and Smith, 1992; Duquet and Marfurt, 1999; Kayran and King, 1983; Linville and Meek, 1995; McCowan et al., 1984; Mulk et al., 1983; Treitel et al., 1967; Yilmaz, 2001).

Random Noise

Disturbances in seismic data, which lack phase coherency between adjacent traces, are considered to be random noise. Unlike coherent noise energy, such energy is usually not related to the source that generates the seismic signals. In land seismic records, near-surface scatterers, wind, rain, and instrument are examples of sources generating random noise. They are spatially random. It can be considered repeatable, as in the case of scattering from near-surface irregularities and inhomogeneities. It can also be considered nonrepeatable, like the cases when wind shakes geophones or a person walks near a geophone. Based on the assumption that random noise is an additive white Gaussian noise (AWGN) (Duval and Rosten, 2000; Yilmaz, 2001), it can be attenuated easily in several different ways such as frequency filtering, deconvolution, wavelet denoising (Berkner and Wells, 1998; Cheng et al., 2001; Zhang and Ulrych, 2003), filtering using Gabor representation (Womack and Cruz, 1994), stacking (Kearey et al., 2002; Yilmaz, 2001), and many other methods. As discussed in Chapter 1, stacking usually suppresses most of the incoherent noise and, therefore, improves the SNR by a factor of $\sqrt{M}$, where M is equal to the number of stacked traces. Figure 2.37 shows a real recorded seismic random noise.

Figure 2.37 Recorded random noise.

2.10.4 Examples of Real Seismic Data

Figure 2.38 shows a 2-D land symmetrical split-spread shot record consisting of 240 receivers showing good first arrivals from direct and head waves in the form of high-frequency linear events at the top of the record. Clear primaries in the form of hyperbolic events are also shown in the middle of the section. Ground-roll noise fills the area between the first arrivals and primaries in the form of many low-velocity, low-frequency, and high-amplitude linear events. The events at the top left and right corners are ambient and wind noise amplified by the automatic gain control (AGC) process.

Also, a 2-D land shot record consisting of 48 receivers over a sabkha, Figure 2.39, showing good head-wave arrivals at far offsets in the form of a high-frequency linear event at the top of the record. Ground roll noise fills the left area of the record in the form of many low-velocity, low-frequency, and high-amplitude linear events. The events at the top right corners are generated by a strong wind blowing almost perpendicular to the profile.

Finally, an Example of a 2-D land symmetrical split-spread shot record, seen in Figure 2.40, consisting of 458 receivers in a desert showing good first arrivals from head waves in the form of high-frequency linear events at the top of the record. Ground roll noise fills the central area of the record in the form of low-velocity, low-frequency, and high-amplitude linear events. The high-frequency low-velocity linear event cutting through the central area of the record is the air wave arrival generated by the source. Note the sand-dune effects on the right and left of the shot location in the form of time delays in the first and later arrivals.

2.11 Summary

This chapter presented important seismic reflection surveying background information, where seismic waves are assumed to be of elastic energy. Rock velocities and densities can characterize different subsurface layers. The seismic amplitudes are weak due to many factors. The reflection and transmission coefficients can be calculated knowing the acoustic impedances via Zoepprits equations. Various seismic events have been presented such as direct, reflection, refraction, and diffractions, in a single-layer or a multi-layer model. The seismic events were classified into linear, hyperbolic, and random. Also, seismic noise (unwanted energy) was classified into coherent and random, with various examples.

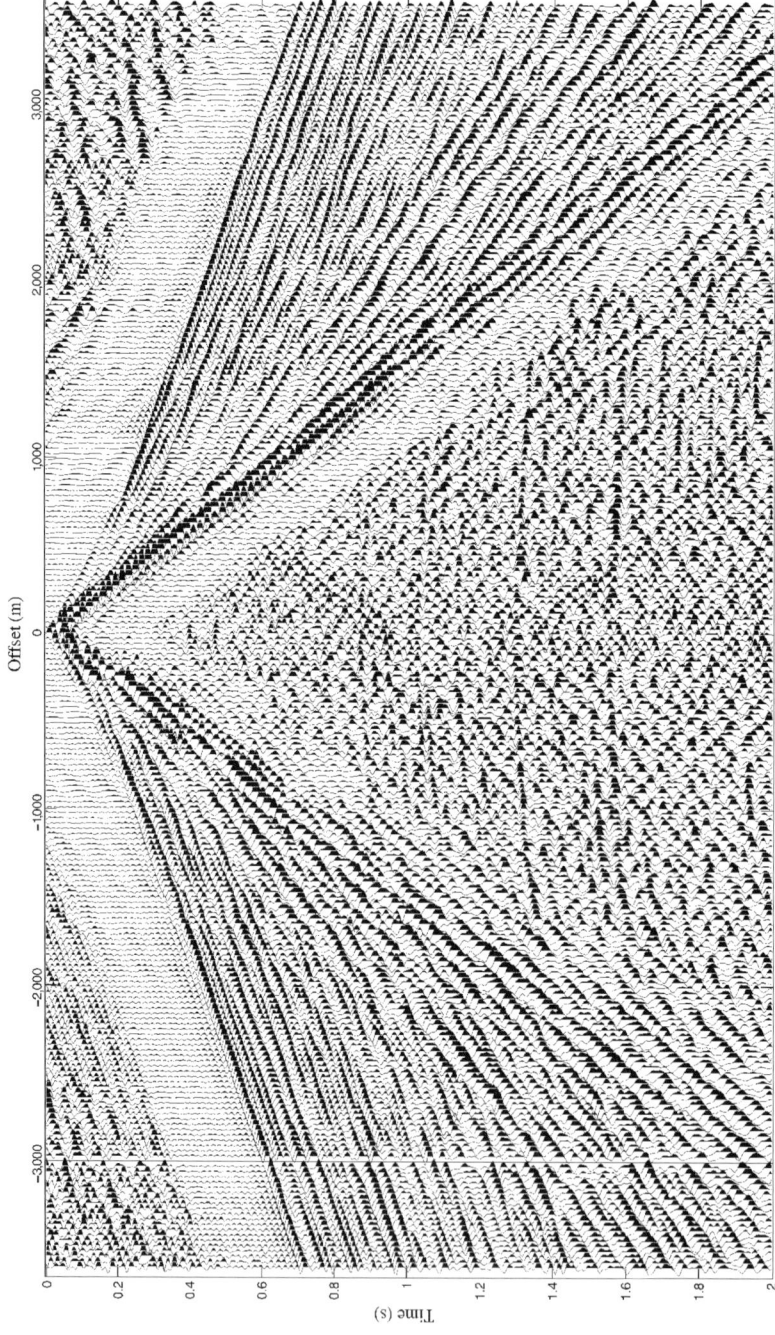

Figure 2.38 Land seismic data.

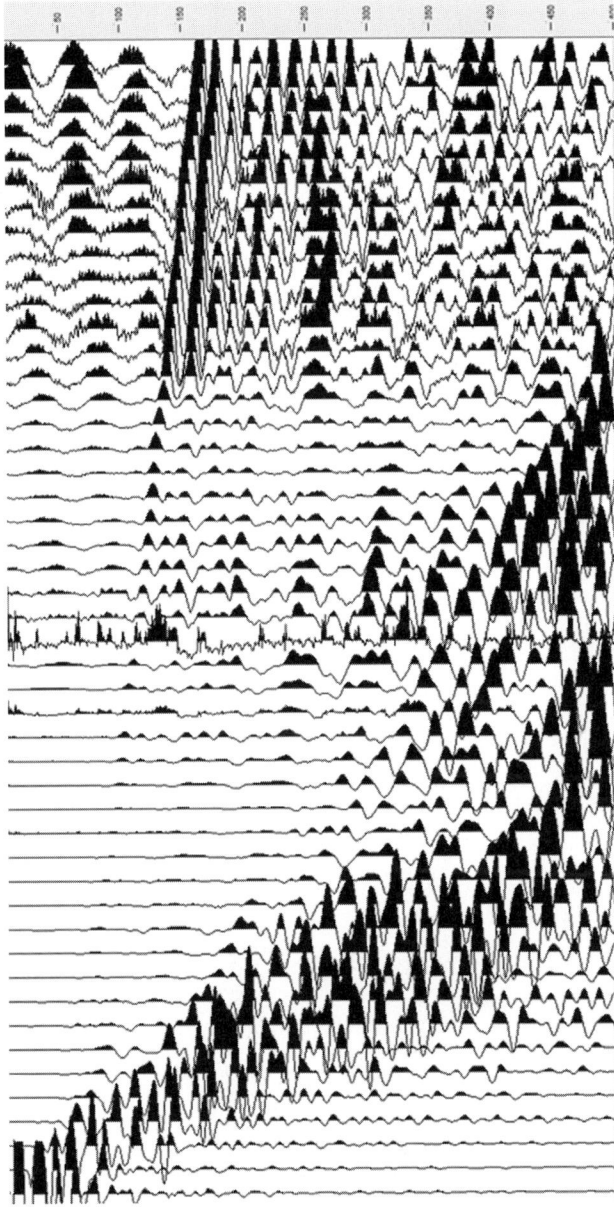

Figure 2.39 An example of a sabkha.

Figure 2.40 An example of land seismic data.

Exercises

2.1 A load of 100 kg is supported by a wire of length 1 m and cross-sectional area of 0.1×10^{-4} m². The wire is stretched by 0.2×10^{-2} m. Find the tensile stress, strain, and Young's modulus for the wire. Note that the gravity acceleration g is equal to 9.8 m/s².

2.2 If a seismic wave is propagating in a dolomite rock with a compressional wave velocity v_p of 3,600 m/s and a shear wave velocity v_s of 2,100 m/s, then determine the dolomite rock Poisson's ratio?

2.3 Consider two seismic waves with frequencies of 5 and 120 Hz to propagate through a rock in which $v_p = 4,000$ m/s and $\beta = 0.4$ dB/λ. By how many dB's will each wave be absorbed?

2.4 A seismic wave is propagating in a sinusoidal periodic manner, where the wiggles recorded by a seismic geophone are 2 ms apart. Obtain the following parameters of the recorded seismic data if the velocity of the propagating seismic wave was 2,300 m/s:

(a) The frequency.
(b) The angular frequency.
(c) The wavelength.
(d) The wavenumber.

2.5 Given the following propagating seismic plane wave:

$$u(x,t) = 5e^{-\beta x}\sin(0.0125\pi x - 25\pi t),$$

and assuming no geometrical (spherical) spreading loss exists. Then, by how much should one multiply the recorded seismic signal in order to restore its original amplitude if the seismic wave travels 1,000 m and the quality factor of the rock was equal to 0.85?

2.6 Given that the group velocity of seismic waves is $v_u = \frac{d\omega}{dk}$, then show that it can be given also by:

$$v_u = v + k\frac{dv}{k}$$

2.7 Calculate the NMO for geophones 600, 1,200, and 3,600 m from the source $v_u = \frac{d\omega}{dk}$ for a reflection at $t_0 = 2.358$ s, given that the compressional velocity = 2,900 m/s. Also, what is the depth z?

2.8 If an incident compressional ray travels with a velocity $v_{p1} = 2,350$ m/s at angle $\theta_{p1} = 35°$, then what is the angle of transmitted shear wave θ_{s1} when $v_{s2} = 590$ m/s?

Table 2.2 *Problem 2.11*

Layer	Depth (m)	Two-way Time (s)	Density (kg/m^3)	Velocity (m/s)	Average Velocity (m/s)	Reflected Ray Amplitude	Transmitted Ray Amplitude	Reflection Coefficient
1	1,000	1	2,100					
2	2,500	2	2,500					
3	4,800	3.1	2,900					

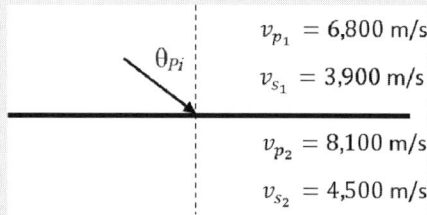

$$v_{p_1} = 6,800 \text{ m/s}$$
$$v_{s_1} = 3,900 \text{ m/s}$$
$$v_{p_2} = 8,100 \text{ m/s}$$
$$v_{s_2} = 4,500 \text{ m/s}$$

θ_{pi}

Figure 2.41 Problem 2.9.

2.9   A compressional wave strikes the interface shown here at angle $\theta_{p_i} = 60°$ to the normal. List all the generated body waves and calculate their angles measured from the normal. Comment on your results (Figure 2.41).

2.10 Consider the model of a single layer dipping reflector in Figure 2.28, where x stands for distance in m, z stands for depth in m, θ the dipping angle, and v stands for the layer velocity in m/s. Show that the time–distance relationship based on this model can be described by the following equation:

$$t(x) = \frac{\sqrt{x^2 + 4z^2 + 4xz \sin\theta}}{v}$$

Hint: Use the Law of Cosines of a triangle with sides a, b, and c and corresponding opposite angles A, B, and C: $a^2 = b^2 + c^2 - 2cb \, \cos(A)$.

2.11 Assuming a normally incident seismic ray with an original amplitude equal to 1 propagates in such a medium, fill Table 2.2 accordingly.

2.12 Using the synthetic shot record shown in Figure 2.42:

(a) Define the direct wave and compute its velocity.
(b) Define the head wave and compute its velocity.
(c) Define the ground roll and compute its velocity.
(d) Define the primary reflection and compute its velocity.
(e) Define the multiple and compute its velocity.

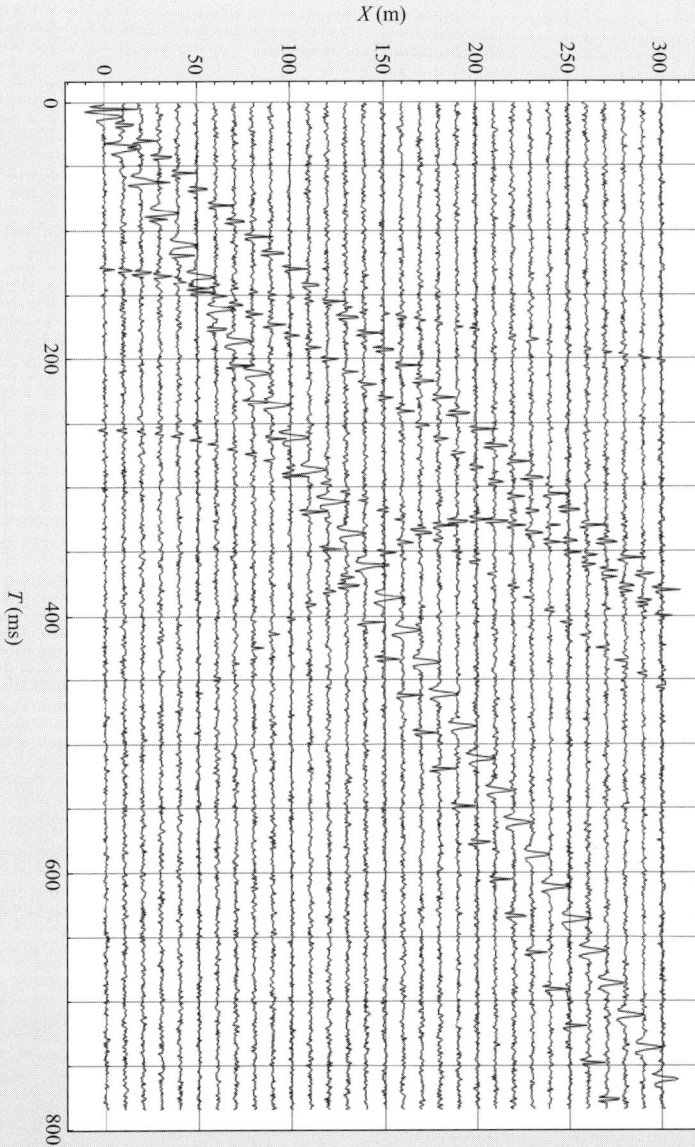

Figure 2.42 Problem 2.12.

  (f) Determine the depth and velocity of the layer.

  (g) Sketch the Earth model of this single horizontal layer and label all the raypaths of all the seismic events.

2.13 Figure 2.43 shows a simple layered Earth model. An explosive shot (S) is fired at the surface at time $t = 0$ and generates a signal with amplitude

Table 2.3 *Problem 2.15*

Material	v_p (m/s)	v_s (m/s)	ρ (kg/m³)
Granite	5,990	3,400	2,700
Sandstone	2,300	720	2,100
Soil	300	140	1,750

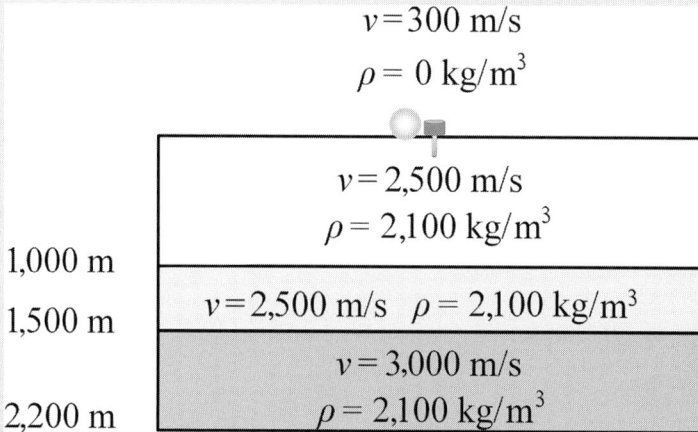

$v = 300$ m/s

$\rho = 0$ kg/m³

$v = 2,500$ m/s

$\rho = 2,100$ kg/m³

1,000 m

1,500 m $v = 2,500$ m/s $\rho = 2,100$ kg/m³

$v = 3,000$ m/s

2,200 m $\rho = 2,100$ kg/m³

Figure 2.43 Problem 2.13.

$A = 1$. A geophone (R) is located very close to the shot point. Assume normal incidence. Ignore geometric spreading and attenuation. Do not consider the direct wave, surface wave, or even multiples.

(a) Sketch the raypath taken by each signal from the shot to the geophone.
(b) Compute the arrival time, reflection coefficients, and amplitude for the first three arrivals to reach the geophone.
(c) Sketch the seismic trace that will be recorded on the geophone.

2.14 Based on the given compressional and shear velocities in Table 2.3, write a MATLAB function that computes Lame's constant λ_r, and the shear modulus μ_r.

2.15 Write a MATLAB code to compensate for amplitude losses based on Equation (2.54). Then, load the seismic data *Yilmaz_data_25_g.mat*. Display the wiggle-variable area for shot gather before and after gain. Also, plot the amplitude envelope for trace number 6 of the shot gather before and after gain correction. Finally, plot the amplitude envelopes for the average trace of the shot gather before and after gain correction. Comment on your results.

Part II

Deterministic Digital Signal Processing for Seismic Data

3

Spectral Analysis of Seismic Data and Useful Transforms

3.1 Introduction

Signal analysis, in the time or space and spectral or other domains, is very important and assists in obtaining a better understanding of signals. Particularly when dealing with seismic studies, it becomes almost standard to analyze seismic data sets, various domains, depending on the dimensions of the acquired data sets. Also, other discrete transforms are very useful for processing seismic data sets, such as the Radon transform, which can be used for seismic wavefield decomposition as well as seismic multiple removal. Of course, this would also require a brief review of the z-transform and the various usages in seismic applications like the minimum phase dipole. This chapter starts by reviewing some of the relavent fundamental background for discrete-time (space) signals and systems. It is very important to understand the basic discrete signals and their interaction with linear systems that can, for example, be digital filters for seismic signals. Afterward, various useful transforms are discussed with examples.

3.2 Discrete-Time(Space) Signals and Systems: A Review

3.2.1 Discrete-Time(Space) Signals

Definition: A discrete-time signal $g[n]$ is a function of an independent variable that is an integer, i.e., $g[n] \in \mathbb{R}$ such that $n \in \mathbb{Z}$, where $\mathbb{R}$ is the set of real numbers and $\mathbb{Z}$ represents the set of integers. Note that $g[n]$ is not defined between two successive samples, and it is incorrect to think that $g[n] = 0$ if n is not an integer, i.e., $g[n]$ is not defined for $n \in \mathbb{Z}$. Also, note that $g[n]$ is called the nth sample of the signal. The following represents a few examples of elementary discrete-time signals:

1. The unit sample sequence (or unit-spike function) or signal $\delta[n]$:

$$\delta[n] = \begin{cases} 1 & \text{for } n = 0, \\ 0 & \text{for } n \neq 0. \end{cases} \tag{3.1}$$

117

2. The shifted unit sample signal $\delta[n - k]$:

$$\delta[n - k] = \begin{cases} 1 & \text{for } n = k, \\ 0 & \text{for } n \neq k. \end{cases} \tag{3.2}$$

3. The unit step signal $u[n]$:

$$u[n] = \begin{cases} 1 & \text{for } n \geq 0, \\ 0 & \text{for } n < 0. \end{cases} \tag{3.3}$$

4. The unit ramp signal $u_r[n]$:

$$u_r[n] = \begin{cases} n & \text{for } n \geq 0, \\ 0 & \text{for } n < 0. \end{cases} \tag{3.4}$$

5. The exponential signal:

$$x[n] = \alpha^n, \tag{3.5}$$

where $\alpha \in \mathbb{R}$.
6. The complex exponential sequence:

$$x[n] = \alpha^n, \tag{3.6}$$

where $\alpha \in \mathbb{C}$.
7. The sinusoidal signal:

$$x[n] = A \cos[\omega_o n + \phi], \tag{3.7}$$

where A is the sinusoid amplitude, ω_o is the angular frequency, and ϕ represents the sinusoid's phase.
8. Dipole: A dipole is a two-length discrete-time signal (or wavelet) and is mathematically defined as (where generally $w[n] \in \mathbb{C}$):

$$w[n] = w[0]\delta[n] + w[1]\delta[n - 1]. \tag{3.8}$$

9. Reverse dipole is given by

$$w_r[n] = w[1] * \delta[n] + w[0] * \delta[n - 1]. \tag{3.9}$$

Figures 3.1 and 3.2 display various elementary 1-D discrete-time signals. In general, discrete-time(space) signals can be classified into:

• Energy Signals: Let $g[n]$ be a real or complex valued discrete-time signal. Then $g[n]$ is called an Energy Signal if its energy E, which is defined as:

$$E = \sum_{n=-\infty}^{\infty} |g[n]|^2, \tag{3.10}$$

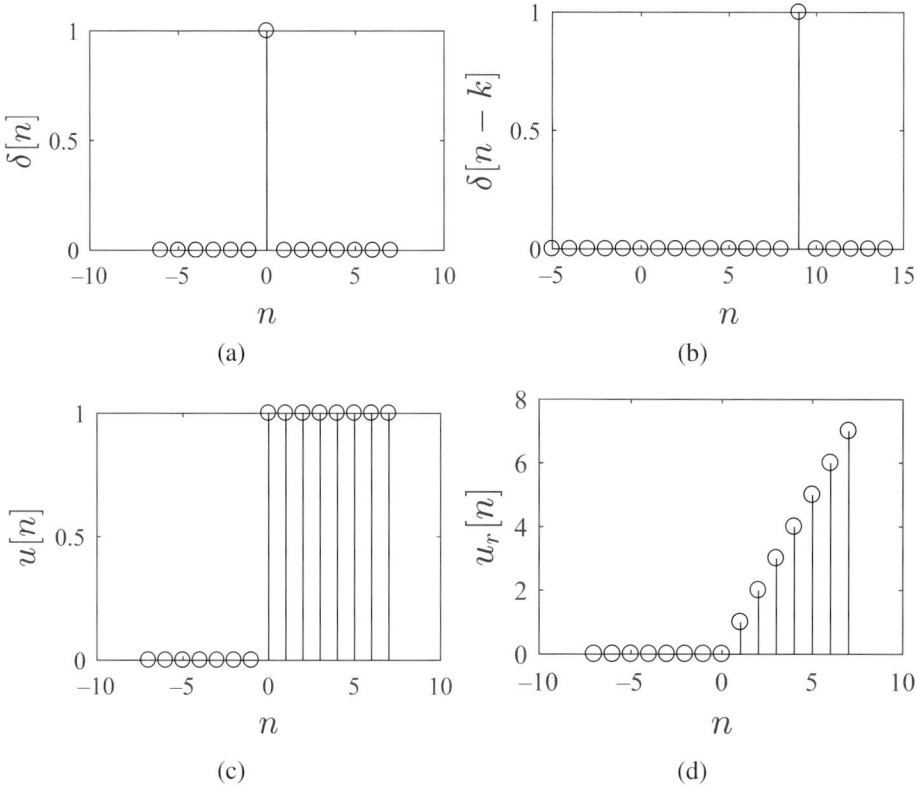

Figure 3.1 Various elementary 1-D discrete-time signals: (a) unit sample signal,
(b) shifted unit sample signal, (c) unit step signal, and (d) unit ramp signal.

is finite, i.e., $0 < E < \infty$. Seismic signals are energy signals, since they are finite
in nature when recorded in computers.

- Power Signals: The average power of a discrete-time signal $g[n]$ is defined as:

$$P = \lim_{N \to \infty} \frac{1}{2N+1} \sum_{n=-N}^{N} |g[n]|^2, \tag{3.11}$$

If E is finite, then $P = 0$. If E is infinite, then P may be finite or infinite. If P is
finite and nonzero, the signal is then called a Power Signal.

- Neither an Energy nor a Power Signal:

Example 3.1 Determine which of the following signals are energy, power, or
neither:

(a) $g[n] = 0.5\delta[n] - 0.25\delta[n]$.
(b) $g[n] = u[n]$.
(c) $g[n] = u_r[n]$.

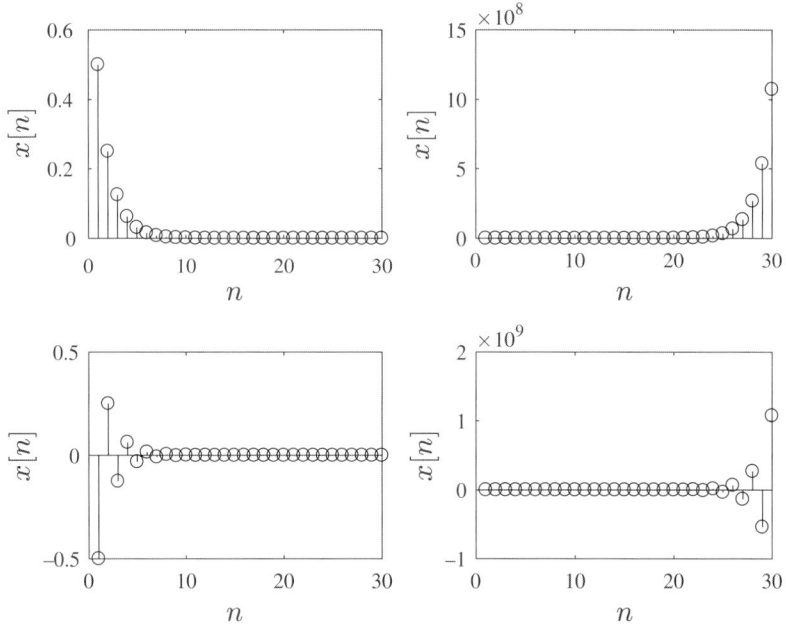

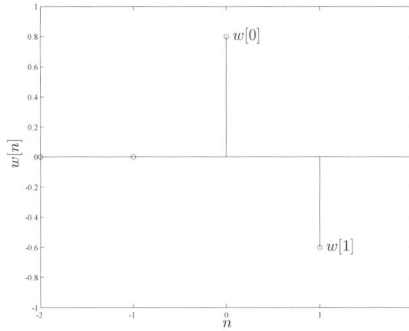

(b)

Figure 3.2 Various elementary 1-D discrete-time signals: (a) exponential signals, and, (b) a dipole wavelet.

Solution: (a) Using Equation (3.10):

$$E = |0.5|^2 + |0.25|^2 = 0.25 + 0.125 = 0.375 < \infty,$$

which is, in this case, an energy signal.

(b) Using Equation (3.10):

$$E = \sum_{n=-\infty}^{\infty} |u[n]|^2 = \sum_{n=0}^{\infty} 1 = \infty$$

Since $E = \infty$, $u[n]$ is not an energy signal. Now, it is desired to check whether it is a power signal or not. Using Equation (3.11):

$$P = \lim_{N \to \infty} \frac{1}{2N+1} \sum_{n=-N}^{N} |u[n]|^2,$$

$$= \lim_{N \to \infty} \frac{1}{2N+1} \sum_{n=0}^{N} 1,$$

$$= \lim_{N \to \infty} \frac{1}{2N+1} \sum_{n=0}^{N} (N+1),$$

$$= \lim_{N \to \infty} \frac{1+1/N}{2+1/N} = \frac{1}{2}.$$

The unit step sequence $u[n]$ is, therefore, a power signal.

(c) Starting by using Equation (3.10):

$$E = \sum_{n=-\infty}^{\infty} |u_r[n]|^2 = \sum_{n=0}^{\infty} n^2 = \infty.$$

Since $E = \infty$, $u_r[n]$ is not an energy signal. Now, using Equation (3.11),

$$P = \lim_{N \to \infty} \frac{1}{2N+1} \sum_{n=-N}^{N} |u_r[n]|^2,$$

$$= \lim_{N \to \infty} \frac{1}{2N+1} \sum_{n=0}^{N} n^2,$$

$$= \lim_{N \to \infty} \frac{1}{2N+1} \sum_{n=1}^{N} n^2,$$

$$= \lim_{N \to \infty} \frac{1}{2N+1} N(N+1)(2N+1)\frac{1}{6} = \infty.$$

Hence, the unit step signal $u_r[n]$ is neither an energy nor a power signal.

In addition, discrete-time signals can be classified into:

• Periodic Signals: A signal $g[n]$ is periodic with period N ($N > 0$) if and only if:

$$g[n+N] = g[n], \tag{3.12}$$

for all n. Note that the smallest value N for which Equation (3.12) holds is called the *Fundamental Period*.

• Aperiodic Signals: If there is no value N that satisfies Equation (3.12), then the signal $g[n]$ is called *Nonperiodic* or *Aperiodic*.

Example 3.2 Determine which of the following signals is periodic or aperiodic:

(a) $g[n] = A \sin[2\pi f_0 n]$
(b) $g[n] = 2 \sin[\frac{\pi}{8} n]$

Solution: (a) $g[n] = A \sin[2\pi f_0 n]$ is periodic when f_0 is a rational number, that is, if $f_0 = k/N$, where $k, N \in \mathbb{Z}$

(b) $g[n] = 2 \sin[\frac{\pi}{8} n]$

$$g[n] = 2 \sin\left[\frac{\pi}{8} n + 2\pi\right] = 2 \sin\left[\frac{\pi}{8}\left(n + 2\pi \frac{8}{\pi}\right)\right]$$

$$= 2 \sin\left[\frac{\pi}{8}(n + 16)\right] \Rightarrow N = 16.$$

Hence, $g[n]$ is periodic with a period $N = 16$.

Furthermore, any discrete-time signal $g[n]$ can be expressed as a weighted sum of shifted unit sample signals (sequences). That is:

$$g[n] = \sum_{k=-\infty}^{\infty} g[k]\delta[n - k]. \tag{3.13}$$

If one takes, for example, the sample value of $g[n]$ at $n = 3$, then based on Equation (3.13):

$$\sum_{-\infty}^{\infty} g[3]\delta[3 - k] = g[3], \tag{3.14}$$

where based on Equation (3.2), $\delta[3 - k]$ is zero except that when $k = 3$, it will be equal to 1.

3.2.2 Discrete-Time(Space) Systems

A discrete-time system is a device or algorithm that operates on a discrete-time signal, called the *Input* or *Excitation*, according to some well-defined rule, to produce another discrete-time signal called the Output or Response of the system (see Figure 3.3). For example, the Earth as a system, can be viewed as an operation or a set of operators performed on the input signal wavelet $w[n]$ to produce the output recorded seismic signal $g[n]$:

$$g[n] = \mathcal{T}\{w[n]\}, \tag{3.15}$$

where $\mathcal{T}\{.\}$ represents an operator. In analysis and design of discrete-time systems, it is desirable to classify the systems based on their general properties they satisfy. If a discrete-time system possesses a given property, then this property must hold for every possible input signal. If that property holds for some and not for others, then

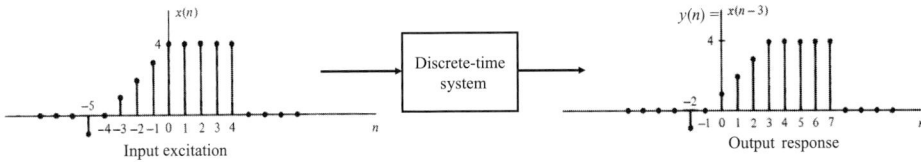

Figure 3.3 A discrete-time(space) system.

the system does not possess that property. In most of the cases, the Earth system can be assumed to be linear and shift-invariant (LSI), when carefully analyzing the corresponding acquired seismic records. So the model that can be followed for seismic filtering, for example, is the use of LSI filters.

Linearity can be stated as follows. Let $T\{.\}$ be an operator that represents, in general, a given system, where $g_1[n] = T\{w_1[n]\}$, $g_2[n] = T\{w_2[n]\}$, and $\alpha_1, \alpha_2 \in \mathbb{R}$, then one can say that the system denoted by $T\{.\}$ is linear if and only if:

$$T\{\alpha_1 w_1[n] + \alpha_2 w_2[n]\} = \alpha_1 g_1[n] + \alpha_2 g_2[n]. \tag{3.16}$$

Example 3.3 Is the system provided by the following relationship

$$g[n] = n^2 w[n],$$

linear or nonlinear?

Solution: Consider $\alpha_1 w_1[n]$ and $\alpha_2 w_2[n]$ as the input of the system, which yields:

$$g_1[n] = n^2 \alpha_1 w_1[n] \qquad g_2[n] = n^2 \alpha_2 w_2[n].$$

Now, consider the following:

$$\alpha_1 T[w_1[n]] + \alpha_2 T[w_2[n]] = n^2 \alpha_1 w_1[n] + n^2 \alpha_2 w_2[n]$$
$$= \alpha_1 g_1[n] + \alpha_2 g_2[n]. \tag{3.17}$$

Also, on the other hand,

$$T[\alpha_1 w_1[n] + \alpha_2 w_2[n]] = n^2[\alpha_1 w_1[n] + \alpha_2 w_2[n]]$$
$$= n^2 \alpha_1 w_1[n] + n^2 \alpha_2 w_2[n]$$
$$= \alpha_1 g_1[n] + \alpha_2 g_2[n], \tag{3.18}$$

which is equal to Equation (3.17). Therefore, the system is linear. In general, the system described by $g[n] = n^\alpha w[n]$, where $\alpha \in \mathbb{R}$, is typically used as a data-independent gain process at the pre-processing stage of seismic data (Mousa and Al-Shuhail, 2011; Yilmaz, 2001). Figure 3.4 shows an example of data from Yilmaz (2001) before and after applying n^2 (or t^2).

(a)

(b)   (c)

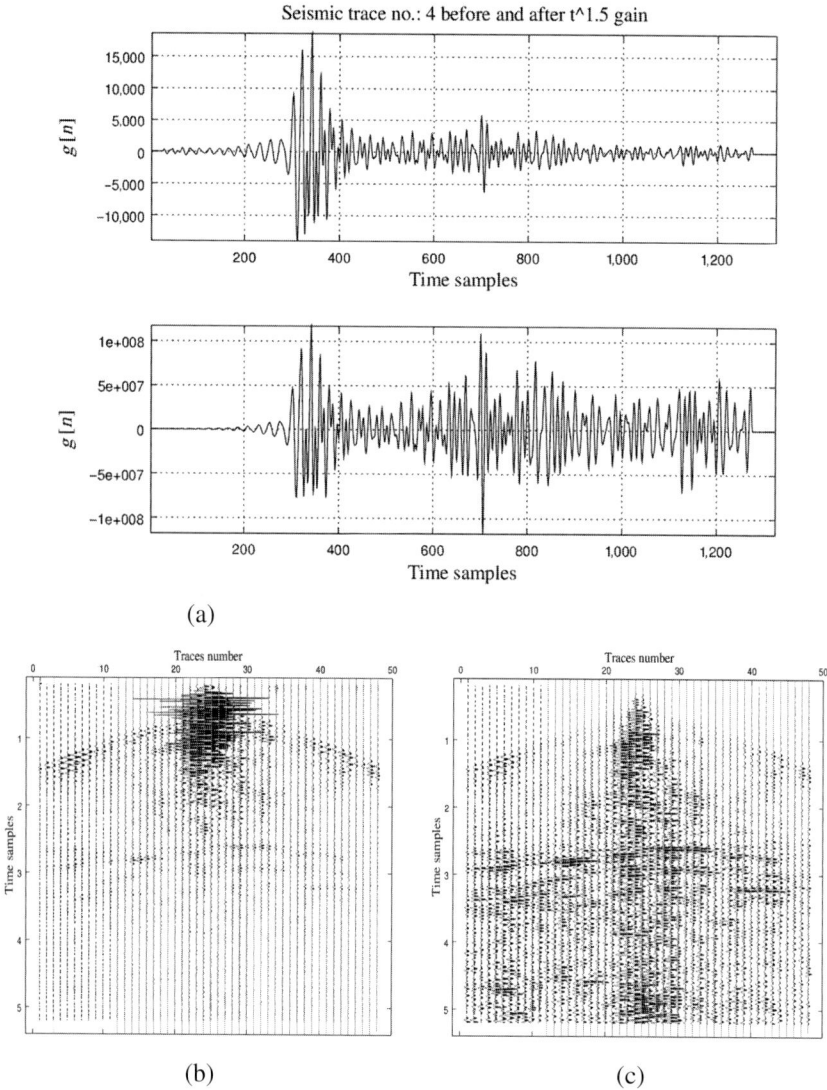

Figure 3.4 This system is known as the t^α (or n^α since n is the independent variable) gain correction method. It is commonly used at the pre-processing stage of seismic exploration data processing (as seen in Section 1.1), where it is applied on a seismic shot record, modified from Yilmaz (2001). (a) Seismic trace 4 before and after applying the gain with $t^{1.5}$. (b and c) The full shot record before and after applying the gain with $t^{1.5}$, respectively.

Another important classification of discrete-time systems is related to whether the system's characteristics change with time (or space). That is, a discrete-time system denoted by $\mathcal{T}\{.\}$ is shift invariant (its input–output characteristics do not change with time) if and only if given $g[n] = \mathcal{T}\{w[n]\}$:

$$g[n - k] = T\{w[n - k]\}. \tag{3.19}$$

To determine if any given system is shift-invariant, excite the system with an arbitrary input sequence $w[n]$, which produces an output, say $g[n]$. Then delay the input sequence $w[n]$ by the same amount k (i.e., excite the system with $w[n - k]$) and re-compute the output. If the output (call it $g[n, k]$ due to an input $w[n - k]$ is equal to $g[n - k]$ for every value of k, then the system is shift-invariant. Otherwise, the system is shift-varying.

Example 3.4 Determine whether the following system

$$g[n] = \frac{1}{3}(w[n] + w[n - 1] + w[n - 2]), \tag{3.20}$$

is shift-invariant or time varying?

Solution: This system is described by the following relationship:

$$g[n] = T[w[n]] = \frac{1}{3}(w[n] + w[n - 1] + w[n - 2]).$$

If one delays the input by k units in time, the output will be:

$$g[n, k] = \frac{1}{3}(w[n - k] + w[n - k - 1] + w[n - k - 2]). \tag{3.21}$$

Now, delay the output $g[n]$ by k units in time, i.e.,

$$g[n - k] = \frac{1}{3}(w[n - k] + w[n - k - 1] + w[n - k - 2]), \tag{3.22}$$

which is equal to Equation (3.21). This implies that the system is shift-invariant. This system is known as the M-point (M here is 3) moving average filter. It is used in seismic data processing to smooth the data (Mousa and Al-Shuhail, 2011).

Note that a discrete-time system that is linear and shift-invariant is called a Linear Shift-Invariant (LSI) system. Also, a system is said to be causal if the output $g[n]$ of the system at a time n depends only on present and past inputs $\{w[n], w[n - 1], w[n - 2], \ldots\}$ but does not depend on future inputs $\{w[n + 1], w[n + 2], w[n + 3], \ldots\}$, i.e.,

$$g[n] = F\{w[n], w[n - 1], w[n - 2], \ldots\}, \tag{3.23}$$

where $F\{.\}$ is an arbitrary function. If the system does not satisfy this definition, it is called *non-causal*.

Example 3.5 The 3-point moving average system in Equation (3.20) is causal, since the output depends on current and past inputs.

Furthermore, unstable systems usually exhibit erratic and extreme behavior and cause overflow in any practical implementation. Hence, it is important to consider the stability property in any practical application of a system. Hence, an arbitrary relaxed system is said to be bounded input-bounded output (BIBO) stable if and only if every bounded input produces a bounded output. Mathematically, some finite numbers exist, say M_w and M_g, such that:

$$|w[n]| \leq M_w < \infty, \quad |g[n]| \leq M_g < \infty, \tag{3.24}$$

for all n. If, for a bounded input sequence $w[n]$, the output $g[n]$ is unbounded, then the system is classified as *Unstable*.

Example 3.6 Check the stability of the M-point moving average system described by:

$$g[n] = \frac{1}{M} \sum_{k=0}^{M-1} w[n-k].$$

Solution: Assume the input is bounded, that is,

$$|w[n]| < B_g.$$

Then, check if the output is bounded or not:

$$|g[n]| = \left| \frac{1}{M} \sum_{k=0}^{M-1} w[n-k] \right|, \text{using the triangular inequaility} \tag{3.25}$$

$$\leq \frac{1}{M} \left| \sum_{k=0}^{M-1} w[n-k] \right|, \text{since the input is bounded} \tag{3.26}$$

$$\leq \frac{1}{M}(MB_g) = B_g. \tag{3.27}$$

Therefore, the system is BIBO stable.

There are two basic methods for analyzing the response of LSI systems to a given input signal, which relate the input to the output of such LSI systems. The first is the Constant Coefficient Difference Equation (CCDE), which is given by:

$$g[n] = -\sum_{k=1}^{M} a_k g[n-k] + \sum_{k=0}^{N} b_k w[n-k], \tag{3.28}$$

where $\{a_k\}$ and $\{b_k\}$ are constant parameters that specify the system and are independent of $w[n]$ and $g[n]$. This expresses the output of the system at time n directly as a weighted sum of past outputs as well as past and present input signal samples. Note that if all the coefficients $a_k = 0$, then the system is called a Finite Impulse

Response. On the other hand, if at least one coefficient $a_k \neq 0$, then the system is called Infinite Impulse Response.

The second is the convolution sum between the system's *Impulse Response*, denoted by $h[n]$, and the input signal $w[n]$. The impulse response of any LSI system, generally speaking, is the response of that system $\mathcal{T}\{.\}$ to an unit sample signal input $\delta[n]$. Note that an LSI system is completely characterized by a single function $h[n]$, i.e., its response to the unit sample signal $\delta[n]$. If Equation (3.13), which represents any discrete-time signal with a weighted sum of shifted unit samples, is substituted in Equation (3.15), then:

$$g[n] = \mathcal{T}\{w[n]\}$$

$$= \mathcal{T}\left\{\sum_{k=-\infty}^{\infty} w[k]\delta[n-k]\right\}, \text{ by linearity}$$

$$= \sum_{k=-\infty}^{\infty} \mathcal{T}\{w[k]\delta[n-k]\}, \text{ since } w[k]\text{'s are constants}$$

$$= \sum_{k=-\infty}^{\infty} w[k]\mathcal{T}\{\delta[n-k]\}, \text{ by shift-invariance}$$

$$= \sum_{k=-\infty}^{\infty} w[k]h[n-k]. \tag{3.29}$$

In other words, the output of an LSI system is a function of the input signal and the system's impulse response where it can be computed through the so-called Convolution Sum via:

$$g[n] = \sum_{k=-\infty}^{\infty} w[k]h[n-k]. \tag{3.30}$$

The operation of convolution is denoted by a $*$, where, in this case,

$$g[n] = w[n] * h[n], \tag{3.31}$$

$$= \sum_{k=-\infty}^{\infty} w[k]h[n-k].$$

Note that the impulse signal $\delta[n]$ is the identity element for the convolution, which means that:

$$g[n] = w[n] * \delta[n] = w[n]. \tag{3.32}$$

If we shift $\delta[n]$ by k, then the convolution sum is also shifted by k units as:

$$w[n] * \delta[n-k] = g[n-k] = w[n-k]. \tag{3.33}$$

Consequently, it can also be shown, using change of variables, that:

$$g[n] = h[n] * w[n], \tag{3.34}$$

$$= \sum_{k=-\infty}^{\infty} w[n-k]h[k].$$

Therefore, the convolution operation is commutative, where:

$$g[n] = w[n] * h[t] = h[n] * w[n]. \tag{3.35}$$

Besides commutative, the convolution also holds following associative and distributive laws, respectively:

$$(w[n] * h_1[n]) * h_2[n] = w[n] * (h_1[n] * h_2[n]), \tag{3.36}$$

$$w[n] * (h_1[n] + h_2[n]) = w[n] * h_1[n] + w[n] * h_2[n]. \tag{3.37}$$

In the case of seismic signals, the Earth's impulse response $h[n]$ can be modeled as the reflectivity signal (sequence) $r[n]$. If one is given an input such as the seismic wavelet sequence $w[n]$, and the Earth's system impulse response $r[n]$, where it is assumed that the Earth system in LSI, then one will be interested to determine the output seismic signal $g[n]$ through convolution.

Example 3.7 Given $h[n] = 4\delta[n] - 4\delta[n-1] - 7\delta[n-2]$ and $w[n] = \delta[n] + 3\delta[n-1]$. Find the convolution $g[n]$ of $h[n]$ and $w[n]$?

Solution:

$$g[n] = \sum_{k=-\infty}^{k=\infty} w[n-k]h[k],$$

$$g[0] = \sum_{k=-\infty}^{k=\infty} w[-k]h[k] = w[0]h[0] = 4,$$

$$g[1] = \sum_{k=-\infty}^{k=\infty} w[1-k]h[k] = w[1]h[0] + w[0]h[1] = 8,$$

Similarly, one can show that $g[2] = -19$ and $g[3] = -21$, so $g[n] = 4\delta[n] + 8\delta[n-1] - 19\delta[n-2] - 21\delta[n-3]$. Note that $g[n] = 0$ for $n \neq 0, 1, 2, 3$. Also, one can compute the convolution using the following matrix form:

$$\begin{bmatrix} 4 & -4 & -7 & 0 \end{bmatrix} \begin{bmatrix} 1 & 3 & 0 & 0 \\ 0 & 1 & 3 & 0 \\ 0 & 0 & 1 & 3 \\ 0 & 0 & 0 & 1 \end{bmatrix} = \begin{bmatrix} 4 & 8 & -19 & -21 \end{bmatrix},$$

or perform the convolution as a non-carry multiplication:

4	−4	−7	
1	3		×
12	−12	−21	
4	−4	−7	+
4	8	−19	−21

All yield the same answer.

Example 3.8 Find the overall impulse response of the system shown in Figure 3.5, where $h_1[n] = \delta[n] + \frac{1}{2}\delta[n-1]$, $h_2[n] = \frac{1}{2}\delta[n] - \frac{1}{4}\delta[n-1]$, $h_3[n] = 2\delta[n]$, $h_4[n] = -2\left(\frac{1}{2}\right)^n u[n]$.

Solution: From Figure 3.5, $h_3[n]$ and $h_4[n]$ are added and cascaded to $h_2[n]$ and the result is added to $h_1[n]$. That is,

$$h[n] = \left(\left[h_3[n] + h_4[n] \right] * h_2[n] \right) + h_1[n]. \tag{3.38}$$

Since convolution is distributive, then,

$$h[n] = h_1[n] + h_2[n] * h_3[n] + h_2[n] * h_4[n]. \tag{3.39}$$

And,

$$h_2[n] * h_3[n] = \delta[n] - \frac{1}{2}\delta[n-1]. \tag{3.40}$$

$$h_2[n] * h_4[n] = -\delta[n]. \tag{3.41}$$

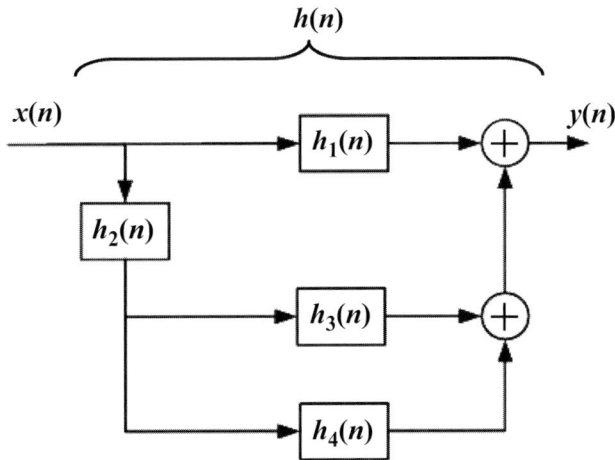

Figure 3.5 Example 3.8 system.

Therefore,

$$h[n] = \delta[n]. \tag{3.42}$$

Example 3.9 Find the step response of the LSI system with impulse response $h[n] = a^n u[n]$, where $|a| < 1$.

Solution: Since the input and impulse response of the system are causal, one can use the last equation as follows:

$$g[n] = \sum_{k=0}^{n} h[k]w[n-k], \tag{3.43}$$

$$= \sum_{k=0}^{n} a^k, \tag{3.44}$$

$$= 1 + a + a^2 + a^3 + \cdots + a^n, \tag{3.45}$$

$$= \frac{1 - a^{n+1}}{1 - a}, \quad \text{for } n \geq 0. \tag{3.46}$$

Additionally, the concepts of signal crosscorrelation and autocorrelation are important operations to consider when processing seismic data. Suppose that we have two real signal sequences $w[n]$ and $h[n]$, each of which has finite energy. The crosscorrelation of $w[n]$ and $h[n]$ is a sequence $R_{wh}[n]$, which is defined as

$$R_{wh}[n] = \sum_{k=-\infty}^{\infty} w[k]h[k+n], \quad n = \ldots, -2, -1, 0, 1, 2, \ldots \tag{3.47}$$

In the special case where $w[n] = h[n]$, the autocorrelation of $w[n]$ is defined as:

$$R_{ww}[n] = \sum_{k=-\infty}^{\infty} w[k]w[k+n], \quad n = \ldots, -2, -1, 0, 1, 2, \ldots \tag{3.48}$$

Both will be heavily used when performing seismic deconvolution and wavelet processing using optimal filters (See Chapters 6–8).

As a direct result of $h[n]$ characterizing any LSI system, it can be shown that $h[n] = 0$ for $n < 0$ is both a *Necessary and Sufficient* condition for causality of the LSI systems. Additionally, it can be shown that a *Necessary and Sufficient* condition for a LSI system to be stable is that its impulse response is absolutely summable. That is:

$$\sum_{k=-\infty}^{\infty} |h[k]| < \infty. \tag{3.49}$$

Example 3.10 Determine the range of values of the parameter a for which the LSI system with the impulse response $h[n] = a^n u[n]$, where $|a| < 1$ is stable.

Solution: This system is causal. Hence, using Equation (3.49):

$$\sum_{k=-\infty}^{\infty} |h[k]| = \sum_{k=0}^{\infty} |a^k|, \tag{3.50}$$

$$= \sum_{k=0}^{\infty} |a|^k,$$

$$= 1 + |a| + |a|^2 + |a|^3 + \cdots,$$

$$= \frac{1}{1 - |a|}, \text{ if } |a| < 1.$$

The system is stable if $|a| < 1$. Otherwise, it is unstable. In effect, $h[n]$ must decay exponentially toward zero as k goes to infinity for this system to be stable.

Example 3.11 Is the following system, $h[n]$, stable or not?

$$h[n] = \begin{cases} a^n, & N_1 \leq n \leq N_2, \\ 0, & \text{otherwise.} \end{cases} \tag{3.51}$$

Solution: Using Equation (3.49), check if the impulse response is absolutely summable:

$$\sum_{k=-\infty}^{\infty} |h[n]| = \sum_{k=-N_1}^{N_2} \left| a^k \right|. \tag{3.52}$$

The above is the summation of finite geometric series. It is always summable irrespective of a as long as a is finite. Therefore, the system is BIBO stable.

Also, a system is called *recursive* if its output $g[n]$ at time n depends on any number of past output values $g[n-1], g[n-2], \ldots$. On the other hand, a system is called *non-recursive* if the output of the system $g[n]$ depends only on the present and past inputs $w[n], w[n-1], w[n-2], \ldots, w[n-M]$.

Last, but not least, we consider a Finite Impulse Response (FIR) system that has an impulse response equal to zero outside some finite interval. On the other hand, an Infinite Impulse Response (IIR) system has an infinite duration impulse response. FIR is non-recursive but can be realized recursively. However, the impulse response of the system will be the same. An IIR system can be realized recursively only. An FIR system described by convolution formula can be implemented (realized). The realization involves multiplications, additions, and finite memory. An IIR system described by the convolution formula cannot be realized because of infinite memory requirement. IIR systems (and FIR) can be realized practically using difference equations. When the coefficients of the difference equation are constant, the system is said to be realized by constant coefficient difference equation (CCDE) (see Equation 3.28). Such systems are implementable even though some of these systems may have an infinite impulse response. FIR and IIR terms determine the type of system one deals with. Recursive and non-recursive terms determine the implementation of such systems. Recursive and non-recursive systems most probably are IIR and FIR systems, respectively.

3.3 The z-Transform

The z-transform is a very useful tool in digital signal processing. The z-transform, in particular, plays the same role in the analysis of discrete-time signals and LTI systems as the Laplace transform does in the analysis of continuous-time signals and LTI systems. It can be applied to obtain analytical expressions for discrete-time (space) signals. Furthermore, linear convolution, for example, can be carried out more easily in the z-transform domain. Also, sometimes discrete systems, like digital filters, can be better analyzed or designed based on their z-transform characteristics. In seismic data processing, digital filters are heavily used and, hence, they require careful design to achieve their goals of processing. Also, as described in Chapter 8, seismic wavelets can sometimes be better analyzed in the z-transform domain.

3.3.1 The Forward z-Transform

The z-transform of a 1-D discrete-time(space) signal (sequence) $g[n]$ is defined as:

$$G(z) = \sum_{n=-\infty}^{\infty} g[n]z^{-n}, \tag{3.53}$$

where n is an integer that represents the time (space) index and z is a complex variable. The convergence of the series in Equation (3.53) depends on $g[n]$ and the value of z. For the direct z-transform, one can write $G(z) = \mathcal{Z}\{g[n]\}$, while for the inverse $g[n] = \mathcal{Z}^{-1}\{G(z)\}$.

> **Example 3.12** The z-transform of $g[n] = \frac{1}{2}\delta[n] - 2\delta[n-1] + 3\delta[n-3]$ can be shown, using Equation (3.53), to be $G(z) = \frac{1}{2} - 2z^{-1} + 3z^{-3}$.

For most of the sequences, a region exists where Equation (3.53) converges in the z-plane, known as the *Region of Convergence* (ROC). That is, the ROC of $G(z)$ is the set of all values of z for which $G(z)$ attains a finite value. The ROC can be determined by finding the values r (where $r \in \mathbb{R}$) for which:

$$\sum_{n=-\infty}^{\infty} |g[n]r^{-n}| \leq c < \infty, \tag{3.54}$$

which means that $g[n]r^{-n}$ is absolutely summable. The proof is simply done by assuming that $z = re^{j\omega}$ and then:

$$|G(z)| = \left| \sum_{n=-\infty}^{\infty} g[n]z^{-n} \right|$$

$$= \left| \sum_{n=-\infty}^{\infty} g[n]r^{-n}e^{-j\omega} \right|$$

$$\leq \sum_{n=-\infty}^{\infty} |g[n]r^{-n}e^{-j\omega}|$$

$$= |g[n]r^{-n}|. \tag{3.55}$$

Therefore, $|G(z)|$ is finite if $g[n]r^{-n}$ is absolutely summable. The problem of finding the ROC for $G(z)$ is equivalent to determining the range of values of r for which the sequence $g[n]r^{-n}$ is absolutely summable. Now, express Equation (3.55) as:

$$|G(z)| \leq \sum_{n=-\infty}^{-1} |g[n]r^{-n}| + \sum_{n=0}^{\infty} |g[n]r^{-n}|,$$

$$= \sum_{n=-\infty}^{-1} |g[n]r^{-n}| + \sum_{n=0}^{\infty} \left| \frac{g[n]}{r^n} \right|,$$

$$= \sum_{n=1}^{\infty} |g[-n]r^{n}| + \sum_{n=0}^{\infty} \left| \frac{g[n]}{r^n} \right|,$$

$$= |g[n]r^{-n}|. \tag{3.56}$$

If $G(z)$ converges in a range of the complex plane, both sums in Equation (3.56) must be finite in that range. If the first sum in Equation (3.56) converges, values of r small enough such that $g[n]r^{-n}, 1 \leq n < \infty$, is absolutely summable must exist. Therefore, the ROC for this sum consists of all points in a circle of radius $r_1 < \infty$. If the second sum in Equation (3.56), values of r large enough such that $g[n]/r^n, 0 \leq n < \infty$, is absolutely summable must exist. Therefore, the ROC for this sum consists of all points outside a circle of radius $r > r_2$. Since the convergence of $G(z)$ requires that both sums in Equation (3.56) must be finite, it follows that the ROC of $G(z)$ is generally specified as the annular region in the z-plane, $r_2 < r < r_1$, which is the common region where both sums are finite. Finally, if $r_2 > r_1$, there is no common region of convergence for the two sums and, hence, $G(z)$ does not exist. In conclusion, a discrete-time signal $g[n]$ is uniquely determined by its z-transform $G(z)$ and the ROC of $G(z)$. Figure 3.6, pictorially depicts the three cases of ROC.

Im(z)

z-plane

r_1

Re(z)

Region of convergence for
$\sum_{n=1}^{\infty} |g[-n]r^n|$

(a)

Im(z)

z-plane

r_2

Re(z)

Region of convergence for
$\sum_{n=0}^{\infty} |\frac{g[-n]}{r^n}|$

(b)

Im(z)

z-plane

r_2

r_1

Re(z)

Region of convergence for
$|G(z)|, r_2 < r < r_1$

(c)

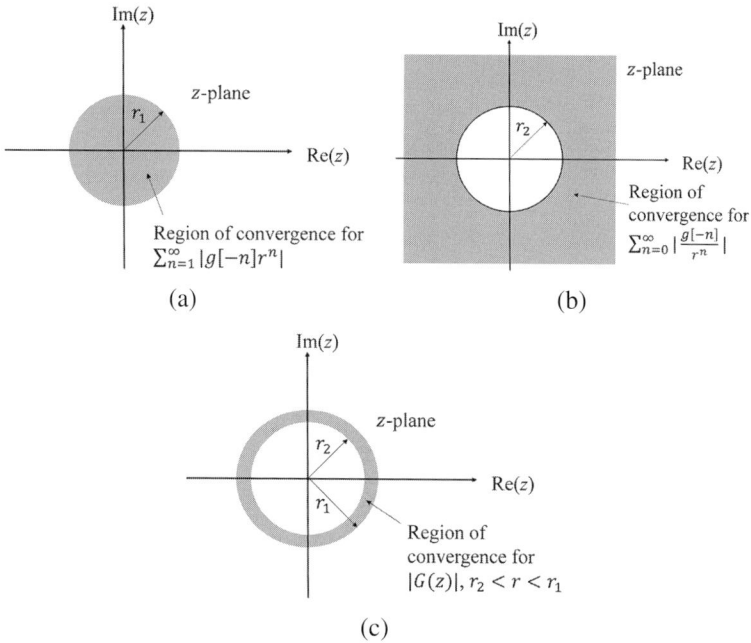

Figure 3.6 Region of convergence (ROC) of a z-transform of $g[n]$. (a) The ROC of the anti-causal part of $g[n]$, (b) the ROC of the causal part of $g[n]$, and (c) the ROC of the $g[n]$ is the intersection of both its causal and anti-causal parts.

Example 3.13 Determine the z-transform of the infinite duration signal $g[n] = (1/2)^n u[n]$.

Solution: Using Equation (3.53), one can obtain:

$$G(z) = \sum_{n=-\infty}^{\infty} g[n]z^{-n}, \tag{3.57}$$

$$= \sum_{n\geq 0} \left(\frac{1}{2}\right)^n z^{-n},$$

$$= \sum_{n\geq 0} \left(\frac{1}{2}z^{-1}\right)^n,$$

$$= \frac{1}{1 - \frac{1}{2}z^{-1}}, \text{ if } \left|\frac{1}{2z}\right| < 1 \Leftrightarrow |z| > \frac{1}{2}.$$

Its ROC can bee seen in Figure 3.17.

Example 3.14 Determine the z-transform of the signal (a and b are real numbers)

$$g[n] = a^n u[n] - b^{-n} u[-n-1] \tag{3.58}$$

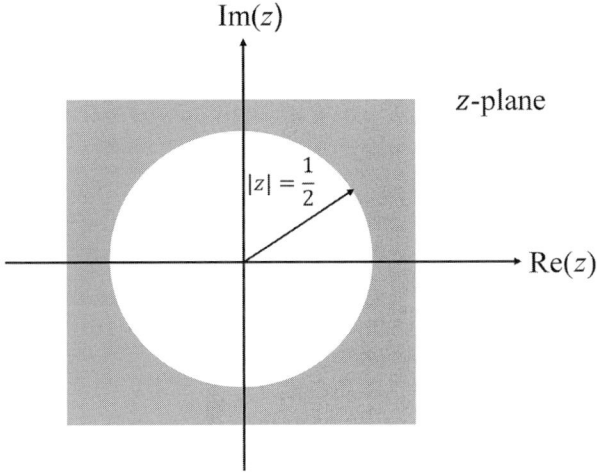

Figure 3.7 The ROC for $G(z)$ in Example 3.13.

Solution:

$$G(z) = \sum_{n=-\infty}^{\infty} g[n]z^{-n},$$

$$= \sum_{n=0}^{\infty} a^n z^{-n} - \sum_{n=-\infty}^{-1} b^n z^{-n},$$

$$= \sum_{n=0}^{\infty} (az^{-1})^n - \sum_{n=-\infty}^{-1} (bz^{-1})^n,$$

$$= \sum_{n=0}^{\infty} (az^{-1})^n - \sum_{n=1}^{\infty} (b^{-1}z)^n. \tag{3.59}$$

The first summation converges if $|az^{-1}| < 1$ or $|z| > |a|$. On the other hand, the second summation converges if $|b^{-1}z| < 1$ or $|z| < |b|$. So two cases to be considered for determining the convergence of $G(z)$ are as follows:

1. Case 1: $|b| < |a|$: $G(z)$ does not exist (DNE), since one cannot find values of z for which both power series converge simultaneously (see Figure 3.8a).
2. Case 2: $|b| > |a|$: There is a ring in the z-plane where both power series converge simultaneously (see Figure 3.8b). Then, one can obtain:

$$G(z) = \frac{1}{1 - az^{-1}} + \frac{1}{1 - bz^{-1}}$$

$$= \frac{2 - (a+b)z^{-1}}{1 - (a+b)z^{-1} + abz^{-2}}, \quad ROC: |a| < |z| < |b|. \tag{3.60}$$

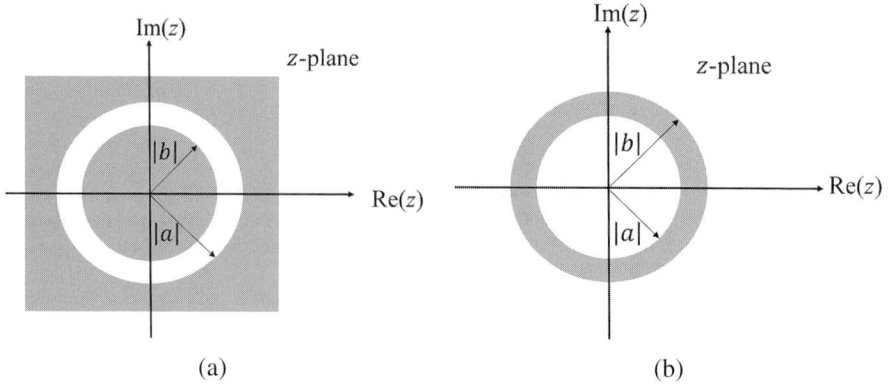

Figure 3.8 The ROC of $G(z)$ in Example 3.14 when: (a) $|a| > |b|$. In this case, the ROC does not exist. (b) The ROC exists when $|a| < |b|$.

The following is a list containing some of the very useful properties of the z-transform that can be helpful in analyzing seismic signals and related problems:

1. Linearity: Given that $g_1[n] \leftrightarrow G_1(z)$, $g_2[n] \leftrightarrow G_2(z)$ and $\alpha_1, \alpha_2 \in \mathbb{R}$, then:

$$\alpha_1 g_1[n] + \alpha_2 g_2[n] \leftrightarrow \alpha_1 G_1(z) + \alpha_2 G_2(z). \tag{3.61}$$

The proof of this relationship follows straight from Equation (3.53). This helps to find the z-transform of a signal by expressing the signal as a sum of elementary signals, for each of which the z-transform is already known.

Example 3.15 Determine the z-transform of the signal $g[n] = \cos[\omega_0 n]u[n]$.

Solution: Using Eulers identity, one can express the signal $g[n]$ as:

$$g[n] = \cos[\omega_0 n]u[n] = \frac{1}{2}[e^{j\omega_0 n} + e^{-j\omega_0 n}]u[n] = \frac{1}{2}e^{j\omega_0 n}u[n] + \frac{1}{2}e^{-j\omega_0 n}u[n].$$

So the z-transform of $g[n]$ can be determined by:

$$G(z) = \frac{1}{2}\mathcal{Z}[e^{j\omega_0 n}u[n]] + \frac{1}{2}\mathcal{Z}[e^{-j\omega_0 n}u[n]]. \tag{3.62}$$

Now, let $a = e^{\pm j\omega_0 n} \Rightarrow |a| = 1$, then,

$$e^{j\omega_0 n}u(n) \xrightarrow{\mathcal{Z}} \frac{1}{1 - e^{j\omega_0}z^{-1}}, \quad ROC: |z| > 1,$$

$$e^{-j\omega_0 n}u(n) \xrightarrow{\mathcal{Z}} \frac{1}{1 - e^{-j\omega_0}z^{-1}}, \quad ROC: |z| > 1.$$

Thus, Equation (3.62) is:

$$G(z) = \frac{1}{2} \frac{2 - z^{-1}(e^{j\omega_0} + e^{-j\omega_0})}{1 - z^{-1}(e^{j\omega_0} + e^{-j\omega_0}) + z^{-2}},$$

$$= \frac{1}{2} \frac{2 - 2z^{-1}\cos\omega_0}{1 - 2z^{-1}\cos\omega_0 + z^{-2}},$$

$$= \frac{1 - z^{-1}\cos\omega_0}{1 - 2z^{-1}\cos\omega_0 + z^{-2}}, ROC : |z| > 1. \tag{3.63}$$

2. Time shifting: Given that $g[n] \leftrightarrow G(z)$, then:

$$g[n - k] \leftrightarrow G(z)z^{-k}. \tag{3.64}$$

So time or space shifting will cause the z-transform of $g[n]$ to be multiplied by a factor z^{-k}. The ROC of $z^{-k}G(z)$ is the same at that of $G(z)$ except for $z = 0$ if $k > 0$ and $z = \infty$ if $k < 0$.

Example 3.16 Find the z-transform of $y[n] = g[n - 2]$ if

$$g[n] = \delta[n] + 2\delta[n - 1] + 5\delta[n - 2] + 7\delta[n - 3] + \delta[n - 5].$$

Solution: Recall that $G(z) = 1 + 2z^{-1} + 5z^{-2} + 7z^{-3} + z^{-5}$, where the ROC is the entire z-plane expected at $z = 0$. $y[n] = g[n - 2] \Rightarrow k = 2$. Using the shifting property in Equation (3.64), yields:

$$G(z) = z^{-2} + 2z^{-3} + 5z^{-4} + 7z^{-5} + z^{-7}.$$

The ROC remains the same, i.e., the entire z-plane expect at $z = 0$.

3. Scaling: If $g[n] \leftrightarrow G(z)$ with ROC: $r_1 < |z| < r_2$, then for a nonzero constant $\alpha \in \mathbb{C}$,

$$\alpha^n g[n] \leftrightarrow G\left(\frac{z}{a}\right), \quad \text{ROC: } |\alpha|r_1 < |z| < |\alpha|r_2. \tag{3.65}$$

Example 3.17 Determine the z-transform of the signal $g[n] = a^n \cos[\omega_0 n]u[n]$.

Solution: Recall from Equation (3.63) that:

$$\cos[\omega_0 n]u[n] \leftrightarrow \frac{1 - z^{-1}\cos\omega_0}{1 - 2z^{-1}\cos\omega_0 + z^{-2}}, \quad \text{ROC: } |z| > 1,$$

then using the scalar property as in Equation (3.65) will yield:

$$a^n \cos[\omega_0 n]u[n] \leftrightarrow \frac{1 - az^{-1}\cos\omega_0}{1 - 2az^{-1}\cos\omega_0 + az^{-2}}, \quad \text{ROC: } |z| > |a|.$$

4. Time inversion: Given that $g[n] \leftrightarrow G(z)$, ROC: $r_1 < |z| < r_2$, then:

$$g[-n] \leftrightarrow G(z^{-1}) \quad \text{ROC: } \frac{1}{r_2} < |z| < \frac{1}{r_1}. \tag{3.66}$$

Example 3.18 Find the z-transform of $u[-n]$.

Solution: Since the z-transform of:

$$u[n] \leftrightarrow \frac{1}{1 - z^{-1}}, \quad \text{ROC: } |z| > 1,$$

then using the time inversion property as in Equation (3.66) yields,

$$\therefore u[-n] \leftrightarrow \frac{1}{1 - z}, \quad \text{ROC: } |z| < 1.$$

5. Differentiation: Given that $g[n] \leftrightarrow G(z)$, then:

$$ng[n] \leftrightarrow -z \frac{dG(z)}{dz}. \tag{3.67}$$

Note that the ROC is the same for both transforms.

Example 3.19 Determine the z-transform of the signal $g[n] = na^n u[n]$.

Solution: Since one knows that:

$$a^n u[n] \leftrightarrow \frac{1}{1 - az^{-1}}, \quad \text{ROC: } |z| > |a|,$$

then,

$$na^n u[n] \leftrightarrow z \frac{d}{dz} \left[\frac{1}{1 - az^{-1}} \right] = \frac{az^{-1}}{(1 - az^{-1})^2}, \quad \text{ROC: } |z| > |a|.$$

6. Convolution: This, probably, is one of the most important z-transform properties, where the convolution in time means multiplication of the z-transform of
the two discrete-time signals in the complex z domain. Consider the following
z-transform pairs, $g_1[n] \leftrightarrow G_1(z)$ and $g_2[n] \leftrightarrow G_2(z)$. The z-transform of the
convolution between $g_1[n]$ and $g_2[n]$ can be stated as follows:

$$g_1[n] * g_2[n] \leftrightarrow G_1(z)G_2(z). \tag{3.68}$$

Example 3.20 Compute the convolution $g[n]$ of the signals:

$$g_1[n] = \delta[n] - 2\delta[n-1] + \delta[n-2],$$

$$g_2[n] = \delta[n] + \delta[n-1] + \delta[n-2] + \delta[n-3] + \delta[n-4] + \delta[n-5].$$

Solution: The z-transform of $g_1[n]$ and $g_2[n]$ can be given as:

$$G_1(z) = 1 - 2z^{-1} + z^{-2}, \qquad G_2(z) = 1 + z^{-1} + z^{-2} + z^{-3} + z^{-4} + z^{-5}.$$

Now, using the convolution property as in Equation (3.68),

$$G(z) = G_1(z)G_2(z) = 1 - z^{-1} - z^{-6} + z^{-7}.$$

Therefore,

$$g[n] = \delta[n] - \delta[n-1] - \delta[n-6] + \delta[n-7].$$

7. The Initial Value Theorem: If $g[n]$ is causal, then:

$$g[0] = \lim_{z \to \infty} G(z). \qquad (3.69)$$

8. Modulation: This property follows directly from Equation (3.65), where $\alpha = e^{j\omega n}$ and, hence:

$$e^{j\omega n} g[n] \leftrightarrow G(e^{j\omega}z). \qquad (3.70)$$

9. Conjugation: Given that $g[n] \leftrightarrow G(z)$, then:

$$g^*[n] \leftrightarrow G^*(z^*). \qquad (3.71)$$

10. Multiplication in the time domain: Let $g_1[n]$ and $g_2[n]$ be real-valued sequences with $G_1(z)$ and $G_2(z)$ being their z-transforms, respectively. Then:

$$g_1[n]g_2[n] \leftrightarrow \frac{1}{j2\pi} \oint_C G_1(v)G_2\left(\frac{z}{v}\right) v^{-1}dv, \qquad (3.72)$$

where $\oint$ is the complex integration.

3.3.2 Rational z-Transforms

An important family of z-transforms are those for which $G(z)$ is a rational function. Recall that for input $w[n]$ the output of the LSI system represented by its impulse response $h[n]$ is $g[n] = w[n] * h[n]$. The convolution property (Equation 3.68) allows expression of this relationship in the z-transform domain as:

$$G(z) = W(z)H(z). \qquad (3.73)$$

So, when $w[n]$ and $h[n]$ are known, then $g[n]$ will simply be:

$$g[n] = \mathcal{Z}^{-1}\{W(z)H(z)\}. \qquad (3.74)$$

If one knows $w[n]$ and we observe the output $g[n]$ of the system:

$$H(z) = \frac{G(z)}{W(z)}, \qquad (3.75)$$

and, therefore, one can obtain the systems impulse response:

$$h[n] = \mathcal{Z}^{-1}\{H(z)\}. \qquad (3.76)$$

$H(z)$ and $h[n]$ are equivalent descriptions of a system in the z-domain and time-domain, respectively. $H(z)$ is called the System Function (or Transfer Function).

The relation in Equation (3.76) is practically useful in obtaining $H(z)$ when the system is described by a linear constant coefficient difference equation of the form:

$$g[n] = -\sum_{k=1}^{N} a_k g[n-k] + \sum_{k=0}^{M} b_k w[n-k]. \tag{3.77}$$

In this case, the system function can be determined directly from Equation (3.77) by computing the z-transform of both sides of Equation (3.77). Thus, by applying the z-transform time-shifting property, one can obtain:

$$G(z) = -\sum_{k=1}^{N} a_k G(z) z^{-k} + \sum_{k=0}^{M} b_k W(z) z^{-k}.$$

Now, rearranging the terms of $G(z)$ yields,

$$G(z)\left[1 + \sum_{k=1}^{N} a_k G(z) z^{-k} \right] = W(z) \sum_{k=0}^{M} b_{k}, z^{-k}.$$

Then, dividing $G(z)$ by $W(z)$ will result in:

$$\frac{G(z)}{W(z)} = H(z) = \frac{\sum_{k=0}^{M} b_k z^{-k}}{1 + \sum_{k=1}^{N} a_k G(z) z^{-k}}. \tag{3.78}$$

Therefore, an LSI system described by Equation (3.77) has a rational system function. Equation (3.78) is the general form for the system (transfer) function of a system described by Equation (3.77). Now, from this general form, we obtain two important special forms:

1. If $a_k = 0$ for $1 \le k \le N$, Equation (3.78) reduces to:

$$H(z) = \sum_{k=0}^{M} b_k z^{-k}. \tag{3.79}$$

   In this case, $H(z)$ contains M zeros, whose values are determined by the system parameters b_k, and an Mth order pole at the origin $z = 0$. This is called an All-Zero system because it contains only trivial poles at $z = 0$, and M non-trivial zeros. It is also an FIR system.

2. If $b_k = 0$ for $1 \le k \le M$, then Equation (3.78) becomes

$$H(z) = \frac{b_0}{1 + \sum_{k=1}^{N} a_k G(z) z^{-k}}. \tag{3.80}$$

   In this case, $H(z)$ contains N poles, whose values are determined by the system parameters a_k, and an Nth order zero at the origin $z = 0$. The system here contains only non-trivial poles and, therefore, is called an All-Pole system. Due to the presence of poles, the impulse response of such a system has infinite duration and, hence, the system is an IIR.

In general, the form in Equation (3.78) contains both poles and zeros and, hence, the corresponding system is called a Pole-Zero system, with N poles and M zeros. Poles and/or zeros at $z = 0$ and $z = \infty$ are implied but are not contained explicitly. Finally, due to the presence of poles, a pole-zero system is an IIR system.

Example 3.21 Determine the transform function of the system described by:

$$g[n] = 0.5g[n-1] + 2w[n]. \tag{3.81}$$

Solution: Taking the z-transform of Equation (3.81) and applying the time-shift property (Equation 3.64) will yield:

$$G(z) = 0.5G(z)z^{-1} + 2W(z).$$

Now, rearranging the terms of $G(z)$ and dividing $G(z)$ by $W(z)$, the transfer function $H(z)$ will, therefore, be:

$$H(z) = \frac{G(z)}{W(z)} = \frac{2}{1 - 0.5z^{-1}} = \frac{2z}{z - 0.5}. \tag{3.82}$$

Definition 3.22 The zeros of a z-transform $G(z)$ are the values of z for which $G(z) = 0$.

Definition 3.23 The poles (Singularities) of a z-transform $G(z)$ are the values of z for which $G(z) = 1$.

If $G(z)$ is a rational function, then:

$$G(z) = \frac{B(z)}{A(z)} = \frac{b_0 + b_1 z^{-1} + b_2 z^{-2} + \cdots + b_M z^{-M}}{a_0 + a_1 z^{-1} + a_2 z^{-2} + \cdots + a_N z^{-N}} \tag{3.83}$$

$$= \frac{\sum_{k=0}^{M} b_k z^{-k}}{\sum_{k=0}^{N} a_k z^{-k}}. \tag{3.84}$$

If $a_0 \neq 0$ and $b_0 \neq 0$, one can avoid negative powers of z by factoring out the terms $b_0 z^{-M}$ and $a_0 z^{-N}$ as follows:

$$G(z) = \frac{B(z)}{A(z)} = \frac{b_0 z^{-M}}{a_0 z^{-N}} \frac{z^M + \frac{b_1}{b_0} z^{M-1} + \frac{b_2}{b_0} z^{M-2} + \cdots + \frac{b_M}{b_0}}{z^N + \frac{a_1}{a_0} z^{N-1} + \frac{a_2}{a_0} z^{N-2} + \cdots + \frac{a_N}{a_0}} \tag{3.85}$$

Since $B(z)$ and $A(z)$ are polynomials in z, they can be expressed in factored form as:

$$G(z) = \frac{b_0}{a_0} z^{-M+N} \frac{(z - z_1)(z - z_2) \cdots (z - z_M)}{(z - p_1)(z - p_2) \cdots (z - p_N)} \tag{3.86}$$

$$= \hat{G} z^{M-N} \frac{\prod_{k=1}^{M}(z - z_k)}{\prod_{k=1}^{N}(z - p_k)} \tag{3.87}$$

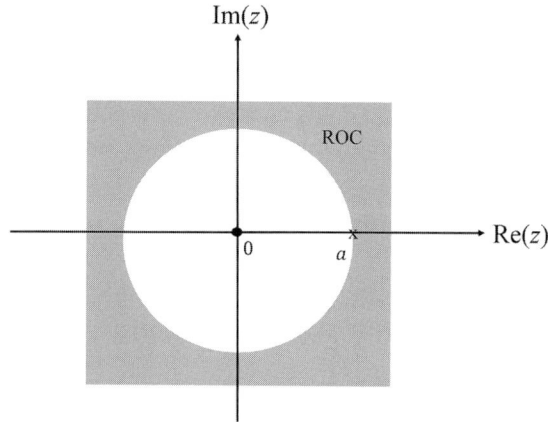

Figure 3.9 The pole-zero map of Example 3.24. $G(z)$ has a pole at $p_1 = a$ and a zero at $z_1 = 0$. Note that this pole is not included in ROC, since the z-transform DNE at a pole.

where $\hat{G} = b_0/a_0$. One can represent $G(z)$ graphically by a Pole-Zero plot (map) in the complex domain. The locations of poles are indicated by crosses (x) and the locations of zeros are indicated by circles (o). The multiplicity of multiple-order poles or zeros is indicated by a number close to the corresponding cross or circle. Note that for a given rational function of $G(z)$, the ROC of a z-transform should not contain any poles.

Example 3.24 Determine the pole-zero plot for the signal $g[n] = a^n u[n], a > 0$.

Solution: The z-transform pair of $g[n]$ can be given by:

$$a^n u[n] \leftrightarrow \frac{1}{1 - az^{-1}} = \frac{z}{z - a}, \quad \text{ROC}: |z| > a,$$

where, clearly, $G(z)$ has a zero at $z = 0$ and a pole at $z = a$.

3.3.3 The Inverse z-Transform

The inverse z-transform, on the other hand, is given by:

$$g[n] = \frac{1}{j2\pi} \oint_C G(z) z^{n-1} dz. \tag{3.88}$$

This integration has to be carried out counter clockwise on a closed contour C in the complex domain, which encloses the origin and lies in the ROC of $G(z)$. In certain cases, the inverse z-transform can be determined in simpler ways, where one might use the residue theorem (Bronshtein et al., 2007), the partial fraction expansion, where $G(z)$ is a rational function of z, or direct expansion of $G(z)$ into a power series in z^{-1}.

Given a z-transform $G(z)$ with its corresponding ROC, one can expand $G(z)$ into a power series of the form:

$$G(z) = \sum_{n=-\infty}^{\infty} c_n z^{-n}, \tag{3.89}$$

which converges in the given ROC. Then, by uniqueness of the z-transform, $g[n] = c_n$ for all n. When $G(z)$ is rational, the expansion can be performed by long division.

Example 3.25 Find the inverse z-transform of $G(z) = \frac{1}{1-\frac{3}{2}z^{-1}+\frac{1}{2}z^{-2}}$ when (a) ROC: $|z| > 1$ and (b) ROC: $|z| < 0.5$.

Solution: (a) Since ROC is an exterior of a circle, one expects $g[n]$ to be causal. Thus, a power series expansion in negative powers of z will be sought. Now, one can obtain the following through long division. But first multiply $G(z)$ by $\frac{z^2}{z^2}$ to obtain:

$$G(z) = \frac{1}{1 - \frac{3}{2}z^{-1} + \frac{1}{2}z^{-2}} = \frac{z^2}{z^2 - \frac{3}{2}z + \frac{1}{2}}.$$

Now, perform long division, which yields the following:

$$G(z) = 1 + \frac{3}{2}z^{-1} + \frac{7}{4}z^{-2} + \frac{15}{8}z^{-3} + \frac{31}{16}z^{-4} + \cdots,$$

and, hence,

$$g[n] = \delta[n] + \frac{3}{2}\delta[n-1] + \frac{7}{4}\delta[n-2] + \frac{15}{8}\delta[n-3] + \frac{31}{16}\delta[n-4] + \cdots$$

(b) In this case, the ROC is the interior of a circle, which implies that $g[n]$ is anti-causal. To obtain a power series expansion in positive powers of z, directly perform long division of $G(z)$ to obtain:

$$G(z) = 2z^2 + 6z^3 + 14z^4 + 30z^5 + \cdots.$$

Now, taking the inverse z-transform of $G(z)$ will result in:

$$g[n] = \cdots + 30\delta[n+5] + 14\delta[n+4] + 6\delta[n+3] + 2\delta[n+2].$$

This example illustrates that the ROC is important. Also, it says that this technique might not be the most convenient and efficient way for determining the inverse z-transform of a rational function. This is very useful if one is interested to know the first few values of $g[n]$. However, it does not provide a closed form, in general, for $g[n]$. Hence, this requires a more convenient method to obtain the inverse z-transform of a rational function. Consider a rational function given in the form:

$$G(z) = \frac{B(z)}{A(z)} = \frac{b_0 + b_1 z^{-1} + b_2 z^{-2} + \cdots + b_M z^{-M}}{a_0 + a_1 z^{-1} + a_2 z^{-2} + \cdots + a_N z^{-N}}, \tag{3.90}$$

$$= \frac{\sum_{k=0}^{M} b_k z^{-k}}{\sum_{k=0}^{N} a_k z^{-k}}.$$

Equation (3.90) is called proper if $a_n \neq 0$ and $M < N$. Otherwise, it is called improper. If one has a proper rational function then this is equivalent to saying that for this rational function, the number of finite zeros is less than the number of finite poles. An improper rational function can always be written as the sum of a polynomial, and a proper rational function as:

$$G(z) = \frac{B(z)}{A(z)} = c_0 + c_1 z^{-1} + c_2 z^{-2} + \cdots + c_{M-N} z^{-(M-N)} + \frac{B_1(z)}{A(z)}, \tag{3.91}$$

where the inverse z-transform of a polynomial can easily be found by inspection. In the following discussion, it will be interesting to determine the inverse z-transform of proper rational functions. This can be done in two steps:

1. Perform a partial fraction expansion of such a proper rational function.
2. Invert each term of the expansion using known z-transforms pairs.

Let $G(z)$ be a proper rational function. For simplifying the following discussion, we eliminate negative powers of z by multiplying both the numerator and denominator of Equation (3.90) by z^N as follows:

$$G(z) = \frac{z^N}{z^N} \frac{b_0 + b_1 z^{-1} + b_2 z^{-2} + \cdots + b_M z^{-M}}{1 + a_1 z^{-1} + a_2 z^{-2} + \cdots + a_N z^{-N}}, \tag{3.92}$$

$$= \frac{b_0 z^N + b_1 z^{N-1} + b_2 z^{N-2} + \cdots + b_M z^{M-M}}{z^N + a_1 z^{N-1} + a_2 z^{N-2} + \cdots + a_N}.$$

Since $N > M$, the following function:

$$\frac{G(z)}{z} = \frac{b_0 z^{N-1} + b_1 z^{N-2} + b_2 z^{N-3} + \cdots + b_M z^{M-M-1}}{z^N + a_1 z^{N-1} + a_2 z^{N-2} + \cdots + a_N} \tag{3.93}$$

is always proper. Therefore, the poles of Equation (3.93) will determine the type of partial fraction expansion of $G(z)$ that will lead to calculating the inverse z-transform of a rational function $G(z)$.

Distinct Poles

Suppose that the poles $p_1, p_2, \ldots, p_N$ of a rational function $G(z)$ are all distinct. Then, one seeks an expansion of the form:

$$\frac{G(z)}{z} = \frac{A_1}{z - p_1} + \frac{A_2}{z - p_2} + \cdots + \frac{A_N}{z - p_N}. \tag{3.94}$$

The problem turns to finding the values of the residues of the expansion $A_1, A_2, \ldots, A_N$. This can be obtained using the following formula:

$$A_k = (z - p_k)\frac{G(z)}{z}\Big|_{z=p_k}. \tag{3.95}$$

This works for both real and complex poles that are distinct. In the case when $G(z)$ has poles that are complex-conjugate, they result in complex-conjugate residues also in the partial fraction expansion. From Equation (3.94), it easily follows that:

$$G(z) = A_1\frac{1}{1 - p_1 z^{-1}} + A_2\frac{1}{1 - p_2 z^{-1}} + \cdots + A_N\frac{1}{1 - p_N z^{-1}}. \tag{3.96}$$

The inverse z-transform, $g[n] = \mathcal{Z}^{-1}\{G(z)\}$, can be obtained by inverting each term in Equation (3.96) and taking the corresponding linear combination. The following formula can be used:

$$\mathcal{Z}\left\{\frac{1}{1 - p_k z^{-1}}\right\} = \begin{cases} (p_k)^n u[n], & \text{if ROC: } |z| > |p_k|, \\ -(p_k)^n u[-n-1] & \text{if ROC: } |z| < |p_k|. \end{cases} \tag{3.97}$$

Note that ff the signal $g[n]$ is causal, then its ROC is:

$$|z| > p_{max} = \max\{|p_1|, |p_2|, \ldots, |p_N|\}. \tag{3.98}$$

In this case, all terms of Equation (3.96) result in causal signal components. Hence, the signal $g[n]$ is given by:

$$g[n] = \underbrace{\left(A_1 p_1^n + A_2 p_2^n + \cdots + A_N p_N^n\right) u[n]}_{\text{A linear combination of real signals}}. \tag{3.99}$$

Assume now that all poles are distinct but some of them are complex. This implies that some terms in Equation (3.96) result in complex exponential components. However, if our signal is real, one should be able to reduce these terms into real components. If $g[n]$ is real, then the polynomial appearing in $G(z)$ has real residue coefficients. Hence, if p_j is a pole with a corresponding residue A_j, its complex conjugate p_j^* is also a pole and with a residue A_j^*. Thus, the combination of two complex-conjugate poles is of the form:

$$g_k[n] = (A_k(p_k)^n + A_k^*(p_k^*)^N)u[n], \tag{3.100}$$

or,

$$g_k[n] = 2|A_k|r_k \cos(\beta_k n + \alpha_k)u[n], \tag{3.101}$$

where $\beta_k = \angle p_k, \alpha_k = \angle A_k$, and $r_k = |p_k|$ if the ROC is $|z| > |p_k| = r_k$. It is observed that each pair of complex-conjugate poles in the z-domain results in a causal sinusoidal signal component with an exponential envelope. Note that the distance r_k of the pole from the origin determines the exponential weighting: growing if $r_k > 1$, decaying if $r_k < 1$, and constant if $r_k = 1$. The angle of the poles with respect to the positive real axis provides the frequency of the sinusoidal signal. Also, the zeros of the rational function affect only indirectly the amplitude and the phase of $g_k[n]$ through A_k.

Example 3.26 Assume $g[n]$ is causal. Determine the partial fraction expansion and inverse z-transform of:

$$G(z) = \frac{1}{1 - \frac{3}{2}z^{-1} + \frac{1}{2}z^{-2}}. \tag{3.102}$$

Solution: In order to obtain the partial fraction expansion, one, first, is required to obtain $G(z)/z$ as follows:

$$
\begin{aligned}
G(z) &= \frac{1}{1 - \frac{3}{2}z^{-1} + \frac{1}{2}z^{-2}}, \\
&= \frac{z^2}{z^2 - \frac{3}{2}z + \frac{1}{2}}, \\
&= \frac{z^2}{(z - 1)(z - \frac{1}{2})}, \\
\Rightarrow \frac{G(z)}{z} &= \frac{z}{(z - 1)(z - \frac{1}{2})}, \\
&= \frac{A_1}{z - 1} + \frac{A_2}{z - \frac{1}{2}}.
\end{aligned}
$$

Then, the residue A_1 is computed using:

$$A_1 = \left.\frac{(z - 1)G(z)}{z}\right|_{z=1} = (z - 1) \left.\frac{z}{(z - 1)(z - \frac{1}{2})}\right|_{z=1} = 2,$$

and,

$$A_2 = \left.\frac{(z - \frac{1}{2})G(z)}{z}\right|_{z=1/2} = \left(z - \frac{1}{2}\right) \left.\frac{z}{(z - \frac{1}{2})(z - \frac{1}{2})}\right|_{z=1/2} = -1.$$

Therefore,

$$G(z) = \frac{2z}{z-1} - \frac{z}{z-\frac{1}{2}},$$

$$= \frac{2}{1-z^{-1}} - \frac{1}{1-\frac{1}{2}z^{-1}}.$$

Now, since $g[n]$ is causal (note the ROC is $|z| > \max\{1, \frac{1}{2}\}$), then $g[n] = (2 - (\frac{1}{2})^n)u[n]$.

Multiple Poles

If $G(z)$ has a pole of multiplicity m, that is, it contains in its denominator the factor $(z - p_k)^m$, then a different partial fraction expansion is required. Let $G(z)$ be a rational function, where its poles are of multiplicity m. Then, one can write $G(z)/z$ as follows:

$$\frac{G(z)}{z} = \frac{A_{1,j}}{z-p_k} + \frac{A_{2,k}}{(z-p_k)^2} + \cdots + \frac{A_{m,k}}{(z-p_k)^m}. \qquad (3.103)$$

To find the residues of $G(z)/z$, one can use the following procedure. Start by computing $A_{m,k}$:

$$A_{m,k} = \frac{(z-p_k)^m G(z)}{z}\Big|_{z=p_k}. \qquad (3.104)$$

Then, proceed by computing the residues backward as follows:

$$A_{m-1,k} = \frac{1}{1!}\frac{d}{dz}\left[\frac{(z-p_k)^m G(z)}{z}\right]_{z=p_k},$$

$$= A_{m-2,k} = \frac{1}{2!}\frac{d^2}{dz^2}\left[\frac{(z-p_k)^m G(z)}{z}\right]_{z=p_k},$$

$$\vdots$$

$$A_{m-l,k} = \frac{1}{l!}\frac{d^l}{dz^l}\left[\frac{(z-p_k)^m G(z)}{z}\right]_{z=p_k}, \quad \text{until } l = m-1.$$

In this case, the inverse transform of the terms of the form $A/(z-p_k)^m$ is required. In the case of a pole with multiplicity $m = 2$, the following transform pair is useful:

$$Z^{-1}\left\{\frac{pz^{-1}}{(1-pz^{-1})^2}\right\} = np^n u[n], \qquad (3.105)$$

provided that the ROC is $|z| > p$.

Example 3.27 Determine the partial fraction expansion and the inverse z-transform of:

$$G(z) = \frac{1}{(1 - z^{-1})^2},$$

(3.106)

where $g[n]$ is causal.

Solution: Again, in order to obtain the partial fraction expansion, one, first, is required to obtain $G(z)/z$ as follows:

$$G(z) = \frac{1}{(1 - z^{-1})^2},$$

$$= \frac{1}{1 - 2z^{-1} + z^{-2}},$$

$$= \frac{z^2}{z^2 - 2z + 1},$$

$$= \frac{z^2}{(z - 1)^2},$$

$$\frac{G(z)}{z} = \frac{z}{(z - 1)^2},$$

$$= \frac{A_{1,1}}{z - 1} + \frac{A_{2,1}}{(z - 1)^2}.$$

The first residue $A_{2,1}$ is computed as follows:

$$A_{2,1} = \left. \frac{(z - 1)^2 G(z)}{z} \right|_{z=1} = \left. (z - 1)^2 \frac{z}{(z - 1)^2} \right|_{z=1} = 1.$$

Also, the second residue $A_{1,1}$ is computed as follows:

$$A_{1,1} = \frac{d}{dz} \left[\left. \frac{(z - 1)^2 G(z)}{z} \right|_{z=1} \right] = \frac{d}{dz} \left[\left. (z - 1)^2 \frac{z}{(z - 1)^2} \right|_{z=1} \right] = 1.$$

Therefore,

$$G(z) = \frac{1}{1 - z^{-1}} - \frac{1}{(1 - z^{-1})^2}.$$

Hence, $g[n] = \mathcal{Z}^{-1}\{G(z)\} = (1 - n)u[n]$.

3.3.4 Analysis of LSI Systems in the z-Domain

Recall that a causal LSI system is one whose unit sample response $h[n]$ satisfies the condition $h[n] = 0$, $n < 0$. Also, recall that the ROC of the z-transform of a causal sequence is the exterior of a circle. Hence, an LSI system is causal if and only if the ROC of the system function is the exterior of a circle of radius $r < 1$, including the point $z = \infty$.

Now, consider stability of an LSI system. Recall that a necessary and sufficient condition for an LSI system to be BIBO stable is that its impulse response $h[n]$ is absolutely summable, i.e., $\sum |h[n]| < \infty$. This implies that $H(z)$ must contain the unit circle within its ROC for the system to be BIBO stable. Hence, if the system is BIBO stable, the unit circle is contained in the ROC of $H(z)$. The converse is also true. Therefore, an LSI system is BIBO if and only if the ROC of the system function includes the unit circle. Now, since the ROC cannot contain any poles of $H(z)$, it follows that a Causal LSI system is BIBO stable if and only if all the poles of $H(z)$ are inside the unit circle.

Example 3.28 An LSI system has the following system function:

$$H(z) = \frac{1}{1 - \frac{1}{4}z^{-1}} + \frac{2}{1 - 5z^{-1}}.$$

Specify the ROC of $H(z)$ for the following conditions:

1. The system is stable.
2. The system is causal.
3. The system is anti-causal.

Solution: The system $H(z)$ has poles $p_1 = 1/4$ and $p_2 = 5$.

1. Since the system is stable, its ROC must include the unit circle and, hence, its ROC: $1/4 < |z| < 5$. This means that $h[n]$ is non-causal and is given by:

$$h[n] = \left(\frac{1}{4}\right)^n u[n] - 2(5)^n u[-n-1].$$

2. Since the system is causal, its ROC: $|z| > 5$. This means that $h[n]$ is unstable and its impulse response will be:

$$h[n] = \left(\frac{1}{4}\right)^n u[n] + 2(5)^n u[n].$$

3. Since the system is anti-causal, its ROC: $|z| < 1/4$. This means that $h[n]$ is unstable and, hence,

$$h[n] = -\left[\left(\frac{1}{4}\right)^n + 2(5)^n\right] u[-n-1].$$

Example 3.29 Consider an LSI system with the following system function:

$$H(z) = \frac{z^{-1} - \frac{1}{3}}{1 - \frac{1}{3}z^{-1}}.$$

(a) Find a difference equation to implement this system.
(b) If $H(z)$ is cascaded with another system represented by the system function $F(z)$ so that the overall system function is unity, i.e., $H(z)F(z) = 1$. If $F(z)$ is to be stable, then obtain $f[n]$.
(c) Is $f[n]$ casual, anti-causal, or non-causal?

Solution: (a) Consider the transfer function $H(z)$, which can be written as

$$H(z) = \frac{G(z)}{W(z)} = \frac{z^{-1} - \frac{1}{3}}{1 - \frac{1}{3}z^{-1}}.$$

Now,

$$\Leftrightarrow G[z] - \frac{1}{3}z^{-1}G[z] = z^{-1}W(z) - \frac{1}{3}W(z).$$

Hence, $g[n] = \frac{1}{3}g[n-1] - \frac{1}{3}w[n] + w[n-1]$.

(b) $H(z)F(z) = 1$, which implies that:

$$F(z) = \frac{1}{H(z)} = \frac{1 - \frac{1}{3}z^{-1}}{z^{-1} - \frac{1}{3}} = \frac{(-3)(1 - \frac{1}{3}z^{-1})}{1 - 3z^{-1}}.$$

$F(z)$ has a zero at $z = 1/3$ and a pole at $z = 3$. Hence, for $F(z)$ to be stable, the unit circle must be included in the ROC of $F(z)$ and the ROC of $F(z)$ is $|z| < 3$. Therefore, one can show that $f[n] = 8(3)^{n-2}u[-n-1]$.

(c) $f[n]$ is an anti-causal sequence.

3.4 The Fourier Transform

The Fourier transform is one of several mathematical tools that are useful in the analysis and design of LSI systems. Such representations basically involve the decomposition of signals in terms of complex exponential/sinusoidal components. The signal with such decomposition is said to be represented in the Frequency (or wavenumber) Domain. Seismic signals are examples of signals that are heavily analyzed and processed in the frequency (or wavenumber) domain (or both frequency and wavenumber domains). Frequency/wavenumber analysis of a signal (which is a function of time/space) involves the resolution of the signal into its frequency/wavenumber (sinusoidal) components. Frequency/wavenumber synthesis involves the recombination of the sinusoidal components to reconstruct the original signal. Spectrum is used when referring to the frequency/wavenumber content of a signal. For the sake of analysis, the discrete-time Fourier transform will be discussed, followed by the discrete Fourier transform.

The Fourier transform of a finite-energy discrete-time seismic signal $g[n]$ (or DTFT of $g[n]$) is defined as:

$$G(e^{j\omega}) = \sum_{n=-\infty}^{\infty} g[n]e^{-j\omega n}, \tag{3.107}$$

where $G(e^{j\omega})$ represents the frequency content of the signal $g[n]$. In order for Equation (3.107) to exist, $g[n]$ must be absolutely summable. The inverse Fourier transform can be obtained using:

$$g[n] = \frac{1}{2\pi} \int_{-\pi}^{\pi} G(e^{j\omega}) e^{j\omega n} d\omega. \tag{3.108}$$

The frequency range for a discrete-time signal is unique over the frequency interval $(-\pi, \pi)$ or, equivalently, $(0, 2\pi)$, where $G(e^{j\omega})$ is continuous and periodic with period 2π. The inverse DTFT is simply the inverse z-transform evaluated on the unit circle. Note that the z-transform $G(z)$ of $g[n]$ exists based on its ROC, and if its ROC contains the unit circle, then the DTFT $G(e^{j\omega})$ exists. However, the existence of $G(e^{j\omega})$ does not necessarily ensure the existence of the z-transform $G(z)$.

Example 3.30 Determine the DTFT of the following signals: (a) $g[n] = 3\delta[n] - 2\delta[n-1] + \delta[n-2]$, and (b) $g[n] = (\frac{1}{2})^n u[n+3]$.

Solution: (a) Using Equation (3.107), the DTFT of $g[n]$ can be computed as follows:

$$G(e^{j\omega}) = \sum_{n=-\infty}^{\infty} g[n] e^{-j\omega n} = 3 - 2e^{-j\omega n}, \tag{3.109}$$

$$= [3 - 2\cos(\omega) + \cos(2\omega)] + j[-2\sin(\omega) + \sin(2\omega)],$$

where the real part $\Re\{G(e^{j\omega})\}$ is $[3 - 2\cos(\omega) + \cos(2\omega)]$ and an imaginary part $\Im\{G(e^{j\omega})\}$ is $[-2\sin(\omega) + \sin(2\omega)]$. Hence, the magnitude and phase of $G(e^{j\omega})$ is given by, respectively:

$$|G(e^{j\omega})| = \sqrt{(3 - 2\cos(\omega) + \cos(2\omega))^2 + (-2\sin(\omega) + \sin(2\omega))^2},$$

and

$$\angle G(e^{j\omega}) = \arctan\left\{ \frac{3 - 2\cos(\omega) + \cos(2\omega)}{-2\sin(\omega) + \sin(2\omega)} \right\}.$$

(b) The DTFT of $g[n]$ can be computed as follows:

$$G(e^{j\omega}) = \sum_{n=-\infty}^{\infty} g[n] e^{-j\omega n} = \sum_{n=-\infty}^{\infty} \left(\frac{1}{2}\right)^n u(n+3) e^{-j\omega n}, \tag{3.110}$$

$$= \sum_{n=-3}^{\infty} \left(\frac{1}{2} - j\omega\right)^n,$$

$$= \frac{8e^{j3\omega}}{1 - \frac{1}{2}e^{-j\omega}}.$$

For the direct Fourier transform, one can adopt the notation:

$$G(e^{j\omega}) = \mathcal{F}\{g[n]\} = \sum_{n=-\infty}^{\infty} g[n]e^{-j\omega n}, \qquad (3.111)$$

while for the inverse Fourier transform:

$$g[n] = \mathcal{F}^{-1}\{G(e^{j\omega})\} = \frac{1}{2\pi} \int_{-\pi}^{\pi} G(e^{j\omega})e^{j\omega n} d\omega. \qquad (3.112)$$

$g[n]$ and $G(e^{j\omega})$ are a Fourier transform pair denoted as:

$$g[n] \leftrightarrow G(e^{j\omega}).$$

Here is a list of Fourier Transform Theorems and Properties that are useful for analysis of systems and signals.

1. Linearity: Given that $g_1[n] \leftrightarrow G_1(e^{j\omega})$, $g_2[n] \leftrightarrow G_2(e^{j\omega})$ and $\alpha_1, \alpha_2 \in \mathbb{R}$, then:

$$\alpha_1 g_1[n] + \alpha_2 g_2[n] \leftrightarrow \alpha_1 G_1(e^{j\omega}) + \alpha_2 G_2(e^{j\omega}). \qquad (3.113)$$

2. Time Shifting: If $g[n] \leftrightarrow G(e^{j\omega})$, then:

$$g[n-k] \leftrightarrow e^{-j\omega k} G(e^{j\omega}). \qquad (3.114)$$

3. Time Reversal: Given $g[n] \leftrightarrow G(e^{j\omega})$, then:

$$g[-n] \leftrightarrow G(e^{-j\omega}). \qquad (3.115)$$

4. Convolution Theorem: Consider the DTFT $g_1[n] \leftrightarrow G_1(\omega)$ and $g_2[n] \leftrightarrow G_2(\omega)$. The DTFT of the convolution between $g_1[n]$ and $g_2[n]$ can be stated as follows:

$$g_1[n] * g_2[n] \leftrightarrow G_1(e^{j\omega})G_2(e^{j\omega}). \qquad (3.116)$$

5. Frequency Shifting: If $g[n] \leftrightarrow G(e^{j\omega})$, then:

$$e^{-j\omega_o n} g[n] \leftrightarrow G(e^{j\omega} - e^{j\omega_o}). \qquad (3.117)$$

6. Modulation: Given $g[n] \leftrightarrow G(e^{j\omega})$, then:

$$g[n]\cos[\omega_o n] \leftrightarrow \frac{1}{2}[G(e^{j\omega} + e^{j\omega_o}) + G(e^{j\omega} - e^{j\omega_o})]. \qquad (3.118)$$

7. Parserval's Theorem: Consider the DTFT $g_1[n] \leftrightarrow G_1(e^{j\omega})$ and $g_2[n] \leftrightarrow G_2(e^{j\omega})$. Then:

$$\sum_{n=-\infty}^{\infty} g_1[n]g_2^*[n] = \frac{1}{2\pi} \int_{-\pi}^{\pi} G_1(e^{j\omega})G_2^*(e^{j\omega})d\omega. \qquad (3.119)$$

Note that when $g_1[n] = g_2[n] = g[n]$, then Equation (3.119) reduces to:

$$\sum_{n=-\infty}^{\infty} |g[n]|^2 = \frac{1}{2\pi} \int_{-\pi}^{\pi} |G(e^{j\omega})|^2 d\omega, \qquad (3.120)$$

which basically states that the energy of $g[n]$ can be computed in the time(space) domain as well as in the frequency domain.

8. Windowing Theorem: Given the DTFT pairs $g_1[n] \leftrightarrow G_1(e^{j\omega})$ and $g_2[n] \leftrightarrow G_2(e^{j\omega})$, then DTFT of the multiplication between $g_1[n]$ and $g_2[n]$ can be given as follows:

$$g_1[n]g_2[n] \leftrightarrow \frac{1}{2\pi} \int_{-\pi}^{\pi} G_1(e^{j\Omega})G_2(e^{j\omega} - e^{j\Omega})d\Omega. \qquad (3.121)$$

9. Differentiation in Frequency Domain: If $g[n] \leftrightarrow G(e^{j\omega})$, then:

$$ng[n] \leftrightarrow j\frac{dG(e^{j\omega})}{d\omega}. \qquad (3.122)$$

10. Conjugation: Given $g[n] \leftrightarrow G(e^{j\omega})$, then:

$$g^*[n] \leftrightarrow G^*(e^{-j\omega}). \qquad (3.123)$$

Example 3.31 Find the inverse DTFT of the system described by the following frequency response:

$$H(e^{j\omega}) = \begin{cases} -j & \text{for } 0 < \omega < \pi, \\ j & \text{for } -\pi < \omega < 0. \end{cases} \qquad (3.124)$$

Solution: The given system shows that the magnitude spectrum is unity for all the values of ω. The phase is equal to $\frac{-\pi}{2}$ for $0 < \omega < \pi$ and is $\frac{\pi}{2}$ for $-\pi < \omega < 0$. Now, using Equation (3.108), one can show that:

$$h[n] = \frac{1}{2\pi} \int_{-\pi}^{0} je^{jn\omega} d\omega - \frac{1}{2\pi} \int_{0}^{\pi} je^{jn\omega} d\omega \qquad (3.125)$$

$$= \frac{1}{n\pi}[1 - (-1)^n],$$

which is equal to:

$$h[n] = \begin{cases} \frac{2}{\pi} \frac{\sin^2(\frac{n\pi}{2})}{n} & \text{for } n \neq 0, \\ 0 & \text{for } n = 0. \end{cases} \qquad (3.126)$$

This is known as a 90° phase shifter system.

A seismic shot record is shown in Figure 3.10a, while trace number 1 of the same record is shown in both time (Figure 3.10b) and frequency (Figure 3.10c and d) domains. From the magnitude spectrum of the 1-D seismic trace, most of

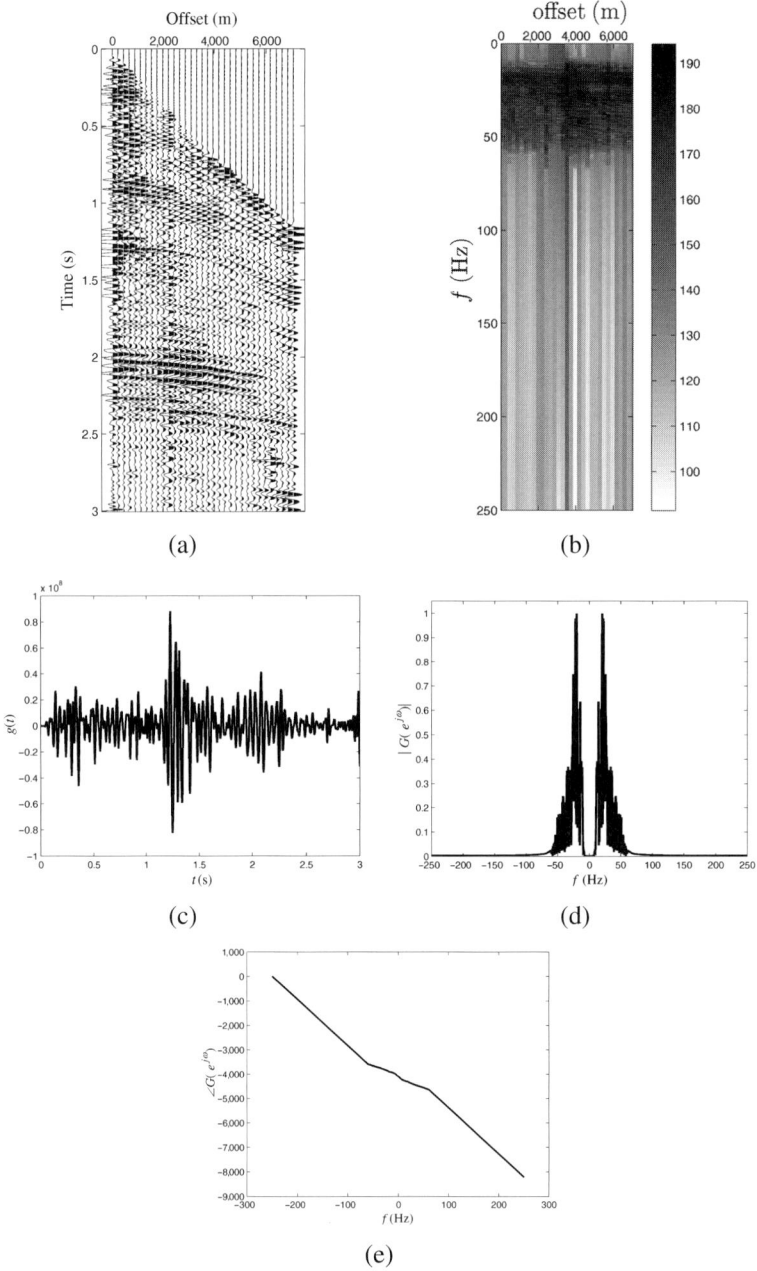

(a)

(b)

(c)

(d)

(e)

Figure 3.10 (a) A seismic data shot gather from Mousa and Al-Shuhail (2011). (b) Seismic trace number 1 of (a). (c) The magnitude spectrum of (b), where the frequencies are within the range 10–70 Hz and the magnitude is even. (d) The phase spectrum of (b), where the phase is odd. (e)The frequency-space magnitude spectrum in dB of (a).

the seismic signal frequencies are within the range 10–70 Hz where the bandwidth is about 60 Hz. Also, since the signal is real, its magnitude spectrum is even, while its phase spectrum is odd. Figure 3.10e shows the frequency-space $(f-x)$ representation, where the magnitude spectrum in dB of every seismic trace in Figure 3.10a, is displayed. Again, one can notice that most of the frequencies are within the range 10–70 Hz.

The DTFT is useful for analyzing signals and systems. However, $G(e^{j\omega})$ is continuous and, hence, cannot be represented in computers. Hence, the Discrete Fourier Transform is discussed next.

3.4.1 The Discrete Fourier Transform

The frequency analysis of discrete-time (space) signals is a continuous function of frequency, which is computationally not a convenient representation, since the computation is performed on digital processors. The representation of a discrete-time signal by samples of its spectrum is considered, which leads to the discrete Fourier transform (DFT). DFT is a powerful computational tool for performing frequency analysis of discrete-time signals. Continuous time aperiodic signals, such as seismic, have frequency response that is continuous and aperiodic. Sampling of continuous time aperiodic signals results in a discrete-time aperiodic signal, which has frequency response that is continuous and periodic. One can reconstruct the discrete-time aperiodic signal from its continuous and periodic frequency signal provided that it is sampled properly equal to the sampling frequency in the time domain (no aliasing of the periodic continuous frequency signal). The reconstruction is achieved using a low pass filter of cutoff frequency.

Also, recall the duality of time and frequency. Continuous frequency periodic signals have a discrete-time aperiodic sequence. Sampling of continuous frequency periodic signals results in a discrete frequency periodic signal that has a time sequence that is discrete and periodic. Hence, one can reconstruct the discrete-time aperiodic signal from its discrete frequency periodic signal, provided that it is sampled well enough in the frequency domain (no overlapping or aliasing of the discrete-time periodic signal). The reconstruction is achieved by taking only N samples of the periodic discrete-time signal.

With this in mind, the DFT will mathematically be discussed. Consider, a finite duration sequence $g[n]$ of length L has a DTFT:

$$G(e^{j\omega}) = \sum_{n=0}^{L-1} g[n]e^{-j\omega n}, \tag{3.127}$$

where $0 \leq \omega \leq 2\pi$. Sampling $G(e^{j\omega})$ at N equally spaced frequencies, where $N \geq L$, i.e., $\omega_l = 2\pi l/N$ for $k = 0, 1, \ldots, N-1$, will result in the following equation:

$$G[l] = G(e^{j\omega})|_{\omega_l = 2\pi l/N} = \sum_{n=0}^{L-1} g[n] e^{\frac{-j2\pi nl}{N}}, \tag{3.128}$$

or, simply,

$$G[l] = \sum_{n=0}^{N-1} g[n] e^{\frac{-j2\pi nl}{N}}, \tag{3.129}$$

where for convenience the upper index is increased from $L-1$ to $N-1$. Note that Equation (3.129) is the N-point DFT of $g[n]$. To recover $g[n]$ from the frequency domain samples $G[l]$, one can use the so-called N-point inverse DFT (IDFT):

$$g[n] = \frac{1}{N} \sum_{l=0}^{N-1} G[l] e^{\frac{j2\pi nl}{N}}. \tag{3.130}$$

Note also that when $g[n]$ has a length $L < N$, then the N-point IDFT yields $g[n] = 0$ for $L \leq n \leq N-1$. The DFT is applicable only to finite length signals, such as seismic signals, and the N-point DFT of $g[n]$ uniquely determines $g[n]$ if and only if $g[n]$ is of length $L \leq N$. In other words, one has to sample the frequency response at $N > L$ points. If not, then one will have time aliasing. The DFT is a sequence rather than a function of a continuous variable, and it corresponds to samples equally spaced in frequency of the Fourier transform of the signal. The DFT plays a central part in the implementation of a variety of digital signal processing algorithms to seismic signals and others. Examples include spectral analysis and linear filtering because DFT is efficiently implemented using the fast Fourier transform (FFT).

Example 3.32 Determine the DFT of (a) $g[n] = \delta[n]$, and (b) $g[n] = \delta[n-m]$.

Solution: Using the DFT definition in Equation (3.129), one obtain:

$$G[l] = \sum_{n=0}^{N-1} \delta[n] e^{-j2\pi nl/N} = 1, \quad 0 \leq l \leq N-1.$$

Similarly,

$$G[l] = \sum_{n=0}^{N-1} \delta[n-m] e^{-j2\pi kn/N} = e^{-j2\pi lm/N}, \quad 0 \leq l \leq N-1.$$

Example 3.33 Compute the 4-point DFT of:

$$g[n] = 2\delta[n] + \delta[n-1] + 2\delta[n-2] + \delta[n-3].$$

Solution: $N = 4$, which implies that $0 \le l \le 3$. Using Equation (3.129):

$$G[l] = \sum_{n=0}^{3} g[n] e^{-j2\pi ln/4} = 2 + e^{-j\pi l/2} + 2e^{-j\pi l} + e^{-j3\pi l/2}.$$

Hence,

$$G[0] = 6,$$
$$G[1] = 0,$$
$$G[2] = 2,$$
$$G[3] = 0,$$

or

$$G[l] = 6\delta[l] + 2\delta[l-2].$$

Example 3.34 Find the 4-point IDFT of:

$$G[l] = 60\delta[l] - 4\delta[l-2].$$

Solution: To obtain the IDFT, Equation (3.130) is followed. Hence,

$$g[n] = \sum_{n=0}^{3} G[l] e^{j2\pi ln/4} = \frac{1}{4}(60 - 4e^{j\pi n}).$$

$$g[0] = 14,$$
$$g[1] = 16,$$
$$g[2] = 14,$$
$$g[3] = 16,$$

or, simply,

$$g[n] = 14\delta[n] + 16\delta[n-1] + 14\delta[n-2] + 16\delta[n-3].$$

Example 3.35 Find the N-point DFT of $g[n]$, where it is given by:

$$g[n] = \begin{cases} 1, & 0 \le n \le L - 1. \\ 0, & \text{otherwise.} \end{cases}$$

Solution: For comparison purposes, $G(e^{j\omega})$ is calculated as follows:

$$G(e^{j\omega}) = \sum_{n=0}^{L-1} g[n]e^{-j\omega n},$$

$$= \sum_{n=0}^{L-1} e^{-j\omega n},$$

$$= \frac{1 - e^{-j\omega L}}{1 - e^{-j\omega}},$$

$$= \frac{\sin(\omega L/2)}{\sin(\omega/2)}e^{-j\omega(L-1)/2}.$$

Now, compuating $G[l]$ should lead to:

$$G[l] = \sum_{n=0}^{L-1} g[n]e^{-j2\pi ln/N},$$

$$= \sum_{n=0}^{L-1} e^{-j2\pi ln/N},$$

$$= \frac{1 - e^{-j2\pi lL/N}}{1 - e^{-j2\pi l/N}},$$

$$= \frac{\sin(\pi lL/N)}{\sin(\pi l/N)}e^{-j\pi l(L-1)/N}, \quad l = 0, 1, \ldots, N-1.$$

Note that if $N = L$, then:

$$G[l] = \begin{cases} L, & l = 0 \\ 0, & l = 1, 2, \ldots, L-1. \end{cases}$$

This is because there is only one nonzero value in the L-point DFT. However, $g[n]$ can be recovered from $G[l]$ by performing the L-point IDFT. Although the L-point DFT is sufficient to uniquely represent the sequence $g[n]$ in the frequency domain, it does not provide sufficient details to see $G(e^{j\omega})$. More points in the frequency domain are needed, say N, where $N > L$. This can be achieved by expanding the sequence length from L points to N points by appending $N-L$ zeros to the sequence $g[n]$. Now, the N-point provides finer interpolation than the L-point DFT. If zero padding increases, a better approximation to DTFT is obtained. It is important to understand that zero padding does not improve frequency resolution because the resolution is proportional to $1/L$ rather than $1/N$. Figure 3.11 illustrates this based on Example 3.35.

Similar to the case of DTFT, the following discussion presents important DFT properties.

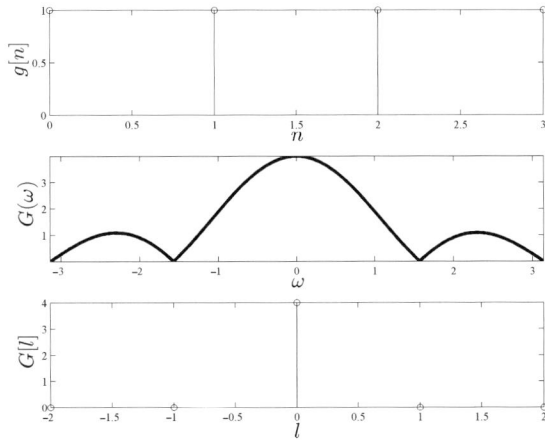

(a)

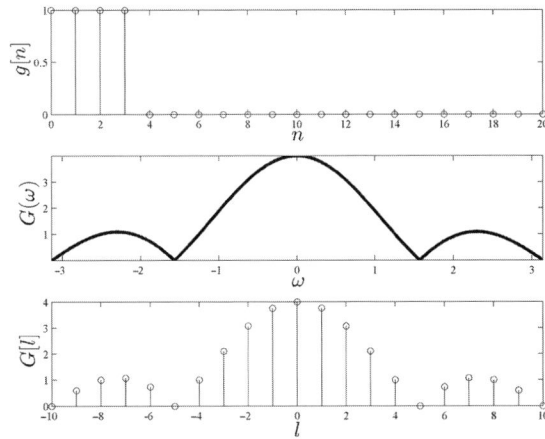

(b)

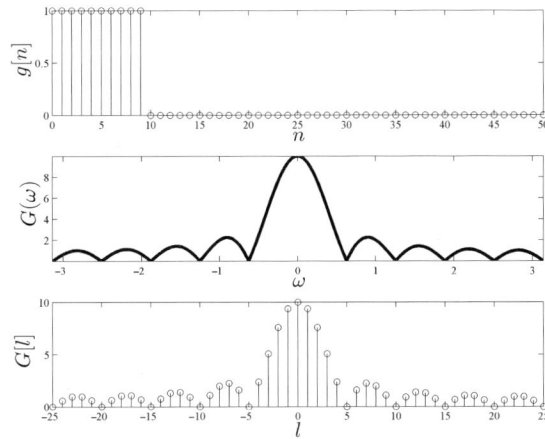

(c)

Figure 3.11 An illustration of the effect of zero padding to better approximate $G(e^{j\omega})$ (in the middle of each sub-figure) by $G[l]$. Various L and N values are presented. (a) $L = N = 4$, (b) $L = 4$ and $N = 20$, and (c) $L = 10$ and $N = 50$.

1. Periodicity: If $g[n]$ and $G[l]$ represent an N-point DFT pair, then:

$$g[n + N] = g[n], \quad \text{for all } n,$$
$$G[l + N] = G[l], \quad \text{for all } l. \tag{3.131}$$

2. Linearity: Given that $g_1[n] \leftrightarrow G_1[l]$, $g_2[n] \leftrightarrow G_2[l]$, and $\alpha_1, \alpha_2 \in \mathbb{R}$, then:

$$\alpha_1 g_1[n] + \alpha_2 g_2[n] \leftrightarrow \alpha_1 G_1[l] + \alpha_2 G_2[l]. \tag{3.132}$$

3. Circular Shift in Time: The circular shift of an N-point signal $g[n]$ by l units to the right produces $g'[n]$, which can be represented as the index modulo N. That is:

$$g'[n] = g[n - l, \text{modulo} N] \equiv g[n - l]_N, \tag{3.133}$$

where $g'[n]$ is simply $g[n]$ shifted circularly by k units to the right in time. Counter-clockwise direction has been arbitrarily selected as the positive direction. Note that a circular shift of an N-point signal is equivalent to a linear shift of its periodic extension, and vice versa.

4. DFT Multiplication: Consider the DFT $g_1[n] \leftrightarrow G_1[l]$ and $g_2[n] \leftrightarrow G_2[l]$, then the multiplication of their DFTs, $G_3[l]$ is:

$$G_3[l] = G_1[l]G_2[l], \quad 0 \leq l \leq N - 1. \tag{3.134}$$

5. Circular Convolution: The time domain $g[n]$ is given by:

$$g[m] = \sum_{n=0}^{N-1} g_1[n]g_2[m - n]_N = \sum_{n=0}^{N-1} g_2[n]g_1[m - n]_N, \quad 0 \leq m \leq N - 1. \tag{3.135}$$

This formula is called the circular convolution. Multiplication of DFTs of two sequences is equivalent to the circular convolution of the two sequences in the time domain. Denoting the circular convolution by $\circledast$ will allow Equation (3.135) to be written as follows:

$$g_1[n] \circledast g_2[n] \longleftrightarrow G_1[l]G_2[l]. \tag{3.136}$$

Example 3.36 Determine the circular convolution of the following two sequences: $g_1[n] = 2\delta[n] + \delta[n - 1] + 2\delta[n - 2] + \delta[n - 3]$ and $g^2[n] = \delta[n] + 2\delta[n - 1] + 3\delta[n - 2] + 4\delta[n - 3]$.

Solution: Using Equation (3.135), one can obtain:

$$g[n] = \sum_{n=0}^{3} g_1[n]g_2[m - n]_4, \quad 0 \leq m \leq 3.$$

Now, for $m = 0$,

$$g[0] = \sum_{n=0}^{3} g_1[n]g_2[-n]_4$$

$$= g_1[0]g_2[0]_4 + g_1[1]g_2[-1]_4 + g_1[2]g_2[-2]_4 + g_1[3]g_2[-3]_4$$
$$= g_1[0]g_2[0] + g_1[1]g_2[3] + g_1[2]g_2[2] + g_1(3)g_2[1]$$
$$= 2 \times 1 + 1 \times 4 + 2 \times 3 + 1 \times 2 = 2 + 4 + 6 + 2 = 14.$$

One can, using the same method, obtain $g[1]$, $g[2]$, and $g[3]$. Therefore,

$$g[n] = 14\delta[n] + 16\delta[n-1] + 14\delta[n-2] + 16\delta[n-3].$$

Now, to verify Equation (3.132) use the DFT and IDFT relations to obtain:

$$G_1[l] = 6\delta[l] + 2\delta[l-2],$$

and,

$$G_2[l] = 10\delta[l] + (-2+2j)\delta[l-1] - 2\delta[l-2] + (-2-2j)\delta[l-3].$$

Multiplying $G_1[l]$ by $G_2[2]$ we obtain:

$$G[l] = G_1[l]G_2[l] = 60\delta[l] - 4\delta[l-2].$$

Hence, one is required to determine the IDFT of $G[l]$, i.e.,

$$g[n] = 14\delta[n] + 16\delta[n-1] + 14\delta[n-2] + 16\delta[n-3],$$

which is equal to $g_1[n] \circledast g_2[n]$.

6. Time Reversal: Suppose that $g[n] \leftrightarrow G[l]$, then,

$$g[-n]_N = G[N-n] \leftrightarrow G[-l]_N = G[N-l].$$

7. Circular Time Shift: Suppose that $g[n] \leftrightarrow G[l]$, then,

$$g[n-l]_N \leftrightarrow G[l]e^{-j2\pi lm/N}.$$

8. Circular Frequency Shift: Suppose that $g[n] \leftrightarrow G[l]$, then,

$$g[n]e^{j2\pi lm/N} \leftrightarrow G[l-m]_N.$$

9. Multiplication of Two Sequences: Suppose that $g_1[n] \leftrightarrow G_1[l]$, and $g_2[n] \leftrightarrow G_2[l]$ then,

$$g_1[n]g_2[n] \leftrightarrow \frac{1}{N}G_1[l] \circledast G_2[l].$$

10. Parsevals Theorem: Suppose that $g[n] \leftrightarrow G[l]$ and $y[n] \leftrightarrow Y[l]$, then,

$$\sum_{n=0}^{N-1} g[n] y^*[n] = \frac{1}{N} \sum_{n=0}^{N-1} G[l] Y^*[l].$$

If $y[n] = g[n]$, then

$$\sum_{n=0}^{N-1} |g[n]|^2 = \frac{1}{N} \sum_{n=0}^{N-1} |G[l]|^2.$$

Linear Filtering Methods Based on DFT

It is known that $G(e^{j\omega}) = W(e^{j\omega}) H(e^{j\omega})$ and $g[n]$ is determined via inverse discrete-time Fourier transform, as stated by Equation (3.134). However, it cannot computationally be performed in a digital computer. DFT can be performed on digital computers.

The DFT will be explored as a computational tool for linear system analysis and, especially, for linear filtering, which is heavily used in seismic data processing. Recall that the output $g[n]$ of an LSI system is the time domain linear convolution of the input $w[n]$ and the system's impulse response $h[n]$. Computing the output using DFT is computationally more efficient than time domain convolution. This is due to the existence of fast algorithms to compute the DFT, namely, the Fast Fourier Transform (FFT).

Suppose that we have a finite duration signal $w[n]$ of length L that excites $h[n]$ of length M. The output sequence $g[n]$ can be expressed in the time domain as the convolution of $w[n]$ and $h[n]$ and the duration of $g[n]$ is $N = M + L - 1$. In the frequency domain, $G(e^{j\omega}) = W(e^{j\omega}) H(e^{j\omega})$, and when expressed using DFT, we need the size of the DFT to be $N = M + L - 1$. Then to recover $g[n]$, one can apply the N-point IDFT. Therefore, one needs to zero pad $w[n]$ and $h[n]$ to the size of $g[n]$. Then, the circular convolution of the $w[n]$ and $h[n]$ will provide the same results as would have been obtained with linear convolution.

Example 3.37 Determine the linear convolution of the following two signals using circular convolution: $g_1[n] = 2\delta[n] + 1\delta[n-1] + 2\delta[n-2] + 1\delta[n-3]$ and $g_2[n] = \delta[n] + 2\delta[n-1] + 3\delta[n-2] + 4\delta[n-3]$.

Solution: When applying the linear convolution directly, one will obtain:

$$g[n] = g_1[n] * g_2[n],$$
$$= 2\delta[n] + 5\delta[n-1] + 10\delta[n-2] + 16\delta[n-3] + 12\delta[n-4]$$
$$+ 11\delta[n-5] + 4\delta[n-6].$$

Note that the size of $g[n]$ is 7. In this case, a 7-point DFT and then IDFT or circular convolution of $g_1[n]$ and $g_2[n]$ can be used to obtain:

$$
\begin{aligned}
g[n] &= g_1[n] \circledast g_2[n], \\
&= 2\delta[n] + 5\delta[n-1] + 10\delta[n-2] + 16\delta[n-3] + 12\delta[n-4] \\
&\quad + 11\delta[n-5] + 4\delta[n-6] \\
&= g_1[n] * g_2[n].
\end{aligned}
$$

If one uses, instead, a 4-point DFT and the IDFT or circular convolution of $g_1[n]$ and $g_2[n]$, then:

$$
\begin{aligned}
g[n] &= g_1[n] \circledast g_2[n], \\
&= 14\delta[n] + 16\delta[n-1] + 14\delta[n-2] + 16\delta[n-3], \\
&\neq g_1[n] * g_2[n].
\end{aligned}
$$

The circular convolution is not the same as the linear convolution unless the discrete-time signals to be convolved are zero-padded to the length of the output signal. If the signals to be convolved were of size M, while the output signal was of size N, where $N > M$, the aliasing effect will take place when the DFT and IDFT of size M are used. The aliasing affects the first $N-M$ samples.

Last, but not least, the DFT is computationally expensive, where it requires N^2 multiplications and N^2 additions. The Fast Fourier Transform (FFT) (Proakis and Manolakis, 2006) represents an efficient way of computing the DFT and its inverse with a cost order of $N \log N$.

3.5 Spectral Analysis of 2-D Seismic Data

Seismic data are already stored digitally, and, when displaying shot records, they come in 2-D, i.e, the offset or distance verses the two-way travel time. 1-D DFT was discussed in the previous section. Here, 2-D DFT will briefly be discussed for the sake of understanding how 2-D seismic records look in the frequency-wavenumber domain. In this case, consider, without loss of generality, a discrete-time space seismic shot record $g[n_t, n_x]$ of size $N_t \times N_x$. Its 2-D DFT $G[n_\omega, n_{k_x}]$ is given by:

$$
G[n_\omega, n_{k_x}] = \sum_{n_t=0}^{N_t-1} \sum_{n_x=0}^{N_x-1} g[n_t, n_x] e^{-j2\pi \left(\frac{n_\omega n_t - n_{k_x} n_x}{N_t N_x} \right)}, \tag{3.137}
$$

for $n_\omega = 0, \ldots, N_t$ and $n_{k_x} = 0, \ldots, N_x$. At the same time, the inverse 2-D DFT $g[n_t, n_x]$ of size $N_t \times N_x$ can be obtained using $G[n_\omega, n_{k_x}]$ and:

$$
g[n_t, n_x] = \frac{1}{N_t N_x} \sum_{n_t=0}^{N_t-1} \sum_{n_x=0}^{N_x-1} G[n_\omega, n_{k_x}] e^{j2\pi \left(\frac{n_\omega n_t - n_{k_x} n_x}{N_t N_x} \right)}, \tag{3.138}
$$

for $n_t = 0, \ldots, N_t$ and $n_x = 0, \ldots, N_x$. Note that both Equations (3.137) and (3.138) comprise the 2-D DFT pair, where n_t and n_x represent the temporal and spatial variables of the seismic shot record, respectively. Also, n_ω represents the frequency variable and n_{k_x} represents the wavenumber variable.

The magnitude spectrum of $G[n_\omega, n_{k_x}]$ is given by:

$$|G[n_\omega, n_{k_x}]| = \sqrt{\Re\{G[n_\omega, n_{k_x}]\}^2 + \Im\{G[n_\omega, n_{k_x}]\}^2}, \qquad (3.139)$$

where $\Re$ stands for the real part and $\Im$ is the imaginary part of a given complex function. In addition, the phase spectrum of $G[n_\omega, n_{k_x}]$ is:

$$\angle G[n_\omega, n_{k_x}] = \arctan\left\{ \frac{\Im\{G[n_\omega, n_{k_x}]\}}{\Re\{G[n_\omega, n_{k_x}]\}} \right\}. \qquad (3.140)$$

It is worth mentioning here that the exponential kernels in the 2-D DFT in Equation (3.137) (or its inverse in Equation 3.138) are separable. This means that in order to compute the 2-D DFT or its inverse, one can apply 1-D DFT in the n_t (or n_ω) (see Equation 3.129) direction followed by another 1-D DFT in the n_x (or n_{k_x}) direction to obtain $G[n_\omega, n_{k_x}]$ (or $g[n_t, n_x]$). Therefore, 1-D FFT can be used in each direction in a cascaded format to obtain the 2-D DFT or 2-D IDFT.

Just as the temporal frequency f (or ω) of a given sinusoid is determined by counting the number of peaks within a unit time, the wavenumber $\grave{k}$ of a seismic dipping event, like a ground-roll or direct wave, can be determined by counting the number of peaks within a unit distance, e.g., (1/m). Note in Chapter 2, the angular wavenumber $k = 2\pi/\lambda$ in rad/distance was used. Because seismic signals are real, it is sufficient for the frequency-wavenumber (whether $f-\grave{k}$ or $\omega-\grave{k}$) domain of seismic sections to consider only two quadrants of the spectrum, which include the positive frequency. Without any loss of generality, and to follow the industry convention when displaying the frequency-wavenumber spectra of seismic data, the spectra will be referred to as the $f-\grave{k}$ representation. Next, a few interesting cases in the $f-\grave{k}$ domain are going to be discussed as follows:

- **Linear (dipping) Seismic Events**: In the $t - x$ domain, linear events can be described in terms of their horizontal velocity or their dip in ms/trace (Yilmaz, 2001). Therefore,

$$v = \frac{\Delta x}{Dip}, \qquad (3.141)$$

where Δx is the distance between adjacent traces in (m) and Dip is the dip in (ms/trace). Now, since,

$$v = \frac{f}{k}. \qquad (3.142)$$

Hence,

$$\grave{k} = \frac{Dip \times f}{\Delta x}.\qquad(3.143)$$

From Equation (3.143), note that down-dip seismic events that are to the right are positive and they map toward the quadrant that contain positive wavenumbers. On the other hand, seismic events with up-dip to the right are assigned negative dip and they map to the quadrant containing negative wavenumber values. Note that, for a given frequency, higher dips are assigned to higher wavenumbers. Now, consider the following scenarios,

1. **Single frequency with no dipping**: When seismic sections, as in Figure 3.12, contain a single frequency with flat seismic events (zero dipping). This implies that the magnitude spectrum of the seismic record will have zero wavenumber and map onto a single point on the frequency axis. Figure 3.13a shows a seismic section in the $t - x$ domain with a single frequency of $f = 10$ Hz with no dipping, while Figure 3.13b shows its $f - \grave{k}$ domain magnitude spectrum with a point (an impulse) at $f = 10$ Hz and $\grave{k} = 0$ 1/m.
2. **Single frequency with dipping**: Consider now seismic data with a single frequency and with constant positive dip along one trace. Along a line of constant time (t), the wavenumber is constant. Therefore, on an $f - \grave{k}$ image spectrum, the single frequency data will plot as a single point at, say, $(f_o, \grave{k}_o)$ in the magnitude spectrum (see Figure 3.14 for an illustration). Also, Figure 3.15a shows a seismic section in the $t - x$ domain with a single frequency of $f = 10$ Hz and $\grave{k} = 0.02$ 1/m and with a $Dip = 5$ ms/trace, while Figure 3.15b shows its $f - \grave{k}$ domain magnitude spectrum with a point (an impulse) at $(f_o, \grave{k}_o) = (10, 0.02)$. Figure 3.15c and d shows the same Figure 3.15a and b, but for $(f_o, \grave{k}_o) = (10, 0.04)$, where the $Dip = 10$ ms/trace.

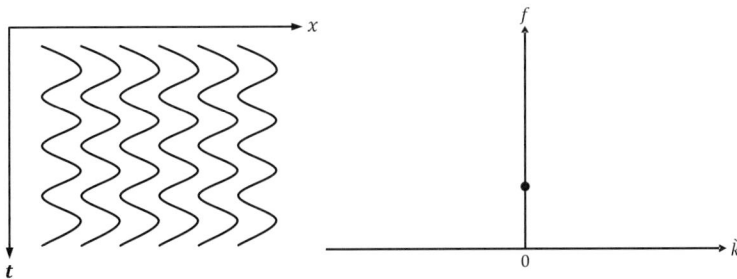

Figure 3.12 A single frequency with no dip in the $t - x$ domain, mapped to a point in the $f - \grave{k}$ domain.

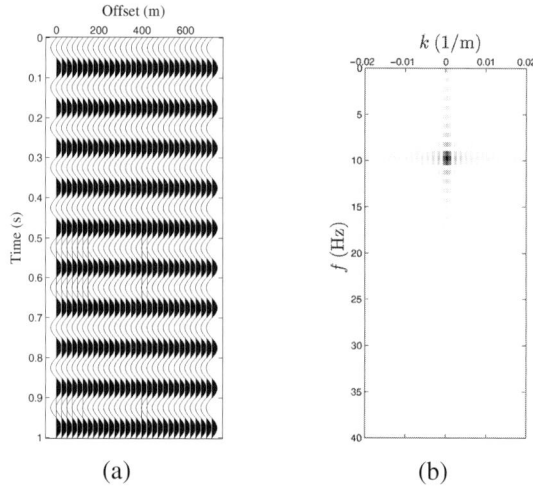

(a)   (b)

Figure 3.13 (a) A synthetic seismic section with a single frequency $f = 10$ Hz with no dip. (b) Its magnitude spectrum, where the wavenumber is zero.

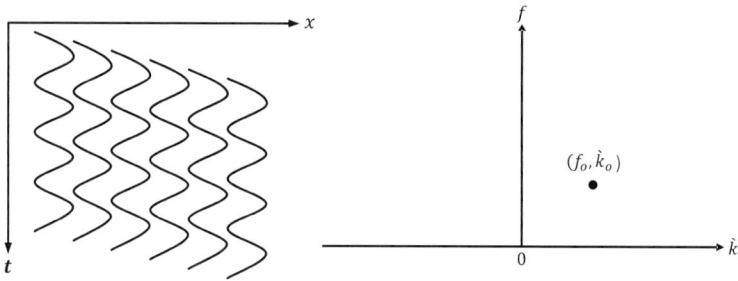

Figure 3.14 A single frequency with a dip in the $t - x$ domain is mapped to a point $(f_o, \grave{k}_o)$ in the $f - \grave{k}$ domain.

3. **Many frequencies with dipping**: Consider dipping seismic data that contains more than one frequency, all with the same dip. Then, since a constant dip = constant horizontal velocity and $v = f/\grave{k}$, a lower f results in a lower $\grave{k}$ and vice versa. As v is constant, the data must have a linear relationship in $f - \grave{k}$. An event with a constant dip in the $t - x$ has a constant dip in the $f - \grave{k}$ for all frequencies. This results in a linear frequency-wavenumber curve in the $f - \grave{k}$ domain, as depicted pictorially in Figure 3.16. Also, Figure 3.13a and b shows synthetic seismic data and its magnitude spectrum, respectively, with $f = [2, 50]$ Hz and $v = 1,500$ m/s. Figure 3.13c and d shows many dipping events and their magnitude spectrum, as well.

• **Hyperbolic Seismic Events**: An event with no dip but with many frequencies will have an infinite horizontal velocity, and infinite wavelength λ, hence $\grave{k} = 0$

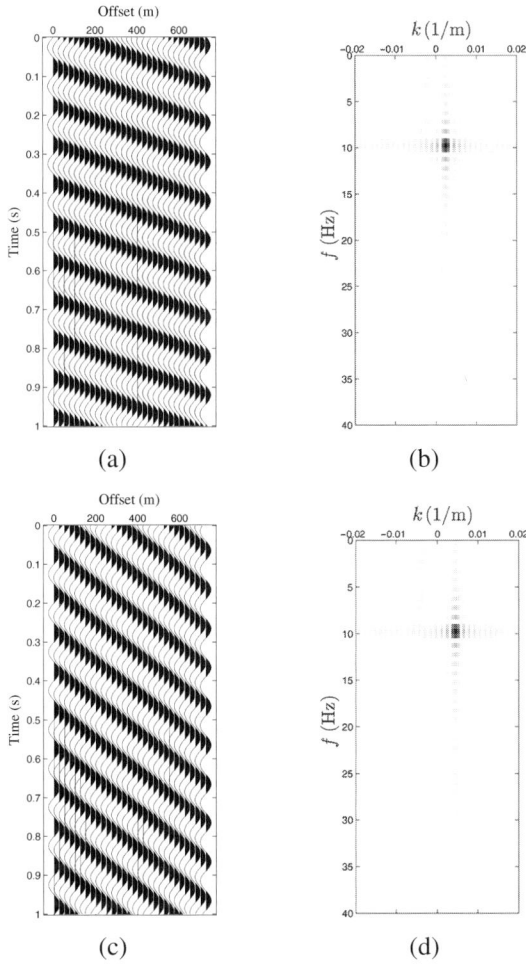

(a)   (b)

(c)   (d)

Figure 3.15 (a) Synthetic seismic section with a single frequency $f = 10$ Hz, $\grave{k} = 0.02$ 1/m and $Dip = 5$ ms/trace. (b) Its magnitude spectrum. (c) Represents (a) but with a single frequency $f = 10$ Hz, $\grave{k} = 0.04$ 1/m, and $Dip = 10$ ms/trace. (d) Its magnitude spectrum.

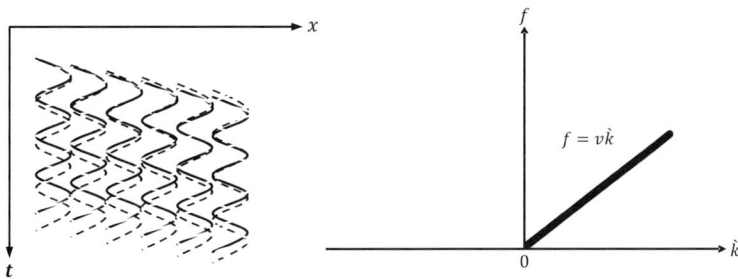

Figure 3.16 Many frequencies with dip in the $t-x$ domain mapped to a point on a linear frequency-wavenumber curve in the $f-\grave{k}$ domain.

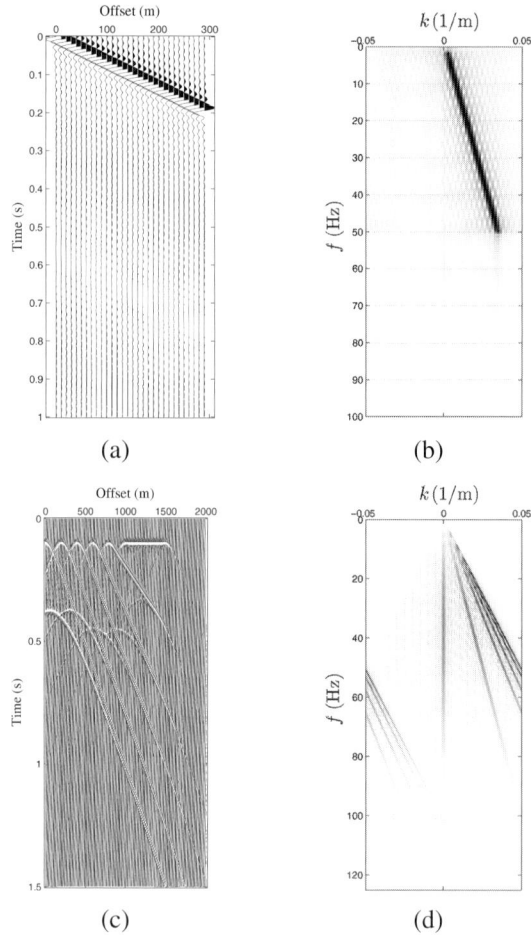

Figure 3.17 (a) A synthetic seismic section with frequency $f = [2, 50]$ Hz with $v = 1,500$ m/s. (b) Its magnitude spectrum. (c) A seismic section containing six dipping events at $0°$, $30°$, $50°$, $60°$, $70°$, and $80°$ with respect to the offset axis. (d) The magnitude spectrum of (c).

(as $\grave{k} = 1/\lambda$). Take, for example, a seismic reflection event largely horizontal in $t-x$ (except at far offsets); they will appear in the region around $k = 0$ in $f-\grave{k}$, while they will take the dip shape in the $f-\grave{k}$, as they are closer to the far offset. That is, a hyperbolic event, in general, will take a fan shape. Figure 3.18 pictorially depicts this, while Figures 3.19 and 3.20 show a few synthetic seismic events that illustrates hyperbolic seismic events in the $t-x$ and $f-\grave{k}$ domains, respectively. Another example is given in Figure 3.21a, where a synthetic seismic section contains a diffraction at $t = 0.4$ s and offset $x = 700$ m, and its magnitude spectrum can be seen in Figure 3.21b.

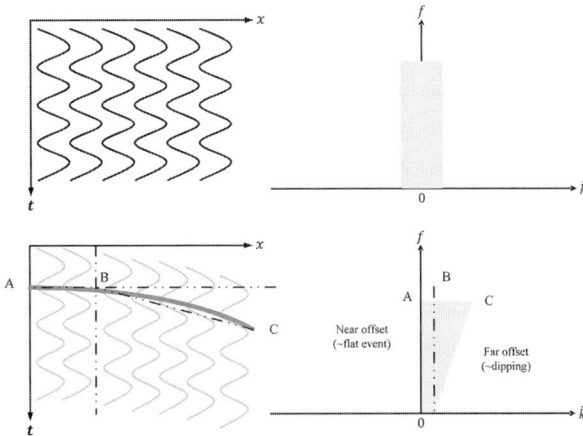

Figure 3.18 An illustration for a seismic event with many frequencies with no dip (top), where the $f-\grave{k}$ domain contains frequencies around $\grave{k} = 0$. A seismic reflection in the $t-x$ domain will appear as a fan shape in the $f-\grave{k}$ domain (bottom). Note the near offset is approximately flat (region A–B), while as the offset increases, the event will approximately become a dipping event (region B–C). Hence, the fan shape in the region A–C exits in the $f-\grave{k}$ domain.

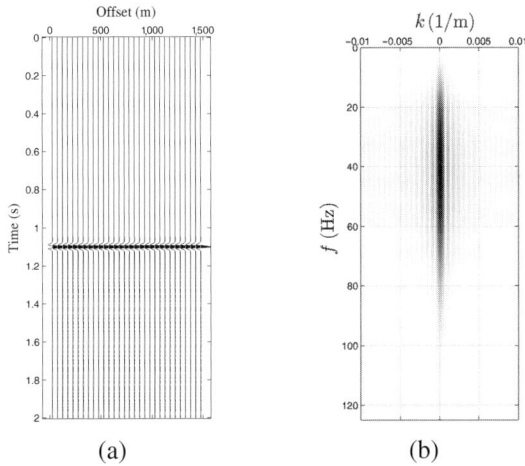

Figure 3.19 (a) A synthetic seismic section with a flat reflection (no dip) and frequency range $f = [4, 80]$ Hz. (b) Its magnitude spectrum.

- **Random Noise**: Recall from Chapter 2 that disturbances in seismic data that lack phase coherency between adjacent traces are considered to be random noise. Some of the sources generating such noise include: wind, rain, and instrument. Figure 3.22a shows a synthetic seismic section containing Gaussian random noise with unit variance and zero mean, where its magnitude spectrum (Figure 3.22b) is white and covers the whole spectrum.

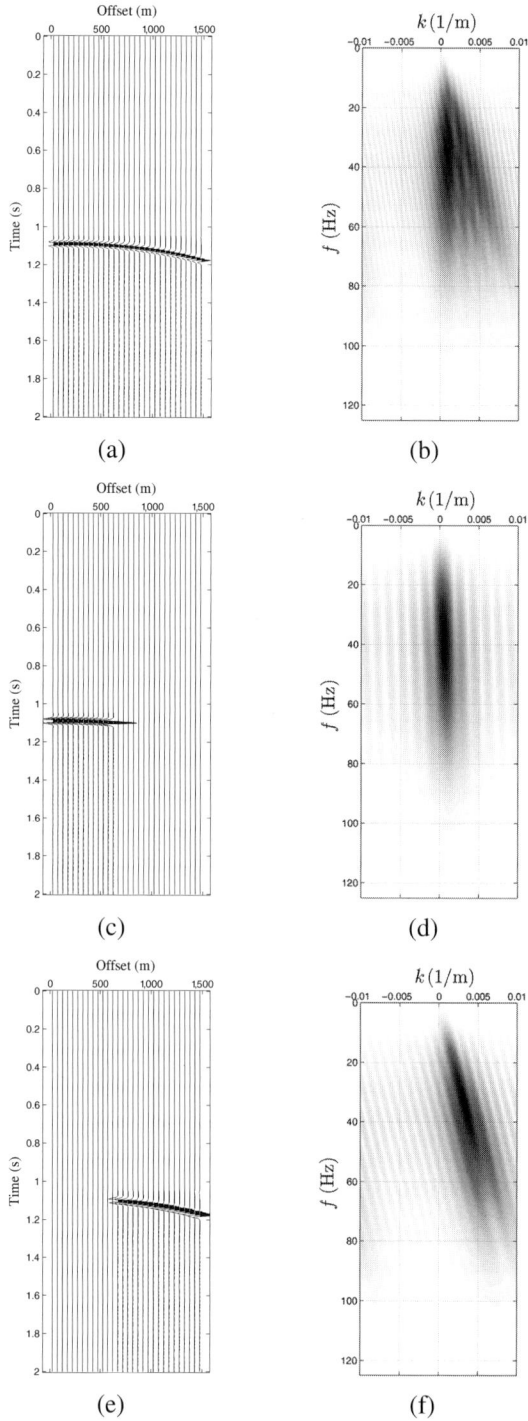

Figure 3.20 (a) A synthetic seismic section containing a reflection (30 traces). (b) Its magnitude spectrum, which appears as a fan shape. (c) Traces 1–13 of (a), and (d) its magnitude spectrum. (e) Traces of (a) from 14–30, and (d) its magnitude spectrum.

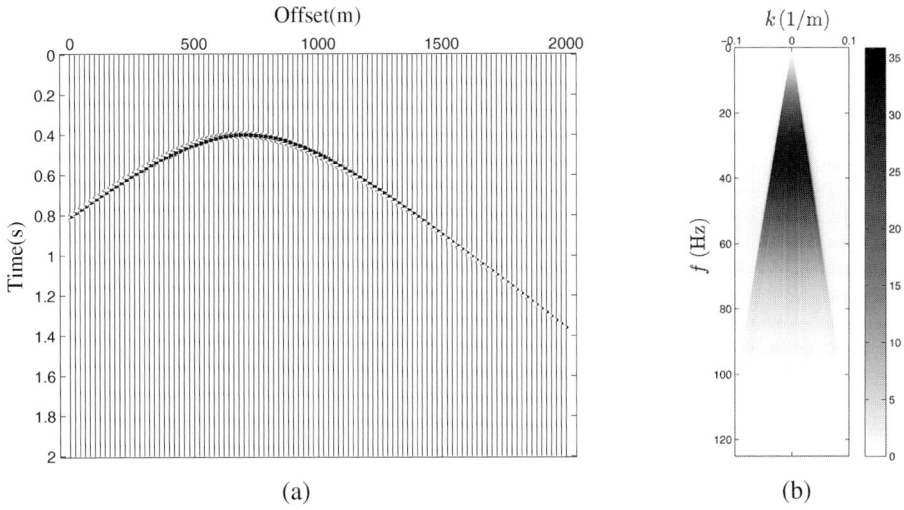

Figure 3.21 (a) A synthetic seismic section containing a diffraction at $t = 0.4$ s and offset $x = 700$ m. (b) Its magnitude spectrum.

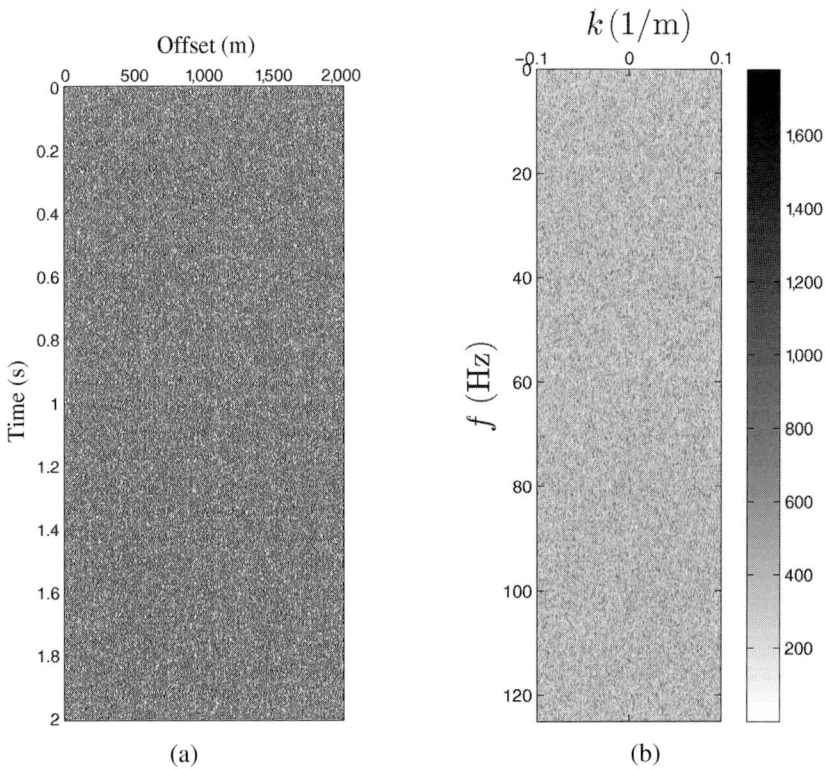

Figure 3.22 (a) A synthetic seismic section containing Gaussian random noise with unit variance and zero mean. (b) Its magnitude spectrum, which shows that the frequency components are white and covers the entire data spectrum.

A practical relationship exists between the $t-x$ and $f-\check{k}$ variables. Stepout can be defined as $\Delta t/\Delta x$. The inverse of the stepout measured in the $t-x$ domain along a constant phase is equal to the ratio of the frequency to the wavenumber associated with the event, that is:

$$\frac{\Delta x}{\Delta t} = \frac{f}{k}. \tag{3.144}$$

So, while retaining a fixed stepout, doubling the frequency f will result in doubling the wavenumber $\check{k}$. Various cases of seismic data $f-\check{k}$ spectra will be shown next.

Examples of the Frequency-Wavenumber Representation of Seismic Data

Figure 3.23 (as seen previously in Chapter 2, Figure 2.38) shows the 2-D land symmetrical split-spread shot record that shows first arrivals from direct and head waves in the form of high-frequency linear events at the top of the record, primary reflections, and ground-roll noise in the form of many low-velocity, low-frequency, and high-amplitude linear events. The magnitude spectrum $f = \check{k}$ is shown in Figure 3.24, where many $f-\check{k}$ dipping events are due to the direct and head waves as well as the dispersive ground-roll noise. The reflection energy is also visible, taking a fan shape starting from $\check{k} = 0$.

Figure 3.25 shows again the 2-D land symmetrical split-spread shot record (seen previously in Chapter 2, Figure 2.40). Recall that the data show good first arrivals from head waves in the form of high-frequency linear events at the top of the record. Ground roll noise fills the central area of the record in the form of low-velocity, low-frequency, and high-amplitude linear events. The high-frequency, low-velocity linear event cutting through the central area of the record is the air-wave arrival generated by the source. Figure 3.26 shows its magnitude spectrum, where most of the seismic reflection energy is centered, while many dipping $f-\check{k}$ events due to the first arrivals, ground-roll, and air waves are visible. Figure 3.27a and b shows a seismic marine data set from Yilmaz (2001) with long-path multiples and its magnitude spectrum.

3.6 Radon Transform

The generalized Radon transform is a tool that is heavily used in many application for image processing (Matus and Flusser, 1993), medical imaging (Cao, 2007; Lehmann, 1999), solution of mathematical problems (Kuchment, 2006; Search et al., 2001), and, most importantly, in the field of reflection seismic data processing (Cao and Bancroft, 2005; Sacchi et al., 2004; Tong et al., 2009; Trad et al., 2003).

The Radon transform is robust in nature and has attracted the attention of seismic data processing scientists and engineers during the last two decades. It has been used for quite some time in different applications, which include seismic

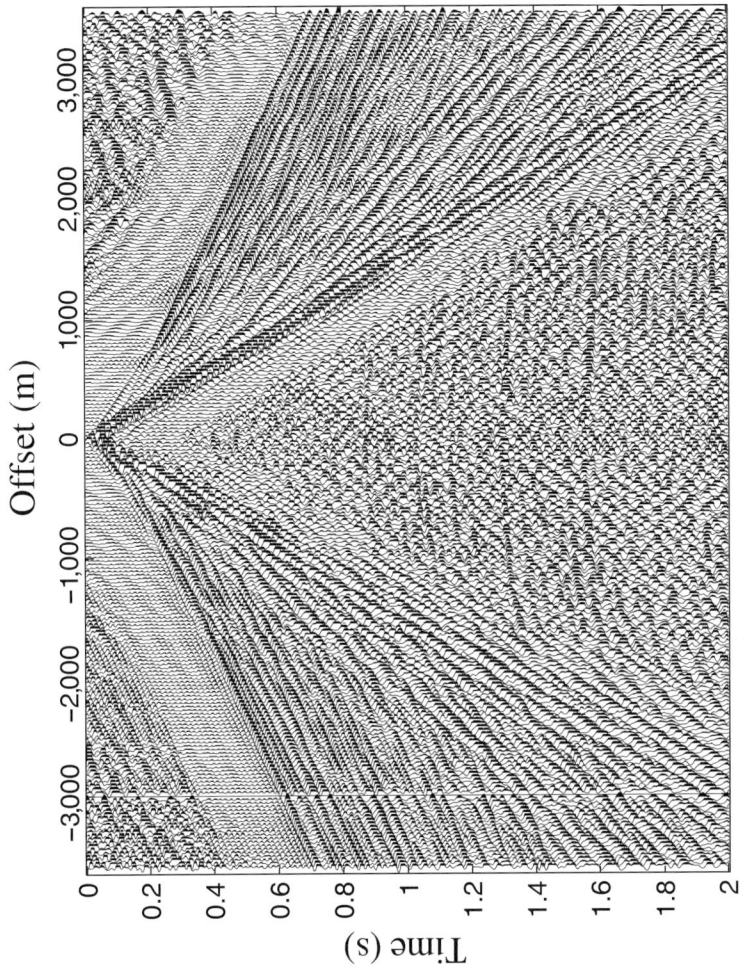

Figure 3.23 Real seismic data seen in Chapter 2, Figure 2.38.

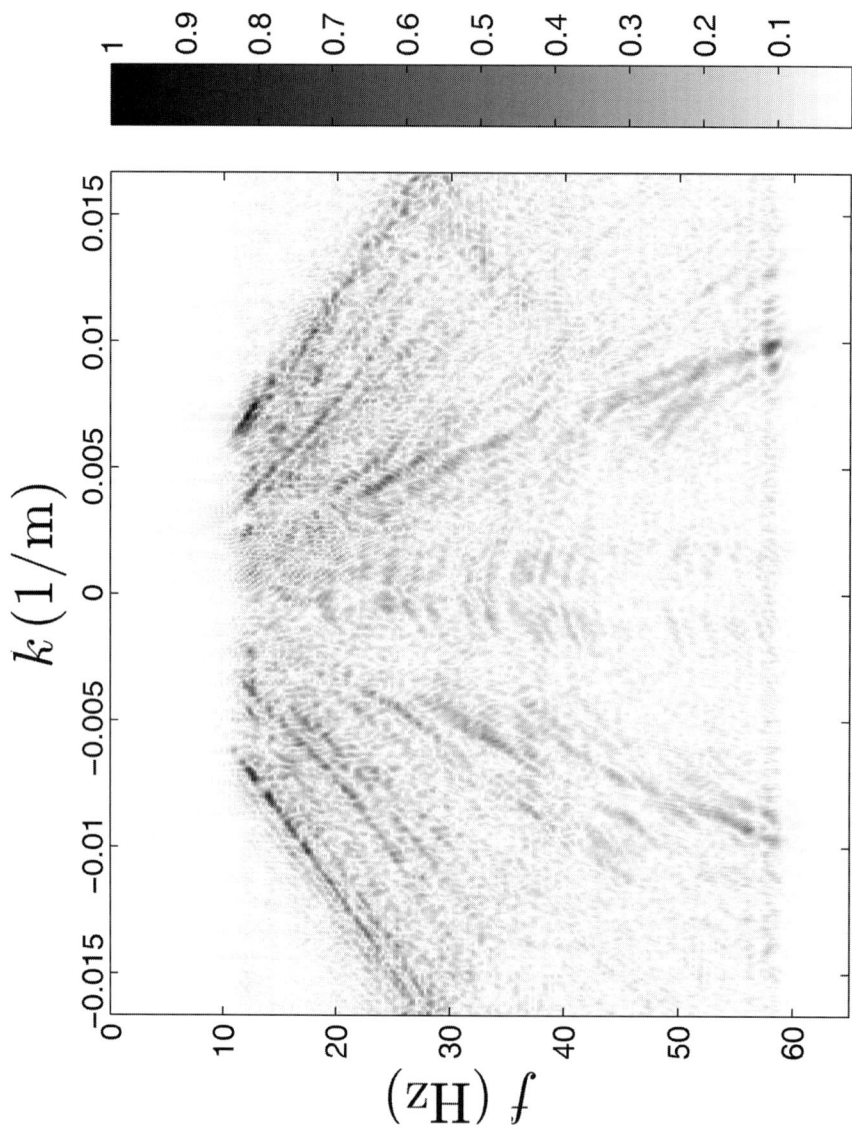

Figure 3.24 The $f - k$ magnitude response of Figure 3.23.

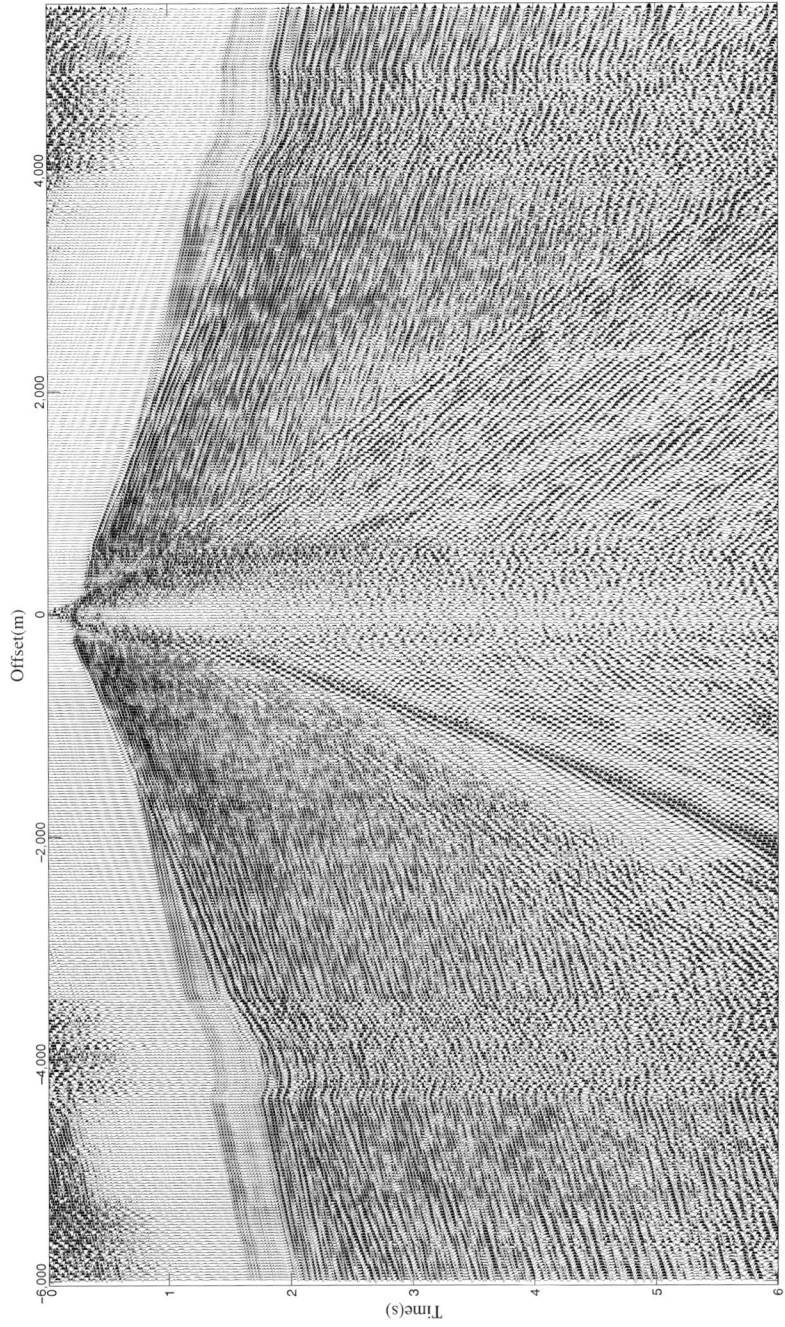

Figure 3.25 Real seismic data seen in Chapter 2, Figure 2.40.

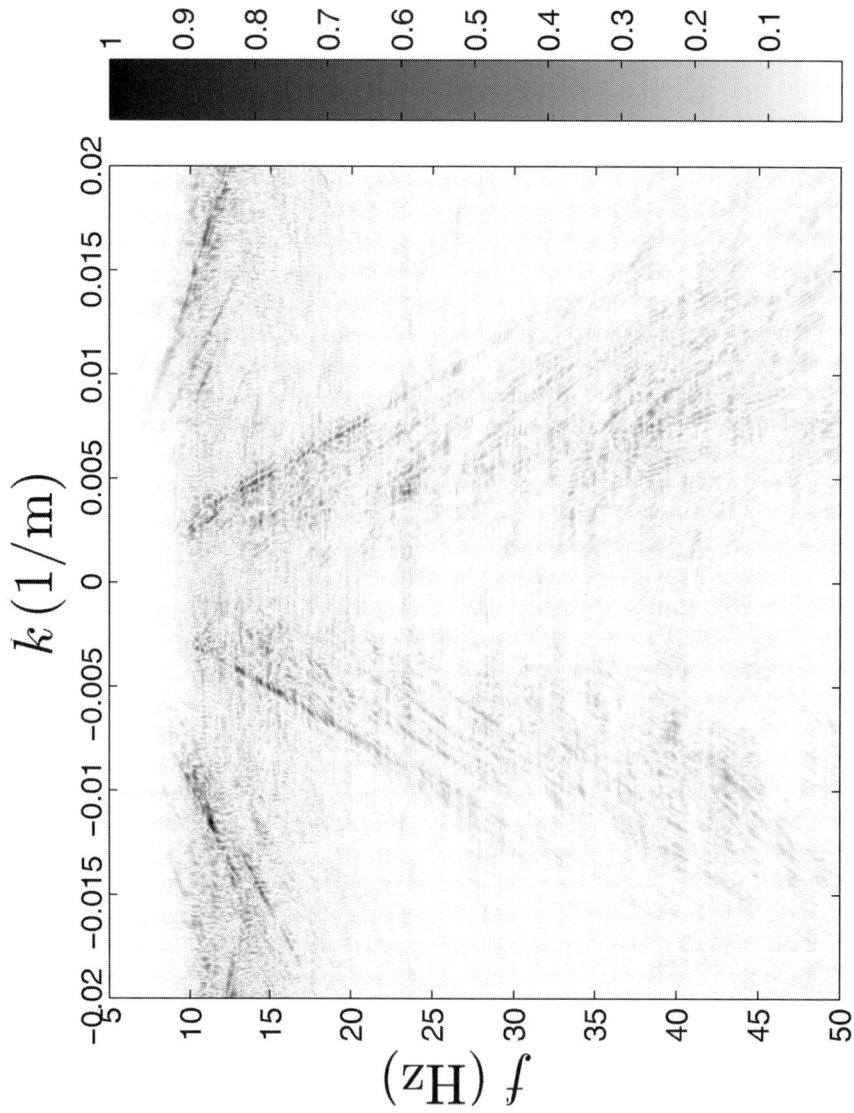

Figure 3.26 The $f-k$ magnitude response of Figure 3.25.

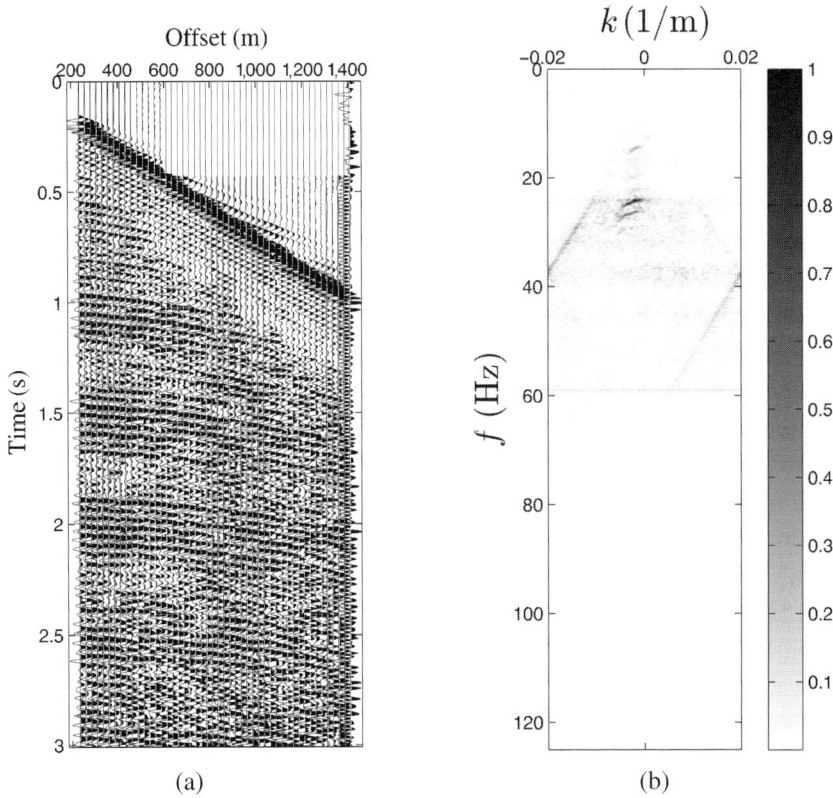

Figure 3.27 (a) Real marine data modified from Yilmaz (2001) with long-path multiples and (b) its magnitude spectrum.

deconvolution (Sacchi et al., 2004), multiple removal as discussed in (Cao and Bancroft, 2005; Minaeian et al., 2009; Thorson, 1985; Tong et al., 2009), and first arrival picking or enhancement (Mousa and Al-Shuhail, 2012).

3.6.1 Radon Transform and Seismic Data Processing

Three main types of Radon transform that are used for seismic data processing exist (Toft, 1996). These transforms are linear Radon transformation (known also as the slant-stack), parabolic Radon transform, and hyperbolic Radon transform. Parabolic Radon transform is mostly used due to its effectiveness compared to linear Radon transform and low computational cost (Trad et al., 2003). The Radon transform the data from the time-space domain (t, x) to the linear Radon domain $(\tau - p)$ or parabolic or hyperbolic Radon domain $(\tau - q)$. Note that τ represents the time interval, p is the slowness, and q is the parameter that defines the curvature of parabola. One of the distinct features of the Radon transform is that the $\tau - p$ and $\tau - q$ domains provide a sparse representation of the linear and parabolic (or hyperbolic) seismic events, respectively (Toft, 1996). Besides low resolution Radon transform, high resolution along with sparse fast high resolution Radon transform has been proposed and used in the seismic data processing industry. The concept of sparse Radon transform was introduced by Thorson (1985), who presented a low resolution Radon transform using least squares. High resolution Radon transform in frequency domain was proposed by Sacchi and Ulrych (1995), and it utilized the regularization technique. Although the regularization method enhanced the focusing power of the transform, due to the non-Toeplitz nature of the matrix the inversion was computationally expensive. Since then, a different version of fast high resolution radon transform has been proposed (Latif and Mousa, 2017; Ng and Perz, 2003, 2004; Sacchi et al., 1999; Trad et al., 2003).

Inverse Radon transform is used for the reconstruction of the actual data. The inverse transform can be obtained by several available techniques, but the most common inversion techniques are the Fourier Slice Theorem and Filtered Back Projection. One of the techniques used for the inverse Radon transform is the use of the Fourier Slice theorem. The Fourier Slice Theorem shows that the function $g(x,t)$ can be reconstructed by taking the 1-D Fourier transform of the Radon transform, which will result in the 2-D Fourier spectrum of $g(x,t)$ (Yilmaz, 2001).

Mathematically, let $g(x,t)$ represent the seismic data in offset x and two-way travel time t. The generalized Radon Transform (Chapman, 1981) pair for this seismic data $g(x,t)$ is given as follows:

$$u(q,\tau) = \int_{-\infty}^{\infty} g(x, t = \tau + q\phi(x))dx, \qquad (3.145)$$

$$\hat{g}(x,t) = \int_{-\infty}^{\infty} u(q, \tau = t - q\phi(x))dq, \qquad (3.146)$$

where q and τ represent a parameter of the curve. Similarly, $\phi(x)$ is the function of the offset parameter x and depends on the path of integration. The discrete form of the generalized Radon transform can be represented by:

$$u[n_q, n_\tau] = \sum_{n_x} g[n_x, n_t = n_\tau + n_q\phi[n_x], n_t = n_\tau + n_q\phi[n_x]], \qquad (3.147)$$

$$\hat{g}[n_x, n_t] = \sum_{n_q} u[n_q, n_\tau = n_t - n_q\phi[n_x]]. \qquad (3.148)$$

3.6.2 Linear Radon Transform

The linear Radon transform integrates the data along planar surfaces. It transforms a linear event to a single point in the transformed Radon panel. Forward linear

Radon transform was developed by Beylkin (1987). Kostov (1990) used Toeplitiz structure for least squares solution of the Radon transform (Beylkin, 1987). To achieve better resolution, Zhou and Greenhalgh (1994) suggested applying the ρ filter in the forward Radon transform (Zhou, 1994). The linear Radon transform is obtained by replacing $\phi(x) = x$ and $q = p$ in generalized Radon transform (Equation 3.145). The forward linear Radon transform equation in the continuous domain is:

$$u(p,\tau) = \int_{-\infty}^{\infty} g(x,t = \tau + px)dx, \qquad (3.149)$$

while the inverse is given by

$$\hat{g}(x,t) = \int_{-\infty}^{\infty} u(p,\tau = t - px)dp. \qquad (3.150)$$

Similarly, the discrete form of these equations is obtained from Equation (3.147).

$$u[n_p, n_\tau] = \sum_{n_x} g[n_x, n_t = n_\tau + n_p n_x], \qquad (3.151)$$

$$\hat{g}[n_x, n_t] = \sum_{n_p} u[n_p, n_\tau = n_t - n_p n_x]. \qquad (3.152)$$

Note also that the linear Radon transform can be calculated via the frequency domain. So, in the frequency domain, the linear Radon transform can be represented as:

$$U(p,\omega) = \int_{-\infty}^{\infty} G(x,\omega)e^{i\omega px}dx, \qquad (3.153)$$

$$\hat{G}(x,\omega) = \int_{-\infty}^{\infty} U(p,\omega)e^{-i\omega px}dp. \qquad (3.154)$$

This provide an efficient way to obtain the linear Radon transform.

Now, the case of a linear and a hyperbolic seismic event and what they will look like in the $(\tau - p)$ or linear Radon transform domain is considered, as presented by Sarajrvi (n.d.). The discussion is based on continuous-domain presentation. Without loss of generality, the case of discrete is assumed to be inferred from its continuous counterpart.

- **Linear seismic events**: The case of a linear seismic event is presented with the following logic. A point source $g(x,t)$ can be modeled as a product of two delta functions. Consider the following wavefield, which formulates a point in the $t-x$ domain:

$$g(x,t) = \delta(x - x_0)\,\delta(t - t_0). \qquad (3.155)$$

Its linear Radon transform using Equation (3.149) is found as follows:

$$g(p, \tau) = \int_{-\infty}^{\infty} g(x, \tau + px) dx, \tag{3.156}$$

$$= \int_{-\infty}^{\infty} \delta(x - x_0) \delta(\tau + px - t_0) dx = \delta(\tau + px_0 - t_0).$$

Clearly, this constitutes a straight line on the form:

$$\tau = t_0 - px_0. \tag{3.157}$$

Hence, a point in the $t-x$ domain is a straight line, the linear Radon domain, and, by duality, a linear seismic event in the $t-x$ domain will constitute a point in the linear Radon domain. Figure 3.28 shows the linear Radon transform of synthetic seismic data containing a linear event.

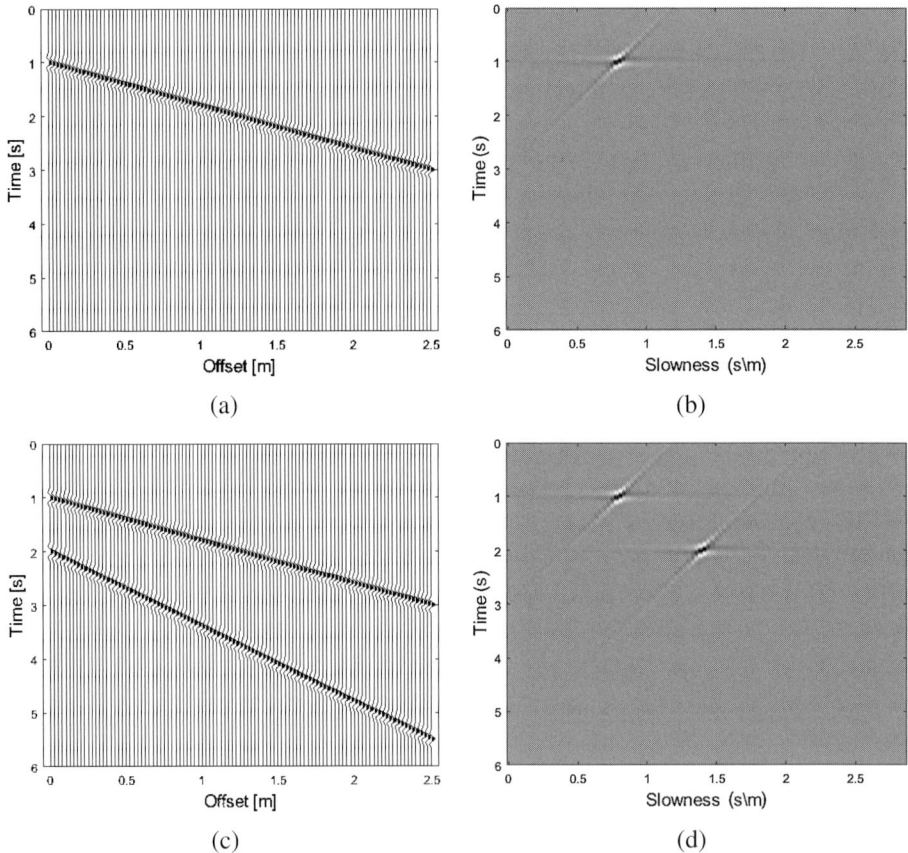

Figure 3.28 Linear Radon Transform of linear events: (a) a single linear event and its (b) Radon transform as a point, (c) two linear events and, (d) their Radon transform as two points.

- **Hyperbolic seismic events**: Consider the following $t-x$ curve $\mathcal{L}$ that is defined by:

$$\tau = t - \phi(x). \tag{3.158}$$

Now, an envelope ϵ is found by solving the following set of equations:

$$\phi(p, \tau; x) = t(x) - \phi(x) - \tau = 0, \tag{3.159}$$

and,

$$\frac{\partial \phi(p, \tau; x)}{\partial x} = \frac{dt}{dx} - \frac{\phi(x)}{dx} = 0. \tag{3.160}$$

Let $\phi(x) = px$ and find the envelope of the hyperbolic curve:

$$t^2(x) = a + bx^2, \tag{3.161}$$

where a and b are constants ($a = 4z^2/v^2$ and $b = 1/v^2$ based on Equation 2.86). Now, based on Equation (3.161), one can obtain:

$$\frac{dt}{dx} = \frac{bx}{t}, \tag{3.162}$$

and,

$$\frac{\phi(x)}{dx} = p. \tag{3.163}$$

Hence, the envelope of the $t-x$ hyperbolic curve in the linear Radon domain can be determined using:

$$\phi(p, \tau; x) = \sqrt{a + bx^2} - px - \tau = 0, \tag{3.164}$$

and,

$$\frac{\partial \phi(p, \tau; x)}{\partial x} = \frac{bx}{t} - p = 0. \tag{3.165}$$

Based on these equations, t and x can be eliminated to obtain a representation in terms of τ and p in relationship with a and b, so that:

$$\tau^2 = t^2 - 2tx + p^2x^2 = x^2(p^2 - b) + a, \tag{3.166}$$

where τ is a function of x and p. To eliminate x, consider:

$$p = \frac{dt}{dx} = \frac{bx}{\sqrt{a + bx^2}}. \tag{3.167}$$

This is equivalent to saying that:

$$x^2(p^2 - b) = -p^2\frac{a}{b}, \tag{3.168}$$

or

$$\tau^2 = -p^2\frac{a}{b} + a, \tag{3.169}$$

where,

$$\frac{\tau^2}{a} + \frac{p^2}{b} = 1, \tag{3.170}$$

represents an ellipse. Substituting the values of a and b in the last equation, one will obtain:

$$\frac{\tau^2}{4z^2} + p^2 = \frac{1}{v^2}. \tag{3.171}$$

So a hyperbolic seismic event in the $t-x$ domain will be represented as an ellipse in the linear Radon domain. Figure 3.29a shows an example of two hyperbolic events, and their Radon transform as two ellipses is shown in Figure 3.29b. Also, Real seismic data from Yilmaz (2001) containing multiples and reverberations is shown in Figure 3.30a. Its linear Radon transform, shown in Figure 3.30b, shows that the primary reflection and the multiples along with their reverberations are in the form of ellipses.

3.6.3 Parabolic Radon Transform

In case of the parabolic Radon transform, the Radon transform integral is taken along a parabolic curve and a parabolic curve in time domain will be mapped to a single point in the $\tau-q$ or Radon domain. On common shot point gathers, seismic events are not linear in nature but they are parabolic or hyperbolic in nature. Refractions and direct waves are linear, reflections and diffractions are hyperbolic. Hampson (1986) proposed an efficient LS-based parabolic transform method. Bradshaw and Ng (1987) proposed the semblance-weighted Gauss-Seidel parabolic Radon transform. The t^2-stretching of the time axis was proposed by Yilmaz (1989). The improved Hampsons' frequency method was improved by Sacchi and Ulrych (1995), Trad et al. (2002), Sacchi et al. (1999), and Sacchi (1999). They incorporated the a priori information for the formulation of high resolution RT. Cary (1998) posed the problem in the TX domain (Ng and Perz, 2004). Sacchi and Porsani (1999) utilized conjugate gradients for their version of high resolution RT. Trad (2003) proposed the robust high-resolution parabolic RT by posing the problem in the FX domain. (Trad et al., 2003; Gholami and

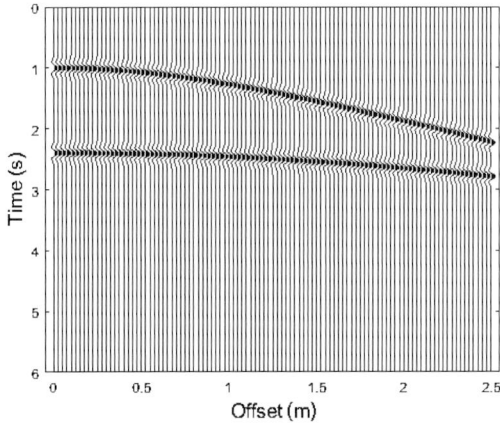

(a)

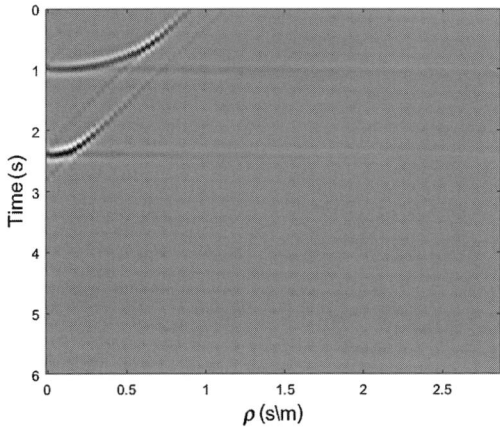

(b)

Figure 3.29 Linear Radon Transform of hyperbolic seismic events in the: (a) time-space domain and (b) linear Radon transform domain.

Siahkoohi, 2009). Some variant of weighted Parabolic Radon transforms are discussed in Cao and Bancroft (2005), Luo et al. (2009), and Kaplan et al. (2010).

In case of parabolic Radon transform $\phi(x) = x^2$, in the forward and inverse parabolic Radon transform Equations (3.145) and (3.146), respectively. Now, for a parabola $\phi(x) = qx^2$, the envelope of a certain curve can be found by solving the set of equations:

$$\psi(p, \tau; x) = t(x) - \phi(x) - \tau = 0, \qquad (3.172)$$

$$\frac{\partial \psi(p, \tau; x)}{\partial x} = \frac{dt}{dx} - \frac{d\phi(x)}{dx} = 0. \qquad (3.173)$$

(a)

(b)

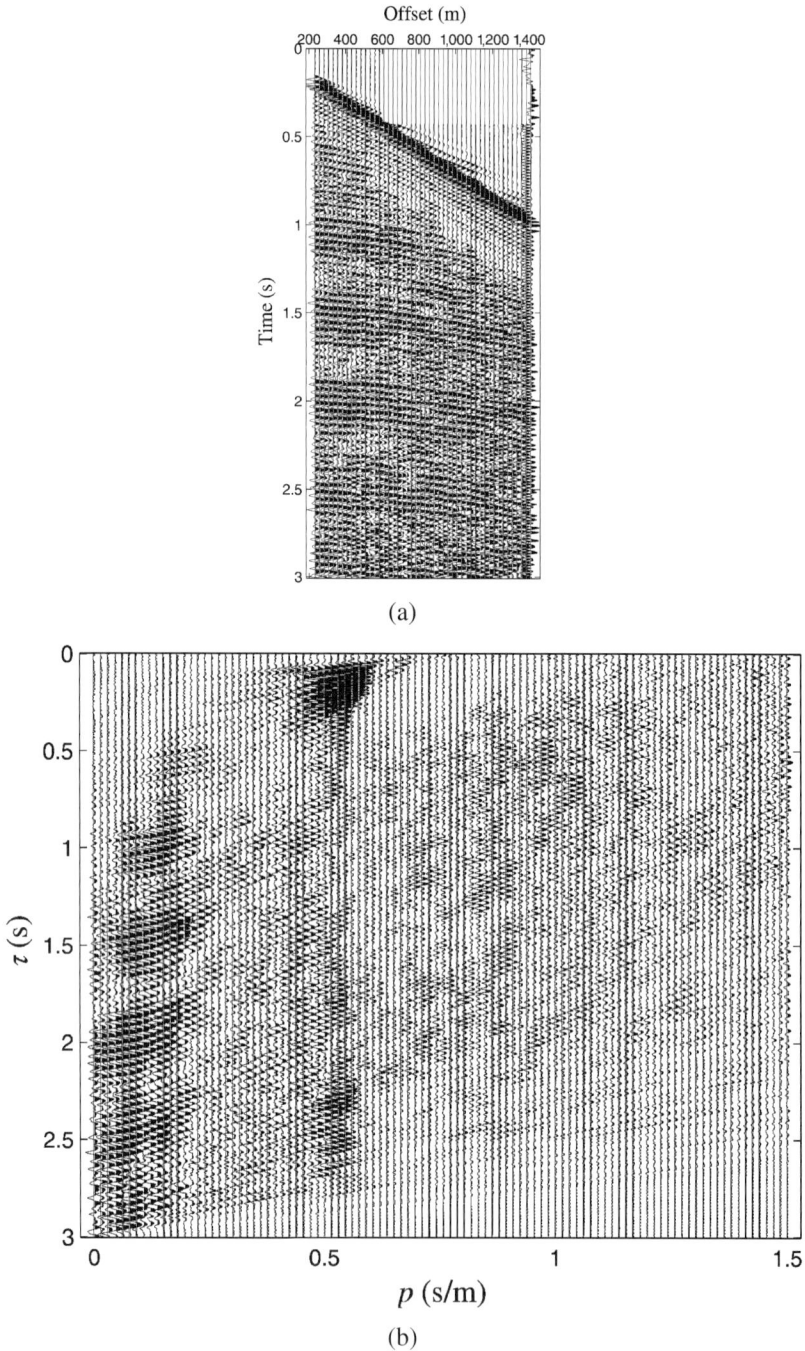

Figure 3.30 (a) Real seismic data modified from Yilmaz (2001) containing multiples and reverberations. (b) Its linear Radon transform, which shows that the primary reflection and the multiples along with their reverberations are in the form of ellipses.

Similar to the case of linear Radon transform, consider the following cases (Sarajrvi, n.d.):

- **Linear seismic events**: Consider Linear curve

$$t(x) = \beta x, \tag{3.174}$$

$$\frac{dt}{dx} = \beta. \tag{3.175}$$

So a single point on C is represented as a line in the transform domain. See Figure 3.31a for an example of a seismic section containing a linear event being transformed into a point in the parabolic Radon domain (Figure 3.31b).

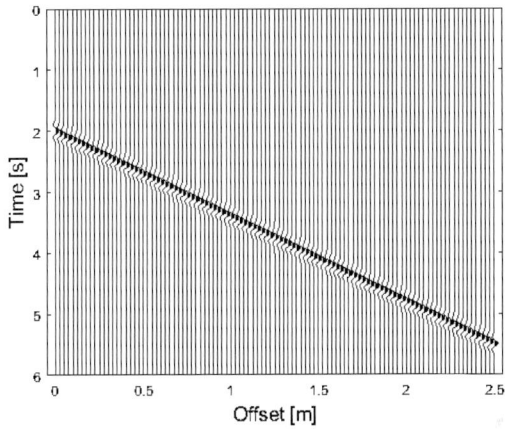

(a)

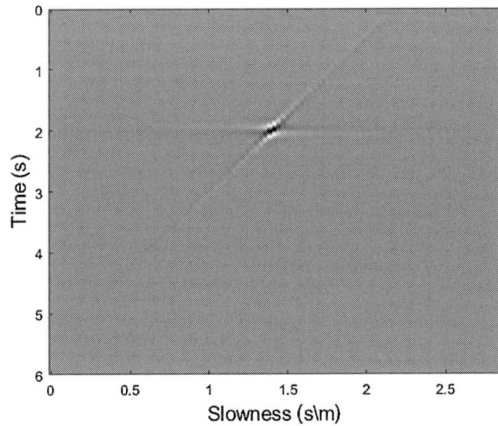

(b)

Figure 3.31 Parabolic Radon transform of (a) a seismic section containing a linear event and a point in the (b) parabolic Radon domain.

- **Parabolic seismic event**: Consider a parabolic curve q. The curvature can be found by solving Equation (3.172):

$$q = \frac{\beta}{2x},$$
(3.176)

$$\tau = \frac{\beta x}{2}.$$
(3.177)

Eliminating x, one can obtain:

$$q\tau = \frac{\beta^2}{4}.$$
(3.178)

Using the hyperbolic seismic event shown in Equation (3.161), one can obtain:

$$q(x) = \frac{b}{2t(x)},$$
(3.179)

$$\tau(x) = \frac{t(x)}{2} + \frac{a}{2t(x)},$$
(3.180)

and eliminating t and rearranging the terms will result in the following:

$$\tau = \frac{1}{4v^2 q} + 4z^2 q.$$
(3.181)

So in the case of a seismic hyperbolic curve C, its parabolic Radon transform is represented as another (short) hyperbola, with more focus to approximate point rather than a hyperbola, as seen in Figure 3.32. Figure 3.33 shows real seismic data from Yilmaz (2001) containing multiples and reverberations and its parabolic Radon transforms, which show a more focused primary reflection and the multiples along with their reverberations. Thus, such unwanted energy can be muted.

3.7 Summary

This chapter introduced discrete-time (space) signals and systems, with a focus on seismic signals. Linear Shift Invariant (LSI) systems are assumed to model many systems. LSI systems can be characterized using their impulse responses and/or through difference equations, where both relate inputs to outputs. For LSI systems, the impulse response provides a means to know whether the system is causal and/or stable. Also, due to their usefulness, such signals and systems were analyzed in other domains such as the z-transform, the DTFT, and DFT domains. Such domains represent alternative views of signals and systems that would lead to better analysis and processing. The LSI system transfer function (the z-transform under zero initial conditions of the LSI system impulse response) and its frequency response provide important means for the analysis and design of such systems. In addition, 2-D seismic data sets were explained in the frequency-wavenumber as well as the Radon transform domains. Such representations of seismic data in an other domain than

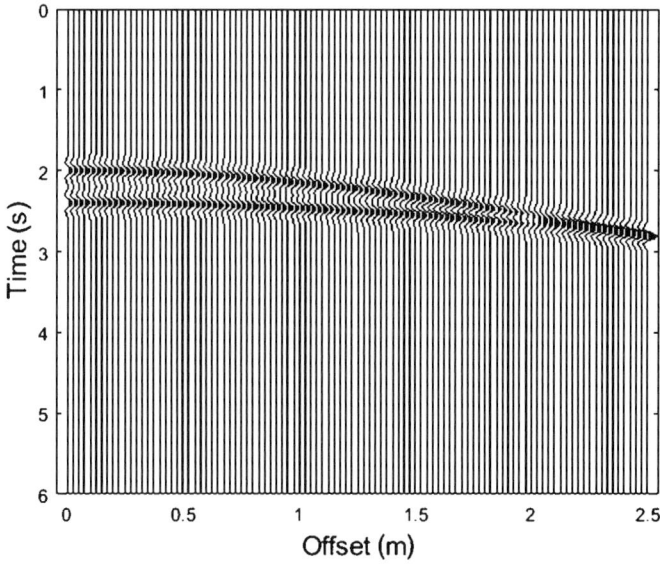

(a)

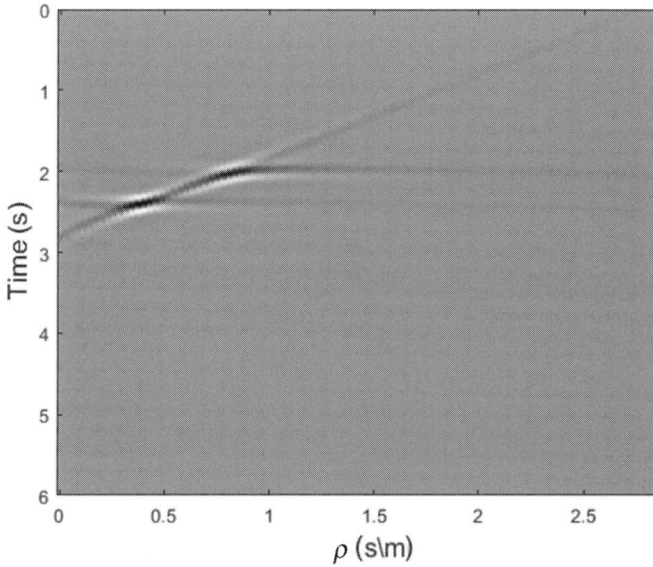

(b)

Figure 3.32 Parabolic Radon Transform of (a) two hyperbolic seismic events and their (b) parabolic Radon transforms, which show a more focused and shorter hyperbolic events that almost approximate points.

the $t-x$ provide an advantage of better signal/unwanted energy separation. Hence, the seismic data primary reflections can be extracted and further enhanced, so that one cannot only identify seismic events in the $t-x$ but also in other transform domains, where these events could be better separated and processed.

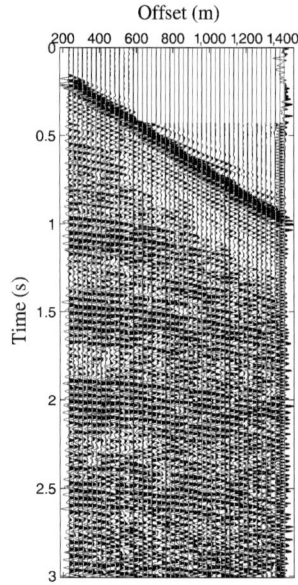

(a)

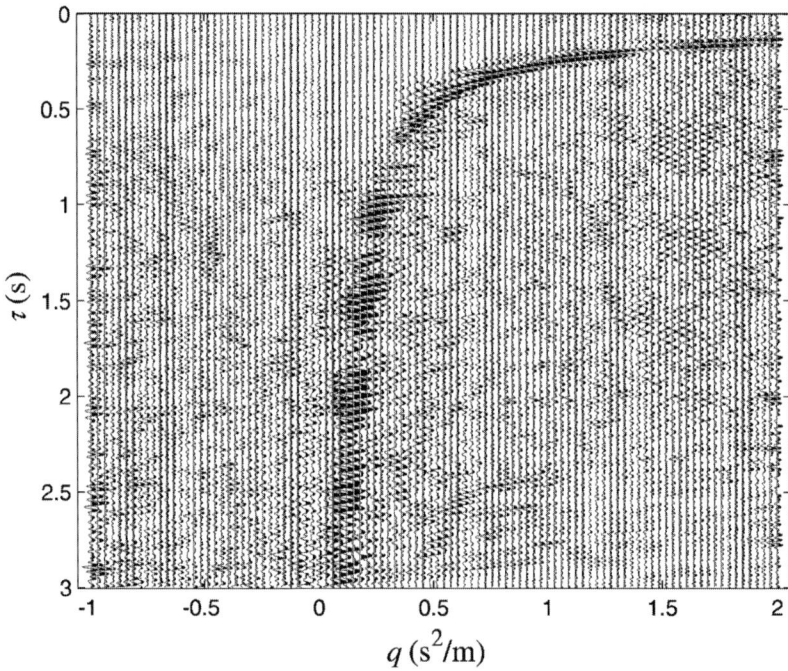

(b)

Figure 3.33 (a) Real seismic data modified from Yilmaz (2001) containing multiples and reverberations. (b) Its parabolic Radon transforms, which show a more focused primary reflection and the multiples along with their reverberations.

Exercises

3.1 A seismic signal $g[n]$ is to pass through an LSI system with the following input–output relationship:

$$y[n] = \frac{1}{3}(g[n+1] + g[n] + g[n-1]).$$

(a) Determine the LSI system impulse response $h[n]$.
(b) Is this system stable or not? Why?

3.2 Let us assume an LTI system has the step response $y[n] = [2 - (\frac{1}{2})^n]$, for $n \geq 0$. Determine the system's impulse response.

3.3 Consider the reflectivity sequence:

$$r[n] = 3\delta[n-1] + 2\delta[n-3] + \delta[n-5].$$

The embedded seismic wavelet is given by:

$$w[n] = \delta[n] + 2\delta[n-1].$$

(a) Sketch $r[n]$.
(b) Sketch $w[n]$.
(c) Compute and sketch the resulting seismogram $g[n]$ from $r[n]$ and $w[n]$. Do not use the z-transform.
(d) Compute the energy of $r[n]$, $w[n]$, and $g[n]$.
(e) Use MATLAB to compute $g[n]$, where $r[n]$ runs over $n = -10, -9, \ldots, 40$, while $w[n]$ runs over $n = 0, 1, \ldots, 50$.

3.4 Using the following difference equation:

$$g[n] = 0.8g[n-1] + 5w[n],$$

obtain, using MATLAB, the discrete-time signal output $g[n]$ for $n = -3, -2, \ldots, 15$, where $w[n] = 2\delta[n] - 3\delta[n-1] + 2\delta[n-3]$. Assume that $g[n] = 0$ for $n < 0$. Plot $w[n]$ and $g[n]$.

3.5 Consider an LSI system with the following system function:

$$H(z) = \frac{z^{-1} - \frac{1}{3}}{1 - \frac{1}{3}z^{-1}}.$$

(a) Find a difference equation to implement this system.
(b) If $H(z)$ is cascaded with another system represented by the transfer function $G(z)$ so that the overall system function is unity. If $G(z)$ is to be stable, then obtain $g[n]$.

(c) Is $g[n]$ causal, anti-causal, or non-causal?

(d) Using MATLAB, plot $H(z)$ zero-pole map.

3.6 Using MATLAB, find the zeros and poles of the following transfer functions:

(a) $H(z) = \frac{2+5z^{-1}+z^{-2}}{1+2z^{-1}+3z^{-2}}$.

(b) $H(z) = \frac{1}{(z-1/3)(z-1/2)}$.

Plot their zero-pole maps.

3.7 Given that:

$$G(\omega) = 1 + \cos(\omega),$$

represents the DTFT of a seismic trace, find the following:

(a) The energy of the seismic trace signal?

(b) If it passes through a LSI system described by the following system function:

$$H(z) = \frac{1}{1 - 0.8z^{-1}}.$$

then determine the output signal.

3.8 Find the inverse DTFT of:

$$G(e^{j\omega}) = \frac{1}{1 - \frac{1}{4}e^{-j10\omega}}.$$

Hint: Use the definition of the DTFT and note that $a^n u[n] \leftrightarrow \frac{1}{1-ae^{-j\omega}}$.

3.9 Consider the impulse response of a LSI system $h[n] = \frac{1}{3}(\delta[n]+\delta[n-1]+\delta[n-2])$. Use MATLAB to compute its DTFT (see Equation 3.107) for $n = -5:5$ and $\omega \in [-\pi,\pi]$. Define ω to be of 512 points. Do not use the FFT function.

3.10 Given $g_1[n] = 4\cos[\frac{2\pi n}{N}]$ and $g_2[n] = 2\sin[\frac{2\pi n}{N}]$, for $0 \leq n \leq N-1$, determine the N-point circular convolution between $g_1[n]$ and $g_2[n]$.

3.11 Consider the following discrete-time signal representing the reflectivity function of a certain subsurface structure:

$$r[n] = 4\delta[n] - 2\delta[n-4] + 0.5\delta[n-5].$$

(a) Compute its z-transform.

(b) Compute its DTFT.

(c) Compute its 6-point DFT.

(d) Plot the seismic trace $g[n]$ with respect to the discrete variable n resulting from a seismic source with a wavelet:

$$w[n] = -0.2\delta[n+1] + 0.5\delta[n] - 0.2\delta[n-1],$$

and the above given reflectivity function. Use both linear convolution and circular convolution to obtain the result. Comment on your findings.

(e) Compute the energy of $g[n]$ from the time domain and DTFT domain using Parserval's theorem.

3.12 Find the linear Radon transform of:

(a) $g(x,t) = 3\sin(\pi x/2 - 120\pi t)$.

(b) $g(x,t) = 5\delta(x - 2, t + 3)$.

3.13 Given the $f-k$ magnitude spectrum for the seismic data sets shown in Figure 3.13, relate the seismic events on the $t-x$ domain (Figure 3.13a) to that of the $f-k$ (Figure 3.13b). Sketch its linear/parabolic Radon transforms labeling clearly different seismic events?

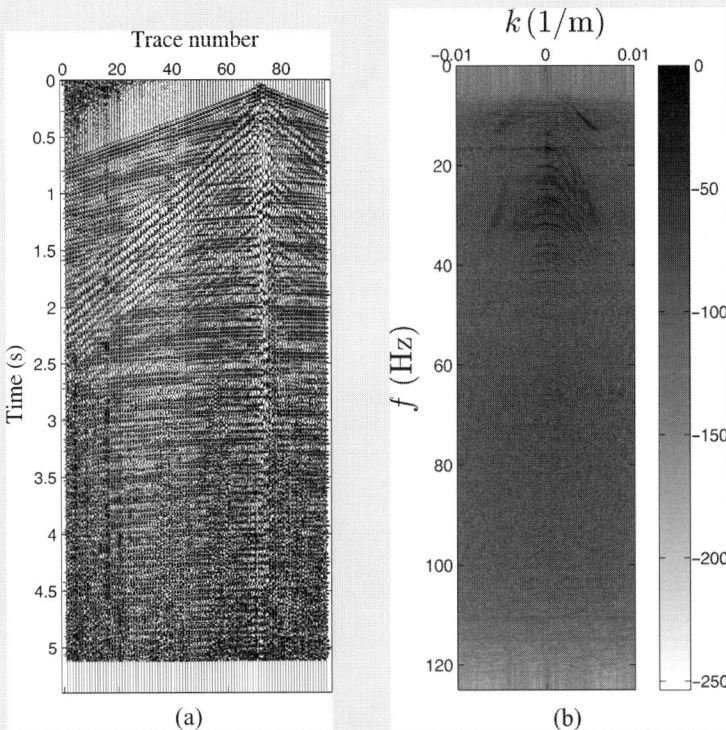

(a)   (b)

Figure 3.34 Problem 3.13 (a) Real land data modified from Yilmaz (2001) (b) its magnitude spectrum.

3.14 Given the synthetic seismic data shown in Figure 3.35, identify the seismic events. Then sketch its: (a) $f-\grave{k}$ magnitude spectrum, (b) linear $\tau-p$ Radon transform, and (c) parabolic $\tau-q$ Radon transform. Label carefully the events in each domain. Note that $\Delta t = 1$ ms and $\Delta x = 25$ m.

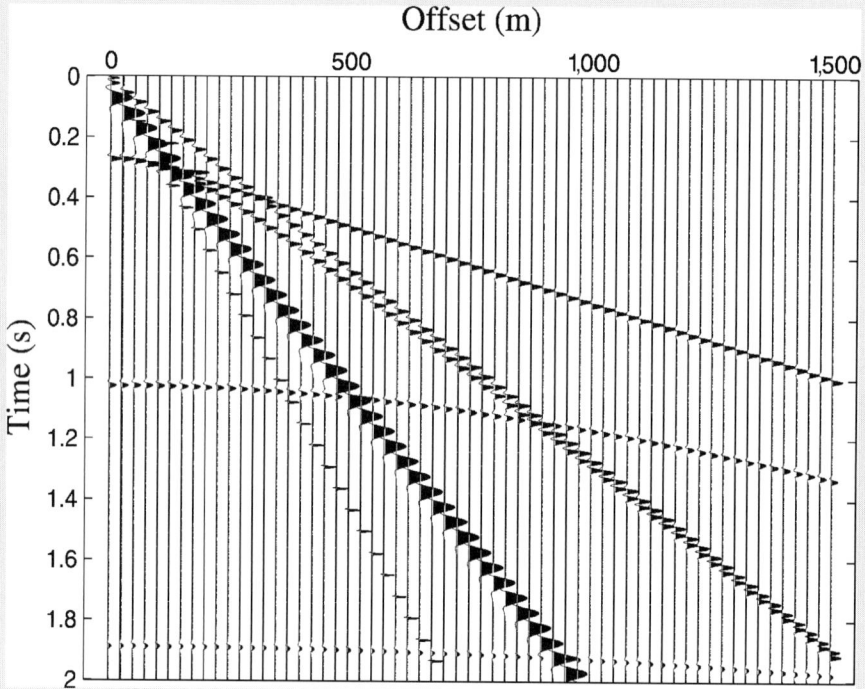

Figure 3.35 Problem 3.14. Synthetic seismic data containing many events.

4

Sampling Theorem for Seismic Data

4.1 Introduction

The goal of DSP is generally to measure and/or filter continuous real-world analog signals and, hence, the first step in DSP is usually to convert the signal from an analog to a digital form, by using an analog to digital converter. More and more aspects of signal processing are being carried out in the discrete-time/digital domain. The decision about whether to go analog or digital depends on many factors such as the task to be carried out, the bandwidths required, the cost, and whether the signal is already digital or not, etc. In addition, whenever one talks generally about digital data and their acquisition systems, Shannon Sampling Theory (Marks, 1991; Shannon, 1949) for sampling continuous time (space) signals must be explained. The seismic data acquisition and processing industry was among the very first to use the digital technology. Gigabytes of two-dimensional (2-D) and terabytes of three-dimensional (3-D) seismic data are acquired in almost every survey nowadays. In this chapter, the reader will go through the sampling theorem. The aliasing effects due to under-sampling of seismic data sets will be explained. This, of course, includes some discussion on how to choose the best parameters for sampling seismic data if the opportunity was given to do so. The theory of compressive sensing (CS) (Candes and Wakin, 2008; Donoho, 2006) is currently considered the state of the art theory of DSP with many applications related to signal and image compression, signal recovery, and many other applications (Candes and Wakin, 2008). It is a recent development in DSP offering the potential of high resolution acquisition of signals from relatively few measured samples under certain conditions. The few measured samples are typically below the minimum number of samples based on Shannon's Sampling Theorem. CS is currently used for various seismic data processing problems. Hence, in this chapter, CS basic principles are introduced. Many of its seismic data processing-related applications will be discussed.

4.2 Sampling Theorem for Time and Spatial Continuous Functions

Multi-dimensional (M-D) processing operations of seismic data can be loosely defined as those processing steps or operations that must operate simultaneously on several seismic traces. M-D processing of seismic data, particularly 2-D and 3-D processing, is useful in discriminating against noise and enhancing seismic signals on the basis of a criterion that can be distinguished from a seismic trace to another, such as moveout or dip. A discrete-time (space or both), such as a seismic wavefield $g(t,x)$, signal is a signal defined by specifying the value of the signal only at discrete-time (space or both), called sampling instants. The implication for sampling a seismic trace $g(t)$, which is a continuous-time signal, is important. If the signal $g(t)$ had frequencies, for example, up to 140 Hz, then sampling at 4 ms, meaning that the maximum acquired frequency components of the signal is at $f = 125$ Hz, will cause a loss of the remaining 15 Hz of the original signal frequency band. The situation is similar for the case of 2-D and 3-D seismic, in general, but may differ when acquiring, say for a 2-D wavefield $g(t,x)$, the maximum possible frequency components, still the spatial sampling interval will have its impact in terms of quality on the acquired data. One-dimensional (1-D) sampling is first discussed, followed by 2-D sampling of seismic data.

4.2.1 1-D Sampling

Without loss of generality, the sampling of a 1-D continuous-time seismic signal will be considered in the following context. In 1-D sampling, the analog seismic signal, say $g_a(t)$, is periodically sampled every Δt seconds to produce a Discrete-Time (DT) seismic signal $g[n]$ given by:

$$g[n] = g_a(n\Delta t), \text{ for } \infty < n < \infty, \tag{4.1}$$

where Δt is the sampling interval $(t = n\Delta t)$. The sampling frequency $f_s = 1/(\Delta t)$ must be selected large enough such that the sampling does not cause any loss of spectral information. If the spectrum of the analog seismic signal can be recovered from the spectrum of the discrete-time seismic signal, there is no loss of information. The seismic signal $g_a(t)$ is aperiodic with a finite energy, and its spectrum is given by:

$$G_a(F) = \int_{-\infty}^{\infty} g_a(t)e^{-j2\pi Ft}dt, \tag{4.2}$$

where F is the frequency of the analog signal $g_a(t)$. The signal $g_a(t)$ can be recovered from its spectrum by:

$$g_a(t) = \int_{-\infty}^{\infty} G_a(F)e^{j2\pi Ft}dF. \tag{4.3}$$

The spectrum of a DT signal $g[n]$ (obtained by sampling $g_a(t)$) is given by:

$$G(e^{jf}) = \sum_{n=-\infty}^{\infty} g[n]e^{-j2\pi fn}, \tag{4.4}$$

and $g[n]$ can be recovered from its spectrum $G(f)$ by:

$$g[n] = \int_{-1/2}^{1/2} G(e^{jf})e^{j2\pi fn} df, \tag{4.5}$$

by using change of variables in Equation (3.108), i.e, $\omega = 2\pi f$. Note that:

$$t = n\Delta t = \frac{n}{f_s}, \tag{4.6}$$

where substituting Equation (4.6) into Equation (4.3) yields:

$$g[n] = g_a(n\Delta t) = \int_{-\infty}^{\infty} G_a(F)e^{j2\pi F \frac{n}{f_s}} dF. \tag{4.7}$$

Comparing Equation (4.7) with Equation (4.5), one concludes that:

$$\int_{-1/2}^{1/2} G(e^{jf})e^{j2\pi fn} df = \int_{-\infty}^{\infty} G_a(F)e^{j2\pi F \frac{n}{f_s}} dF. \tag{4.8}$$

Now, define the relative/normalized frequency as:

$$f = \frac{F}{f_s}. \tag{4.9}$$

Substituting Equation (4.9) into Equation (4.8) yields:

$$\frac{1}{f_s} \int_{-f_s/2}^{f_s/2} G(F)e^{j2\pi \frac{F}{f_s}n} dF = \int_{-\infty}^{\infty} G_a(F)e^{j2\pi n \frac{F}{f_s}} dF, \tag{4.10}$$

which can be divided into an infinite number of intervals of width f_s. That is:

$$\int_{-\infty}^{\infty} G_a(F)e^{j2\pi n \frac{F}{f_s}} dF = \sum_{l=-\infty}^{\infty} \int_{(l-\frac{1}{2})f_s}^{(l+\frac{1}{2})f_s} G_a(F)e^{j2\pi n \frac{F}{f_s}} dF, \tag{4.11}$$

where l is an integer. Note that $G_a(F)$ in the frequency interval $(l - 1/2)f_s$ to $(l + 1/2)f_s$ is identical to $G_a(F - lf_s)$ in the interval $f_s/2$ to $f_s/2$. Hence,

$$\sum_{l=-\infty}^{\infty} \int_{(l-\frac{1}{2})f_s}^{(l+\frac{1}{2})f_s} G_a(F)e^{j2\pi n\frac{F}{f_s}}dF = \sum_{-\infty}^{\infty} \int_{-\frac{f_s}{2}}^{\frac{f_s}{2}} G_a(F - lf_s)e^{j2\pi \frac{n}{f_s}(F+lf_s)}dF, \quad (4.12)$$

$$= \int_{-\frac{f_s}{2}}^{\frac{f_s}{2}} \sum_{-\infty}^{\infty} G_a(F - lf_s)e^{j2\pi \frac{n}{f_s}(F+lf_s)}dF.$$

Comparing Equations (4.10)–(4.12), one can see that:

$$G(F) = f_s \sum_{n=-\infty}^{\infty} G_a(F - lf_s), \quad (4.13)$$

or,

$$G(e^{jf}) = f_s \sum_{n=-\infty}^{\infty} G_a[(f - l)f_s]. \quad (4.14)$$

So the frequency domain of the sampled seismic signal is a periodic repetition of a scaled analogue spectrum with a period equal to f_s.

Finally, it can be shown that, by using interpolation, one can reconstruct, if needed, the original continuous-time signal $g_a(t)$ from its samples $g[n]$ using:

$$g_a(t) = \sum_{n=-\infty}^{\infty} g_a(n\Delta t)\frac{\sin[\frac{\pi}{\Delta t}(t - n\Delta t)]}{\frac{\pi}{\Delta t}(t - n\Delta t)]}, \quad (4.15)$$

which is a sinc function that is shifted every $n\Delta t$ and is weighted by $g_a(n\Delta t)$. The sampling theorem (Proakis and Manolakis, 2006) states that a band-limited continuous-time signal with the highest frequency (Bandwidth) B Hz can be uniquely recovered from its samples provided that the sampling rate $f_s \geq 2B$ samples per second. Note that $f_s = 2B$ is called the Nyquist frequency. Based on Equation (4.15), the recovery of $g_a(t)$ from its samples $g[n]$ requires an infinite number of samples. Practically, one can use a finite number of samples of the signal and deal with finite-duration signals. Finite-duration signals need to be band-limited prior to sampling to avoid spectrum interference, which is known as aliasing. Figure 4.1 illustrates the effect of proper sampling and the effect of improper sampling that causes aliasing. Note that the sampling theorem works for any independent variable such as time, space, etc.

> **Example 4.1** Consider the sinusoids $u_1(t) = \sin(20\pi t)$ and $u_2(t) = \sin(100\pi t)$, where both are sampled at a rate of $F_s = 40$ Hz. Obtain the corresponding discrete-time signals and comment on the results.

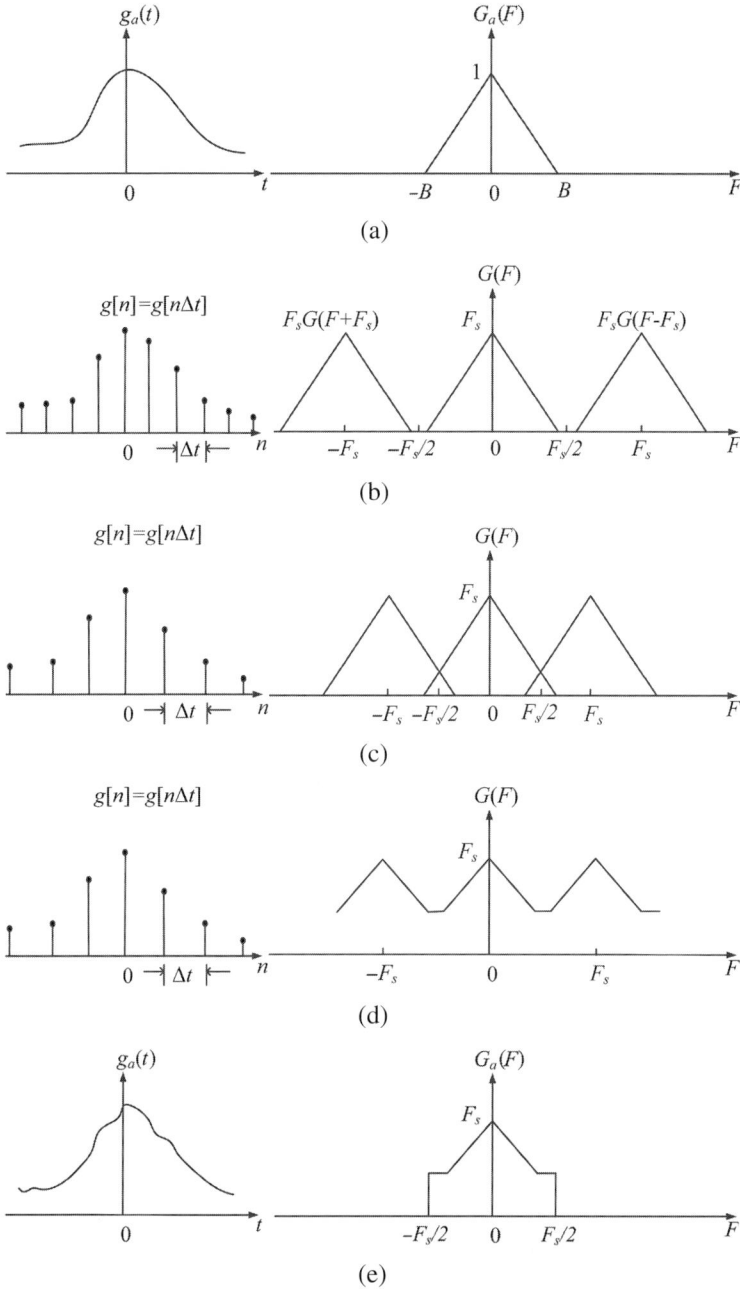

Figure 4.1 This illustrates the impact of sampling (a) a continuous-time domain seismic signal, properly in (b) and improperly (c)–(d). So that when the signal is to be recovered as shown in (e), it will be aliased and not equal to the one in (a).

Solution: The corresponding discrete-time signals are as follows:

$$u_1[n] = \sin(20\pi n \Delta t),$$

$$= \sin\left[\frac{20\pi n}{40}\right], \text{ since } \Delta t = \frac{1}{F_s}$$

$$= \sin\left[\frac{\pi n}{2}\right].$$

Also,

$$u_2[n] = \sin(100\pi n \Delta t),$$

$$= \sin\left[\frac{100\pi n}{40}\right],$$

$$= \sin\left[\frac{5\pi n}{2}\right],$$

$$= \sin\left[\frac{\pi n}{2} + \frac{4\pi n}{2}\right],$$

$$= \sin\left[\frac{\pi n}{2}\right]\cos[2\pi n] + \sin[2\pi n]\cos\left[\frac{\pi n}{2}\right],$$

$$= \sin\left[\frac{\pi n}{2}\right].$$

Hence, $u_1[n] = u_2[n]$ and they cannot be distinguished. So when sampling $u_1[n]$ and $u_2[n]$ at 40 samples/second, the 50 Hz sinusoid in $u_2[n]$ is called an alias of the 10 Hz sinusoid in $u_1[n]$ at $F_s = 40$ Hz.

Finally, Figure 4.2 shows a seismic signal $g(t)$ from Yilmaz (2001) with various sampling rates. Clearly, improper sampling will result in aliasing. This is also witnessed when observing the whole seismic shot gather in Figure 4.3, where some of the events start to fade as the time sampling rates are decreased.

4.2.2 2-D Sampling

Of several ways to generalize 1-D sampling to the 2-D case, the most straightforward is sampling in rectangular coordinates called *Rectangular Sampling*. Let Δx be the spatial sampling interval along the x-axis and Δt be the temporal sampling interval along the t-axis. Now, following the same procedure for sampling 1-D signal $g(t)$, one can show that:

$$G(e^{jf}, e^{jk}) = f_s k_s \sum_{l=0}^{\infty} \sum_{m=0}^{\infty} G_a[(f-l)f_s, (\hat{k}-m)k_s], \qquad (4.16)$$

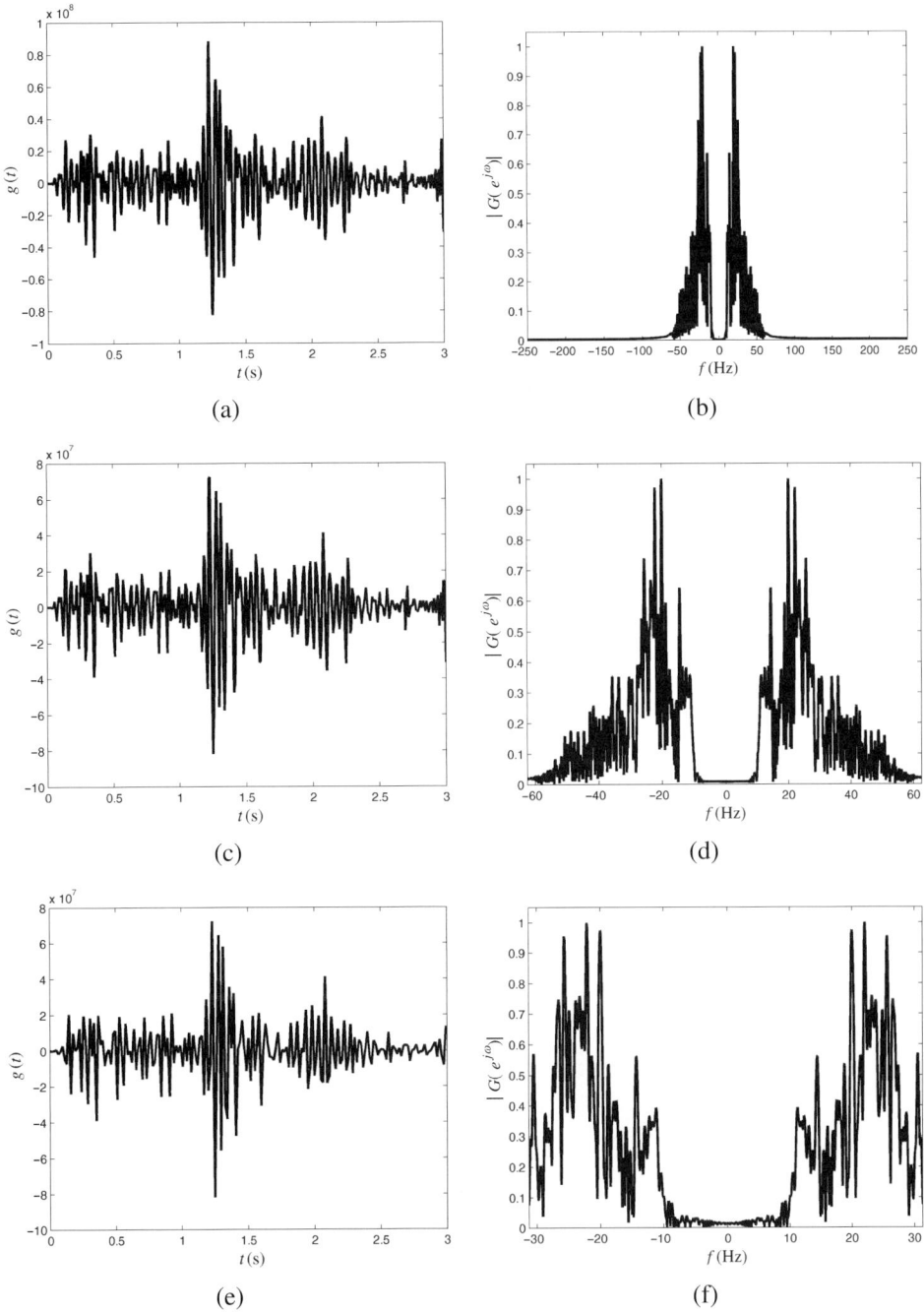

Figure 4.2 Seismic trace number 1 from the first seismic data shot gather by Mousa and Al-Shuhail (2011): (a) at $dt = 2$ ms and (b) its magnitude spectrum. (c) at $dt = 8$ ms and (b) its magnitude spectrum. (a) at $dt = 16$ ms and (b) its magnitude spectrum. Clearly, improper sampling will result in aliasing.

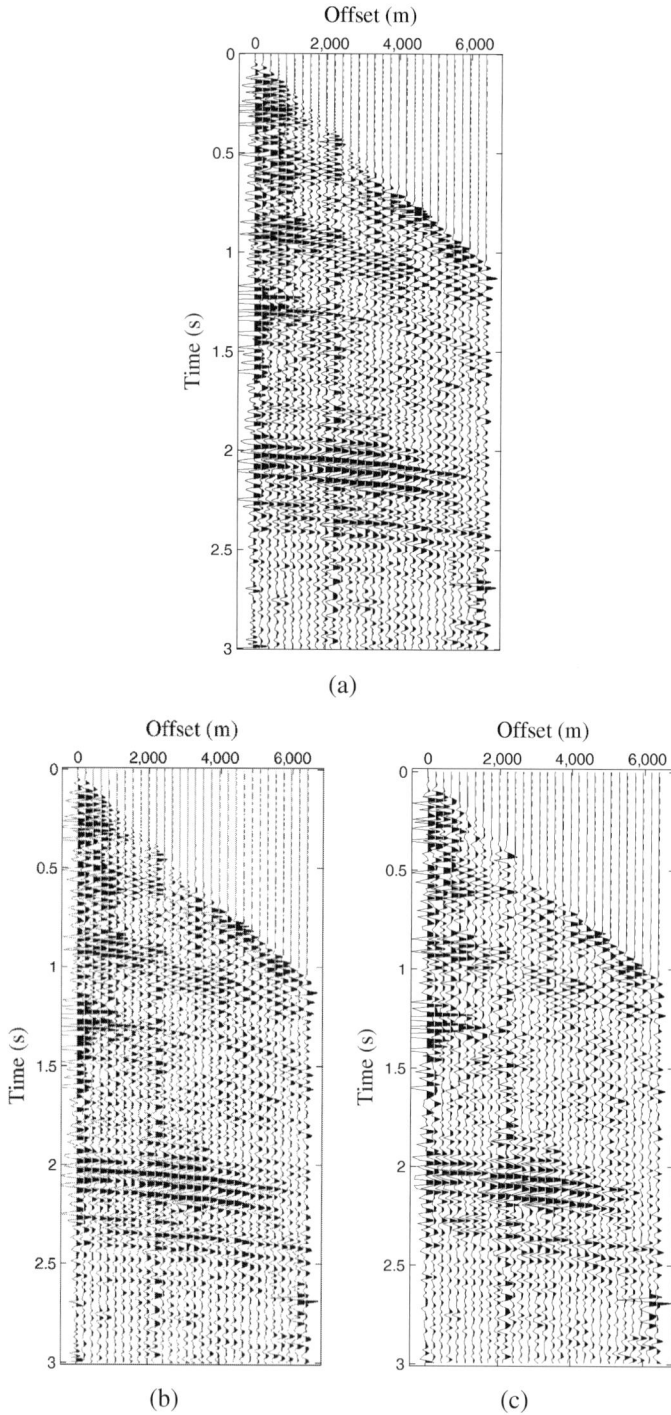

Figure 4.3 The first seismic data shot gather by Mousa and Al-Shuhail (2011) with traces sampled in time at: (a) $dt = 2$ ms. (b) $dt = 8$ ms and (c) $dt = 16$ ms.

where $G_a(\omega, k)$ is the 2-D Fourier transform of the analogue seismic data $g_a(t, x)$, $k_s = 1/\Delta x$, $f_s = 1/\Delta t$. That is, the spectrum of the continuous function $g_a(t, x)$ repeats every $1/\Delta x$ in the wavenumber domain $\hat{k}$ and, at the same time, repeat itself every $1/\Delta t$ in the frequency-domain f.

The 2-D seismic signal $g_a(t, x)$ can be totally recovered from its discrete-time domain $g[n_t, n_x]$ provided that $g_a(t, x)$ is sampled at rates higher then or equal to the Nyquist frequency ($f_s \geq \frac{1}{2\Delta t}$) and the Nyquist wavenumber ($k_s \geq \frac{1}{2\Delta x}$). Note that if the spectrum $G(e^{jf}, e^{jk}) \neq 0$ outside the range $[-\frac{1}{2\Delta t}, \frac{1}{2\Delta t}]$, $[-\frac{1}{2\Delta x}, \frac{1}{2\Delta x}]$, then aliasing occurs.

Example 4.2 Consider Figure 4.4. Recall (from Equation 2.101) that the dip mouveout (DMO) difference between the travel-time of a reflection measured at the surface is:

$$\Delta T_d = \frac{2\Delta x \sin \theta}{v}. \qquad (4.17)$$

The apparent velocity of the reflected wave at the surface is defined as (Kearey et al., 2002; Yilmaz, 2001):

$$v_a = \frac{\Delta x}{\Delta T_d} = \frac{v}{2 \sin \theta}. \qquad (4.18)$$

Now, for a signal with a frequency f, the apparent wavelength of the reflected wave is:

$$\lambda_a = \frac{v_a}{f} = \frac{v}{2f \sin \theta}. \qquad (4.19)$$

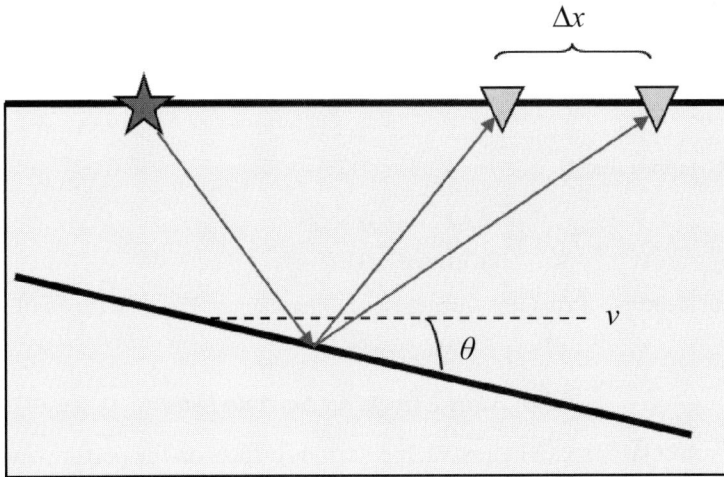

Figure 4.4 Illustrating the concept in Example 4.2.

Hence, the apparent wavenumber is:

$$\grave{k} = \frac{f}{v_a} = \frac{2f \sin \theta}{v}. \tag{4.20}$$

Now, according to the sampling theorem, there must be at least two samples per wavelength. That is,

$$\Delta x \leq \frac{\lambda_a}{2} = \frac{v}{4f \sin \theta}, \tag{4.21}$$

or,

$$\grave{k} \leq \frac{1}{2\Delta x}. \tag{4.22}$$

So, if the velocities and maximum dip of the subsurface are known for the area being surveyed. Also, if the predominant frequencies of the seismic signals are roughly known, then one can use Equation (4.22) to determine the proper distance between the geophones in order to satisfy the sampling theorem and, therefore, avoid spatial aliasing.

Example 4.3 Obtain the minimum spatial sampling interval needed to avoid aliasing for a seismic signal with $f = 70$ Hz, $\theta_{max} = 45°$ in an area of velocities of about $2,000$ m/s.

Solution: Since $f = 70$ Hz, $\theta_{max} = 45°$ and $v = 2,000$ m/s, then a sampling interval of $\Delta x \leq 10$ m is necessary to avoid aliasing.

Example 4.4 Given that the trace spacing $\Delta x = 25$ m and with an event dip $= 20$ ms/trace, at what frequency f will the event become spatially alias?

Solution: The Nyquist wavenumber k_N must be found first. That is $k_N = \frac{1}{2\Delta x} = \frac{1}{50} = 0.02$ 1/m. However,

$$\grave{k} = \frac{dip \times f}{\Delta x}, \tag{4.23}$$

which implies that:

$$f = \frac{\grave{k}\Delta x}{dip} = \frac{0.02 \times 25}{0.02} = 25 \text{ Hz}. \tag{4.24}$$

Hence, if the event contains frequencies higher than 25 Hz, then they will alias.

4.2.3 Alias Effects on Seismic Data

Aliasing, particularly spatial aliasing, has serious effects on the performance of later seismic data processing steps like frequency-wavenumber ($f-\grave{k}$) filtering, which is used to attenuate ground roll noise in land seismic data. In this case, the $f-\grave{k}$

filtering process will perceive seismic events with steep dips at high frequencies as different from how they actually are and, hence, will not process them properly. Another affected step is seismic migration, which depends on the spectrum of the seismic wavefield. It will move the spatially aliased frequency components in the wrong direction, and it generates a dispersive noise that degrades the migrated section quality. Hence, it is important to further discuss the effects of seismic signal aliasing by considering the following cases:

- **The Mono-Frequency Case**: By using a single frequency sinusoid ($f = f_o$), one can see that frequencies above the Nyquist frequency really are not lost after sampling but they appear at frequencies below the Nyquist. This appearance is called folding or aliasing. For example, consider the following sampled zero-phase sinusoid with a sampling time interval Δt:

$$g_1[n] = g_1(n\Delta t) = A \cos[2\pi f_o n \Delta t]. \tag{4.25}$$

Now, consider another sinusoid with the same parameters except that its frequency is equal to $f = f_o + m f_s$, i.e.,

$$g_2[n] = g_2(n\Delta t) = A \cos[2\pi (f_o + m f_s) n \Delta t], \tag{4.26}$$

where $m \in \mathbb{Z}$. Then expanding Equation (4.26) yields:

$$\begin{aligned}
g_2[n] &= A \cos[2\pi (f_o + m f_s) n \Delta t], \\
&= A \cos[2\pi f_o n \Delta t + 2\pi m n f_s \Delta t], \\
&= A \cos[2\pi f_o n \Delta t + 2\pi m n], \\
&= A \cos[2\pi f_o n \Delta t], \\
&= g_1[n]. \tag{4.27}
\end{aligned}$$

Since $g_1[n] = g_2[n]$, then it will be difficult to distinguish them from each other. In addition, let $f = -f_o + m f_s$ in Equation (4.25), then:

$$\begin{aligned}
g_3[n] &= A \cos[2\pi (-f_o + m f_s) n \Delta t], \\
&= A \cos[-2\pi f_o n \Delta t + 2\pi m n f_s \Delta t], \\
&= A \cos[-2\pi f_o n \Delta t + 2\pi m n], \\
&= A \cos[2\pi f_o n \Delta t], \\
&= g_1[n]. \tag{4.28}
\end{aligned}$$

In this case, $g_1[n] = g_3[n]$, which is again difficult to distinguish them from each other. So there are two types of aliasing that occur, namely, one at positive frequencies (called aliasing) and one at negative frequencies (called folding). The locations of the spurious spectral components within the Nyquist interval can

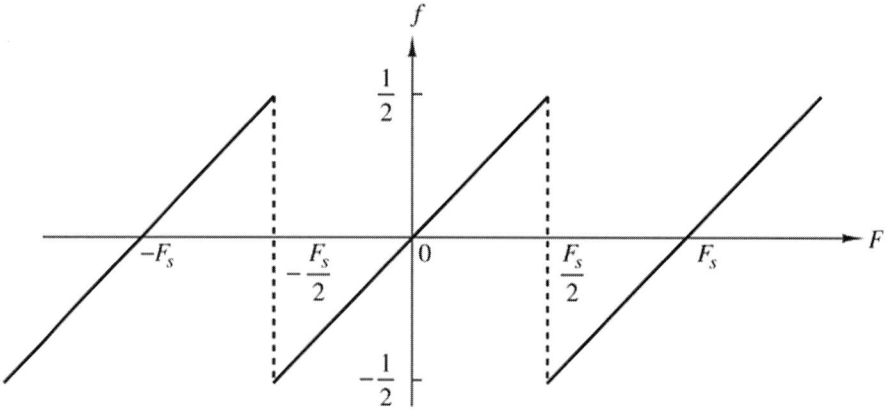

Figure 4.5 An illustration showing when aliasing/folding occur.

be easily found using Figure 4.5. In the case of Equation (4.25), the frequency f_o is mapped by folding on the horizontal lines that pass through integer multiples of f_N.

To compute the alias (or folded) frequency f_a, one can use the following relation (Yilmaz, 2001):

$$f_a = 2mf_N - f_{max}, \qquad (4.29)$$

where f_N is the Nyquist frequency, f_{max}, the maximum signal frequency and m is the minimum integer such that $fa < f_N$. Note that when $f_a > 0$ then this means that it is an alias frequency and if $f_a < 0$ then it is a fold frequency.

> **Example 4.5** Obtain the alias frequency for $g(t) = 2\cos(130\pi t)$ when it is sampled at $\Delta t = 8$ ms.

> **Solution:** $f_N = \frac{1}{8 \times 10^{-3}} = 62.5$ Hz and $f_{max} = 65$ Hz. This implies that $f_a = |2 \times 62.5 - 65| = 60$ Hz $< f_N$. This means that 60 Hz will have the alias frequency.

Consider seismic sections that contain dipping events (at $Dip = 12$ ms/trace) but at different frequencies ($f = 10$, 30, and 50 Hz, like those in Figure 4.6). In each seismic section, the seismic event dips are positive. However, for the 50 Hz section, the dip is ambiguous where the seismic events have a checkboard effect or character that makes it difficult to determine whether the dips are positive or negative. Part of the mapped seismic events have moved from the positive wavenumber quadrant to the negative wavenumber quadrant, which is the wrong quadrant, for this section and, hence, aliasing occurs. Spatial aliasing not only causes mapping to the wrong frequency-wavenumber quadrant but also causes mapping with wrong dip.

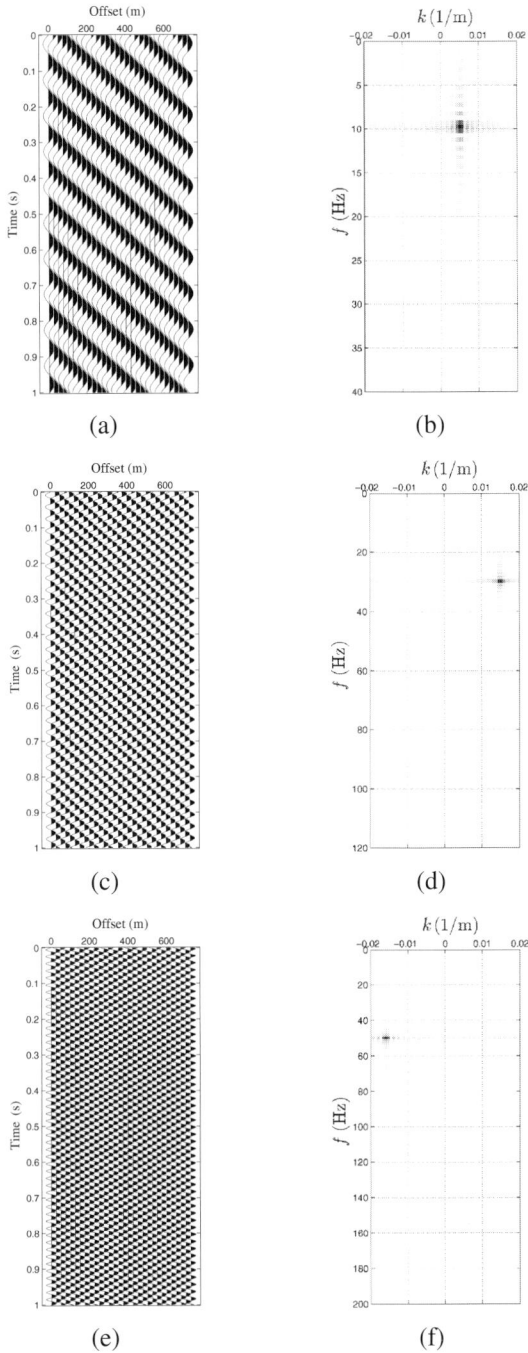

Figure 4.6 A synthetic seismic section with a $Dip = 12$ ms/trace, (a) with $f = 10$ Hz and $\grave{k} = 0.0048$ 1/m. (b) Its magnitude spectrum. (c) $f = 30$ Hz and $\grave{k} = 0.0144$ 1/m and (d) its magnitude spectrum. (e) $f = 50$ Hz and $\grave{k} = 0.024$ 1/m and (f) its magnitude spectrum.

Undersampling the data mainly has two effects (Yilmaz, 2001). First, it will bandlimit the spectrum of the continuous-time seismic signal with the maximum frequency being f_N, and nothing can be done about it, unless one is interested in reacquiring the data. Second, contamination of the digital signal spectrum by the original continuous-time signal high frequencies will occur beyond f_N. In order to keep the recoverable frequency free from aliased frequencies, a low-pass filter, sometimes referred to as an anti-aliasing filter or pre-filter, is applied in the field before the A/D conversion of seismic signals takes place. Typical cut-off values are $0.5-0.75$ of f_N and must roll-off steeply so that the frequencies above f_N become highly attenuated (Yilmaz, 2001).

- **Continuum of Frequency Components**: Suppose that one has seismic sections as in Figure 4.7 with the same dip but with different frequencies. Then, for a given dip, all frequency components map onto the $f-\grave{k}$ plane along a straight line that passes through the origin. The higher the dip, the closer the radial line in the $f-\grave{k}$ domain to the wavenumber axis, where the zero-dip components map along the frequency axis. Spatially aliased frequencies are located along the linear segments that wrap around at k_N to the opposite quadrant in the $f-\grave{k}$ amplitude spectrum. The steeper the dip, the lower the frequency f_a at which the spatial aliasing occurs. Events with the same dip in the $t-x$ domain, regardless of their location, map onto a single radial line in the $f-\grave{k}$ domain.

Finally, a few ways to avoid spatial aliasing exist. some of them are (Yilmaz, 2001):

1. Apply the data through an analog low-pass filter prior to A/D conversion. This is done in the field.
2. Apply spatial anti-alias filtering by array response weather conventional or digital processing, which also can be done in the field via various testing.
3. Select a sufficient small trace spacing Δx, where this may either require a data-dependent interpolation algorithm or modifying the field acquisition geometry. The former is done offline and before processing, while the latter, of course, requires reacquisition but with maybe more sources and more receivers (geophones).
4. Apply wavenumber filtering on the data to remove the seismic energy beyond $0.5k_N$ for all the frequencies. This can be applied after the data is recorded and ready for processing.

4.3 Overview of Compressive Sensing Applications in Seismic Data Processing

The advent of technology and demand for higher resolution data has increased the amount of data acquired for seismic exploration. Nowadays, most of the acquired

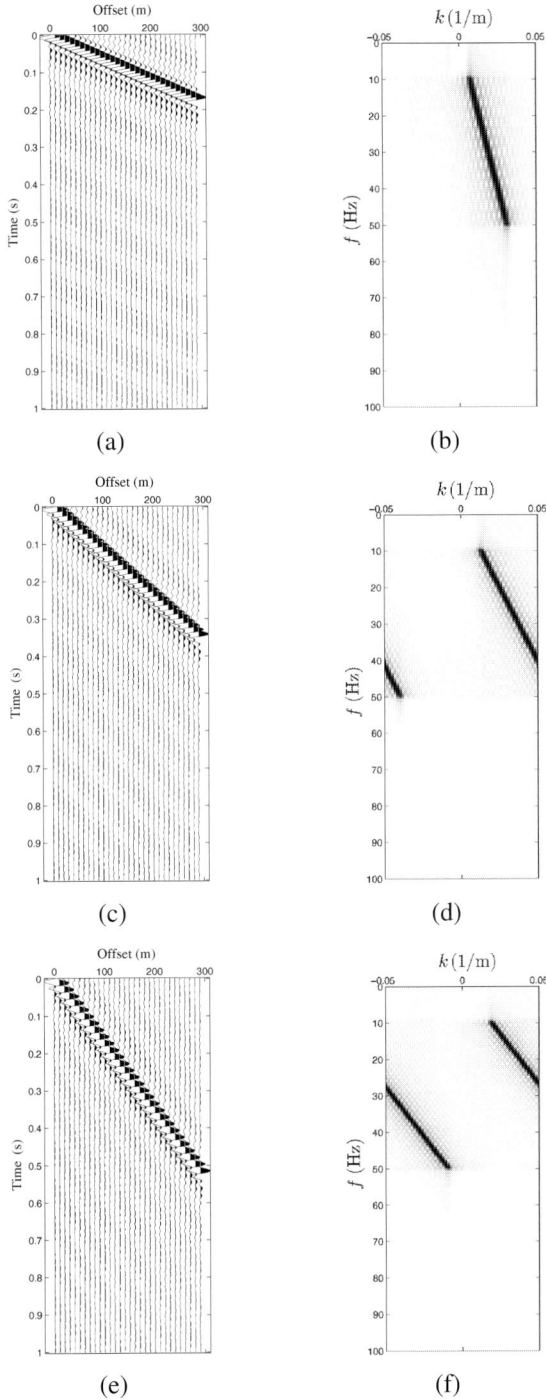

(a)   (b)

(c)   (d)

(e)   (f)

Figure 4.7 A synthetic seismic section with $f = [10, 50]$, (a) with a $Dip = 6$ ms/trace and (b) its magnitude spectrum. (c) A $Dip = 12$ ms/trace and (d) its magnitude spectrum. (e) A $Dip = 18$ ms/trace and (f) its magnitude spectrum.

data is huge, mostly in terabytes, and processing of such huge data takes a lot of processing power and time. Furthermore, most of the seismic exploration techniques depend on acquisition of data, and the existing acquisition techniques do not utilize the structure of the seismic data. By using compressive sensing, one can make use of the fact that seismic data is sparse in a particular domain. This feature allows the use of compressive sensing, for which the sampling rate depends on the sparsity of the data.

Like other fields, compressive sensing has been proposed for different seismic processing applications and seismic data acquisition. The first advancement was the continuous acquisition as instances of compressive sensing and identification of seismic data (Hennenfent and Herrmann, 2005; Herrmann et al., 2009a; Herrmann, 2009b). A lot of progress has been made in the selection of a sparse transform domain and the random sensing schemes for the compressive sensing (Herrmann, 2009a). Curvelet-based seismic data processing was presented by Elad et al. (2005). Seismic data restoration is discussed by Herrmann and Hennenfent (2008) and Cao et al. (2011). Interpolation of seismic data using compressive sensing and curvelet is presented by Lin and Herrmann (2007); Sastry et al. (2007); Hennenfent and Herrmann (2008), and Yang et al. (2013). Model for simultaneous acquisition using compressive sensing is presented by Lin et al. (2009). Separation of multiples using compressive sensing is discussed by Herrmann et al. (2007b). Compressive sensing can also be used for marine data acquisition as discussed in (Herrmann and Wason, 2012; Mansour et al., 2012). The deconvolution problem of seismic data was discussed by Saachi in Gholami and Sacchi (2012).

Compressive sensing exploits the fact that almost all the signals are sparse in some particular domain (Zibulevsky and Pearlmutter, 2001). For example, many captured signals such as seismic data, medical images, photos, and music, can be compressed by representing them in appropriate basis. When appropriate bases are used, then most of the coefficients will be zero or almost zero. The biggest challenge is to find an appropriate transform that can then transform the seismic data in a sparse fashion. For the seismic data, curvelets and wave atoms are extensively used (Herrmann et al., 2009b; Tang et al., 2009). The curvelet transform composes wavefields as superposition of multi-scale and highly localized waveforms. Because of this, one can get needle-like curvelets at small scales. It has been shown by Candès and Demanet (2003) that curvelet transforms are suitable for compressing seismic data, due to their near invariance under wave propagation. For oscillatory wave fronts, wave atoms are more suitable (Candès, 2006).

The basic concept behind compressive sensing is rather simple; if one already has captured enough samples, then increasing the sample rate will not increase the resolution of the signal. According to the theory of compressive sensing, the sampling rate does not depend on the frequency content of the signal; rather, it depends on the information content of the signal. A signal with low information content can be recovered, without any error, only using a small number of measurements. Fortunately, almost all the real world signals have low information content (Zibulevsky and Pearlmutter, 2001). Here, low information content means that the signal of interest is not changing and is sparse in nature. Hence, compressed sensing exploits the fact that most of the real-life signals have low information content to bypass the Shannon theorem (MacKenzie, 2009).

4.3.1 Properties of Compressive Sensing

Compressive sampling is the technique that goes against the common wisdom in data acquisition and enables one to recover signals with fewer samples than the Nyquist. To make this possible, compressive sensing relies on two principles: sparsity and incoherence of the signal in particular domains (Candès, 2008; Needell and Vershynin, 2009). Sparsity uses the fact that the informational content of the signal might be much less than suggested by the bandwidth of the signal. Simply, sparsity deals with the number of nonzero elements in a signal. A signal of size N is said to be K sparse if it has maximum K nonzero elements. Incoherence covers the duality between the transformed domain and the captured domain (e.g., time and frequency). If a signal is sparse in time, then it will be spread out in a frequency domain or vice versa. In other words, incoherence is needed to measure the signal in such a way that the maximum amount of information can be extracted from the signal by using small amounts of measurement (Baraniuk, 2007). Mathematical techniques necessary to implement compressive sensing include the selection of appropriate transforms. Particularly, the l_1 optimization is used for the representation of the signal in the sparse domain. The l_1 minimization concentrates the energy of the signal onto a few nonzero coefficients.

4.3.2 Mathematical Theory

There are many different ways to setup and solve the compressive sensing problem. Mostly, vector-based techniques are used, as they are simple to model and solve. In this section, a compressive sensing problem is solved using l_1-norm minimization. Let $g[n]$ be a 1-D seismic signal, acquired using traditional sampling techniques. Once acquired, the data contain N uniformly spaced samples and can be represented as follows:

$$g[n] = \sum_{i=1}^{N} x_i \psi_i, \tag{4.30}$$

where x_i are the coefficients sequence of $g[n]$, with ψ_i as the basis of a sparsifying transform. To make the formulation easier, the signal $g[n]$ in terms of the matrix is as follows:

$$g = \psi x. \tag{4.31}$$

For the multiplication to hold, the dimension of ψ should be $N \times N$. If ψ is a sparsifying transform, then g is a sparse vector with size N. Let g be K sparse vector, when represented in basis ψ. In other words, the number of nonzero entries in g are K, and remaining $N-K$ entries are zero. As g is a K sparse vector, instead of acquiring all the N elements of g, the signal can be recovered by fewer samples. In simple words, compressive sensing helps to acquire only the nonzero entries of the g. It achieves this by introducing a new matrix known as a measurement matrix (ϕ) with size $M \times N$, where $M \ll N$. The sampled signal g can be represented in terms of a new basis ϕ, as follows:

$$y = \phi g, \tag{4.32}$$

and

$$y = Ax. \tag{4.33}$$

Here $A = \phi\psi$, sometimes referred to as a sensing matrix, is a matrix with dimension $M \times N$. So instead of using ψ for sampling, one can use A for sampling the signal. However, the size of the observed vector y is $M \times N$ and one has reduced the size of the measurements from N to M, where $M \ll N$. To recover the original signal, one has to find the sparsest coefficient vector $\hat{x}$ by solving the following relation:

$$A\hat{x} = b, \tag{4.34}$$

where

$$b = \phi g. \tag{4.35}$$

Here A, a rectangular matrix, forms an underdetermined system of linear equations. In this case, the number of unknowns is greater than the number of equations. Restricted isometric property (RIP) can be utilized for the perfect reconstruction of the undersampled data as shown by Candès (2008). The measurement matrix ϕ satisfies the RIP of order K, if a $\delta_K \in (0, 1)$ exists, such that;

$$(1 - \delta_K)\|x\|_2^2 \leq \|\phi x\|_2^2 \leq (1 + \delta_K)\|x\|_2^2. \tag{4.36}$$

The RIP is a sufficient condition for many sparsifying recovery theorems. Besides RIP, the sensing matrix and representation matrix should have incoherence $\mu(\phi, \psi)$, being:

$$\mu(\phi, \psi) = \sqrt{n} \times \max_{1 \leq K, j \leq n} |\langle \phi_K . \psi_j \rangle|, \tag{4.37}$$

where,

$$\mu(\phi, \psi) \in [1, \sqrt{n}]. \tag{4.38}$$

Once the RIP and incoherence are satisfied, then recovery algorithms are required for the representation of the signal in the sparse domain. Particularly, the l_1-norm can be used to solve the ill-conditioned relation Equation (4.34), which concentrates the energy of the signal onto a few nonzero coefficients. By using the l_1-norm for the reconstruction, the problem becomes:

$$\min \|x\|_1 \text{ subject to } Ax = b. \tag{4.39}$$

Once the coefficients are estimated by the l_1-norm, then the original signal can be recovered by using the following relation:

$$g = \psi \hat{x}. \tag{4.40}$$

Equation (4.39) is only true for a noise-free system of the form $g = \psi x$. However, by following the same procedure, it can be shown that in the presence of noise, Equation (4.39) changes into the following problem:

$$\min \|x\|_1 \text{ subject to } |Ax - b| < \epsilon. \tag{4.41}$$

This relation is true for any system of the form, $g = \psi x - n$, where n is noise in the signal.

Example 4.6 The signal $g(t) = \cos(40\pi t) + \cos(150\pi t)$, is sampled in time with 1,024 samples ($N = 1,024$). The signal and its frequency content are shown in Figure 4.8a and b. From the frequency response it is evident that the signal is sparse in the frequency domain and only has two frequencies. As the signal is sparse, one can use compressive sensing to acquire the same signal with far fewer samples. For this example, $M = 256$ is used along with the discrete Fourier transform (DFT) as its sparsifying transform. The recovered signal and the signal without the compression in frequency response is shown in Figure 4.8a. The signal and recovered signal in the time domain is shown in Figure 4.8a and c. The error between the original signal and the compressed signal is also plotted in Figure 4.8d.

Now, the following subsections discuss the application of compressive sensing for two seismic data processing problems, namely, interpretation of seismic traces and multiple attenuation. The Radon transform is used as the sparsifying domain in the both cases, since it was presented in book (see Chapter 3).

4.3.3 Missing Seismic Trace Interpolation

Interpolation/reconstruction of missing seismic traces is one of the key pre-processing steps, whereby the resolution of the data is increased. During the last few years, the demand for denser seismic traces has significantly increased. In most cases, due to the limitations of hardware, budgets, and computing power, it

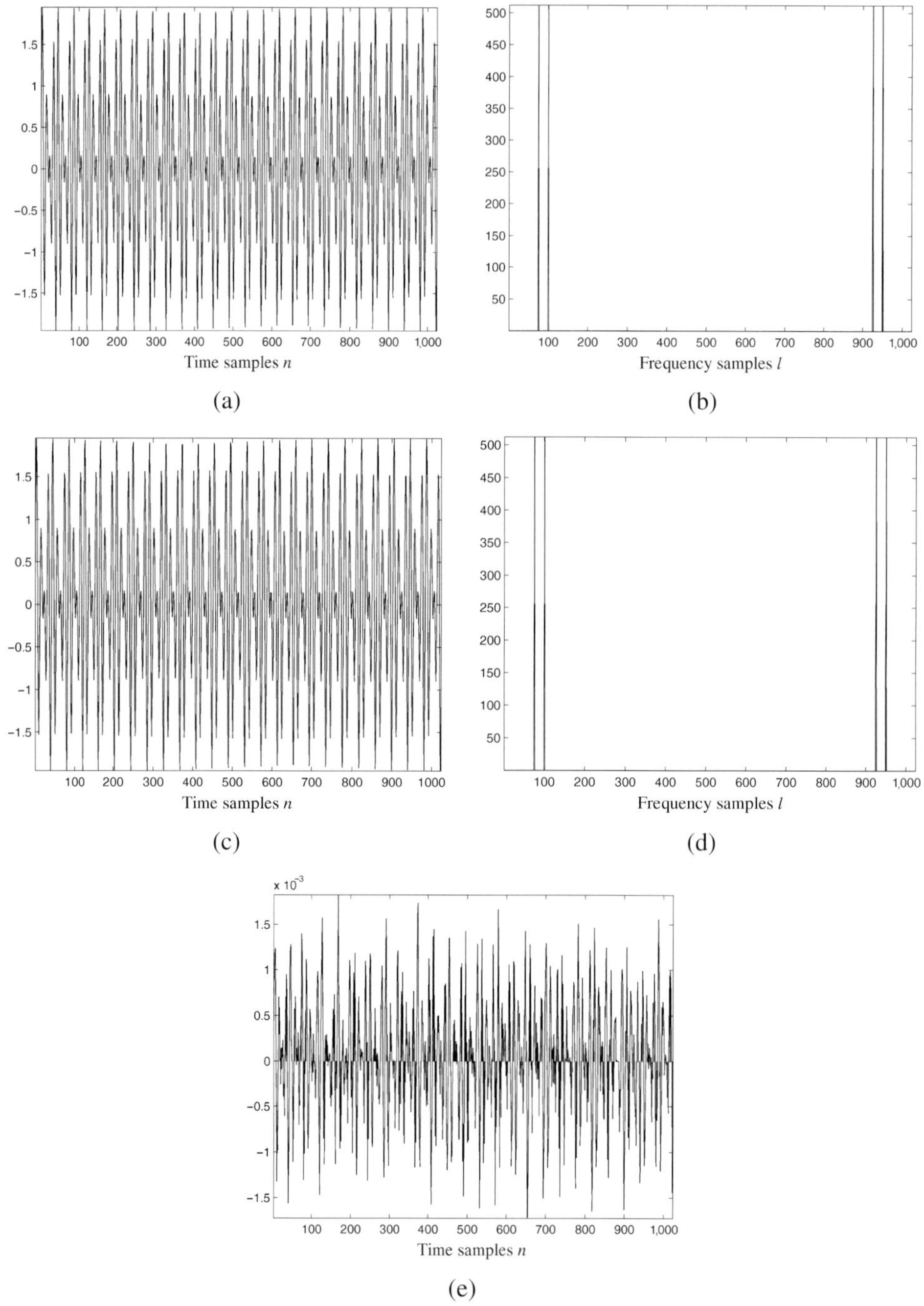

Figure 4.8 A compressive sensing example: (a) The original signal with $N = 1,024$, (b) its sparse frequency representation. (c) The recovered signal in time domain after applying compressive sensing with $M = 256$, (d) the recovered signal in frequency domain after applying compressive sensing with $M = 256$. (e) The error between (a) and (c).

is not possible to collect large samples of seismic traces. Sometimes, due to poor conditions or the topological structures of the Earth's surface, it is not possible to place seismic sensors (geophones), resulting in missing traces in the acquired data. Thus, reducing the amount of acquired data as well as effective reconstruction of complete seismic data from acquired incomplete seismic data are very important issues. Most famous methods are based on transformation of seismic data into the sparse domain by using transforms like wavelet, Radon, or curvelet, as discussed in Trad (2003), Liu and Sacchi (2004), and Hennenfent and Herrmann (2006). Similarly, there are some techniques based on filters, as discussed in Gulunay (2003). Low rank-based interpolation is presented in Recht et al. (2013). Energy of the primary reflections can also be used for the interpolation, as discussed in Berkhout and Verschuur (2006). The same energy concept is extended in Curry and Shan (2010). Compressive sensing-based interpolation, using the curvelet transform, are presented in Cao et al. (2012); Herrmann et al. (2007a).

Now, given a seismic record, the proposed compressive sensing with parabolic Radon transform $\tau-q$ method for interpolation of missing seismic traces is stated as follows:

1. Use the compressive sensing to acquire the seismic data in the compressed fashion.
2. Use the following facts for the thresholding:

   1. The curves are converted to single points in the $\tau-q$ domain.
   2. These points have higher value than the points corresponding to the noise or irrelevant curves.

3. Apply the inverse parabolic Radon transform to the previous step result.

The synthetic data is presented in Figure 4.9a. The total number of traces in the seismic data is 50 ($N = 50$), out of which, 19 are missing. Compressive sensing with parabolic Radon transform is used for the interpolation of missing seismic traces. Figure 4.9b shows the compressed $\tau - q$ domain with $M = 10$. Now, based on the second step of the proposed algorithm, the binary image is presented in Figure 4.9c. Finally, the results after the application of the inverse $\tau-q$ transform are presented in Figure 4.9d. Besides 18 missing traces (33%), with the help of compressive sensing, additional 80% compression is achieved.

To show the robustness of the proposed setup, data with different noise levels were tested. The total number of traces and number of missing traces are the same as the noiseless case. The results after the addition of the noise with 0.2 standard deviation and 60% compression are presented in Figure 4.10. Similar comparisons after the addition of white noise with standard deviation 0.4 are presented in Figure 4.11.

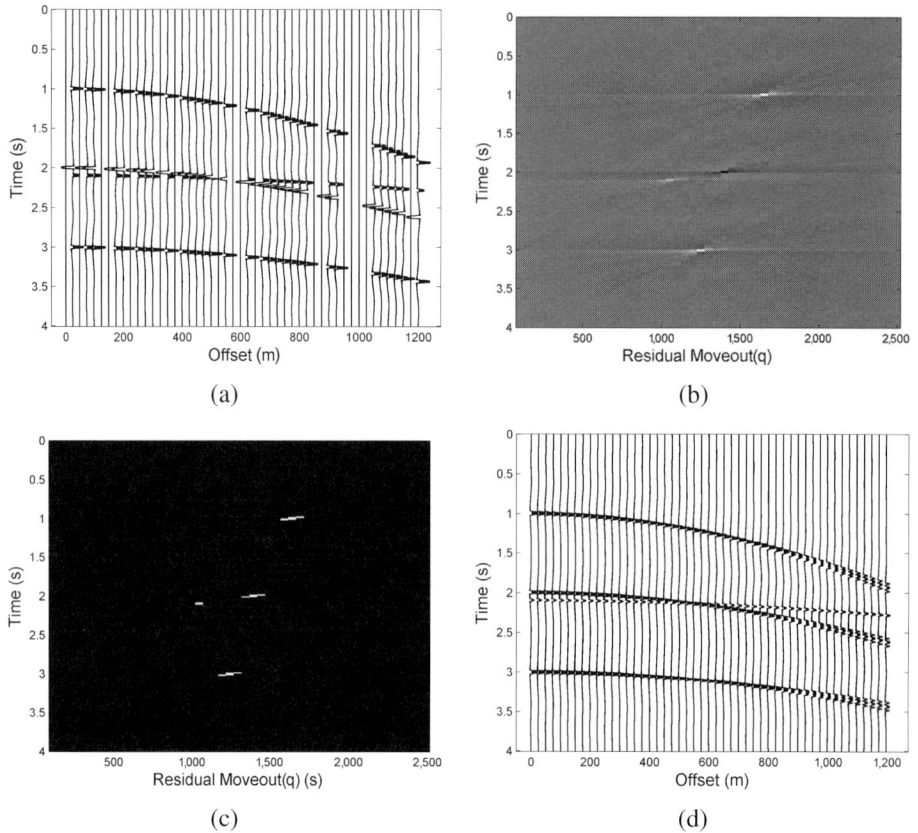

(a)

(b)

(c)

(d)

Figure 4.9 (a) Synthetic seismic reflected data. (b) The compressed $\tau - q$ transform of (a), after the acquisition in compressed fashion with $M = 10$ (80% compression). (c) The $\tau - p$ domain after performing thresholding of (b). (d) The interpolated data after taking the inverse $\tau - q$ transform to (c).

4.3.4 Primary Arrivals and Multiples Separation

The captured data contain multiple reflections from the same Earth layer. Due to the complex geological conditions, different kinds of multiple reflections are generated; these reflections significantly reduce the resolution of recorded seismic data as they get mixed with primary waves. These reflections interfere with the primary reflections, which in turn make the processing of seismic data more difficult. Besides difficulty in identification of primary reflections, we also face difficulties in the data analysis and interpretation. Attenuation of multiple reflections from the recorded seismic data is one of the key issues in the field of seismic data processing, so we can obtain the description of the concerned surface as accurately as possible.

Along with primary reflections, data contain many undesired components such as multiple reflections, refractions, diffractions, etc. The presence of these unwanted

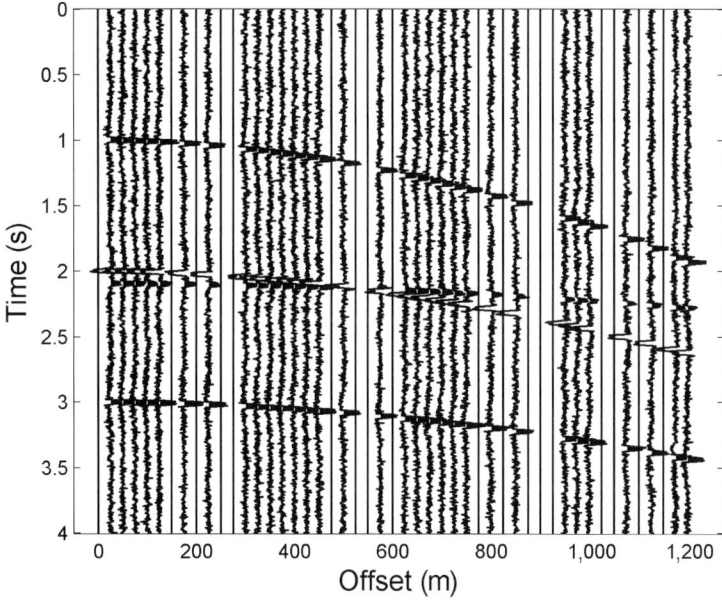

(a)

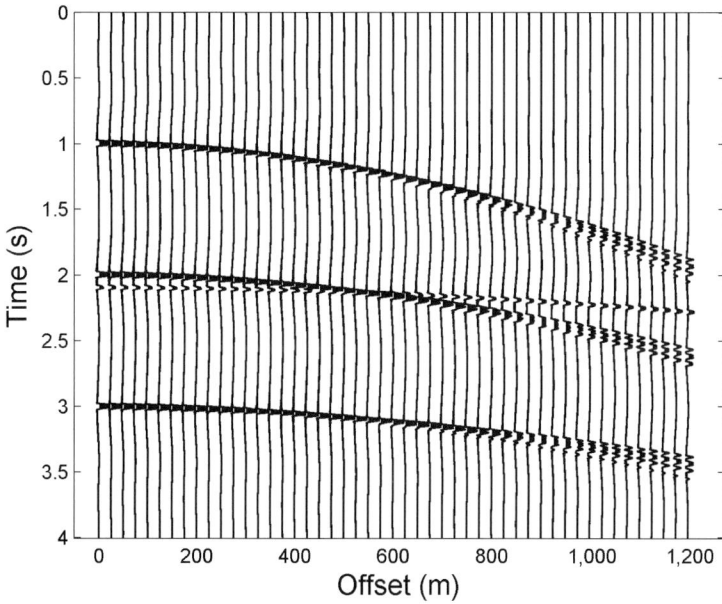

(b)

Figure 4.10 (a) Synthetic seismic reflected data with 20% random Gaussian noise. (b) The interpolated data after taking the inverse $\tau - q$ transform and $M = 10$ (80% compression).

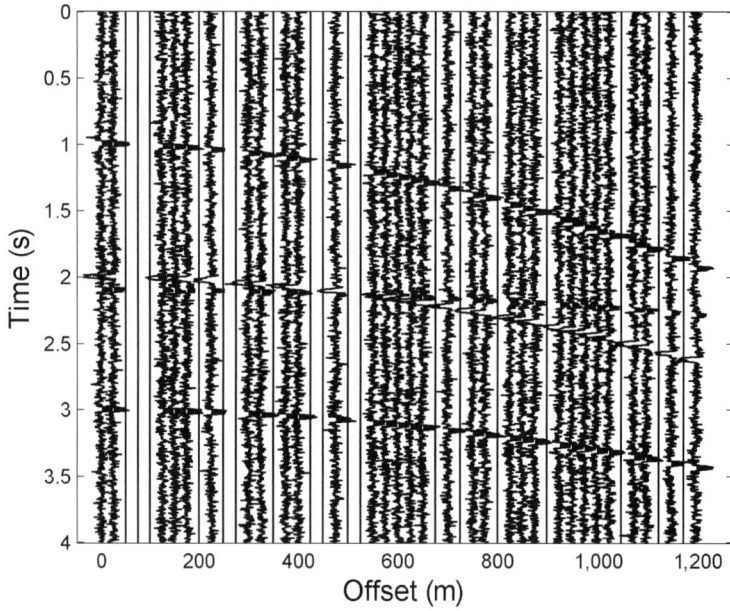

(a)

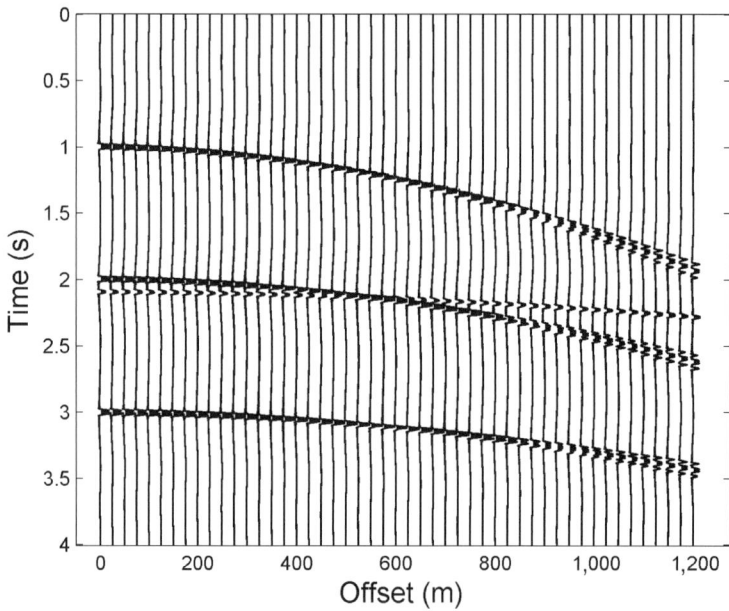

(b)

Figure 4.11 (a) Synthetic seismic reflected data with 40% random Gaussian noise.
(b) The interpolated data after taking the inverse $\tau - q$ transform and $M = 10$ (80%
compression).

components makes the processing more difficult. Migrations and inversion schemes also require seismic data without these multiple reflections. So presence of these multiples limits the performance of the algorithm used for seismic signal processing. Attenuation of multiple reflections is a very important step in seismic data processing. A lot of algorithms have been presented for the multiple attenuation. Methods depending on the moveout and different transforms have been discussed by Berkhout and Verschuur (2006). In a transformed domain, the unwanted regions are muted to separate the primary reflections. Methods based on prediction and extraction of multiples are also presented. Wave-field extrapolation and predictive deconvolution are examples of these techniques. These techniques work well with surface-related multiple removal. These techniques are presented by Verschuur (1991). These methods are time consuming so adaptive algorithms are presented in Verschuur (1997); Berkhout and Verschuur (2006); Ginie (2013). There are some techniques that use independent component analysis for the multiple removal, as discussed in Donno (2011), Kaplan and Innanen (2008), and Lu (2006).

Given a seismic record, the proposed compressive sensing with $\tau-q$ method for multiple reflections removal is stated as follows:

1. Use the compressive sensing with parabolic Radon transform to acquire the seismic data in the compressed parabolic Radon domain (described later).
2. Use the following facts for the muting of multiples (one for the primary reflections, zero for multiple reflections):

   (a) In the $\tau-q$ domain, primary reflection and multiple reflections have the same value of q.
   (b) For that particular value of q, the value with the lowest t corresponds to the primary reflection.

3. Apply the inverse parabolic Radon transform to the previous step result.

Seismic synthetic data were generated using a Ricker wavelet of 10 Hz and three primary reflections (see Figure 4.12a). Figure 4.12b presents the same data set with three multiple reflections. The total number of traces in the data was 61 ($N = 61$) with spatial sampling interval 75 m and sampling interval 4 m/s. A parabolic Radon transform domain ($\tau-q$), for CS-based Radon transform formulation, is presented in Figure 4.12c. To test the efficiency of the proposed algorithm, it was compared with existing methods for Radon transform. The comparison was performed with both low resolution and fast high resolution Radon transforms. For low resolution Radon transform, the adjoint method and least square method (Thorson, 1985), solved by the Levinson solver, was used (see Figure 4.12d and e, respectively). For the high resolution Radon transform, the Chloskey and complex gradient with fast Fourier transform (CG-FFT) is utilized (Sacchi et al., 1999; Trad et al., 2003).

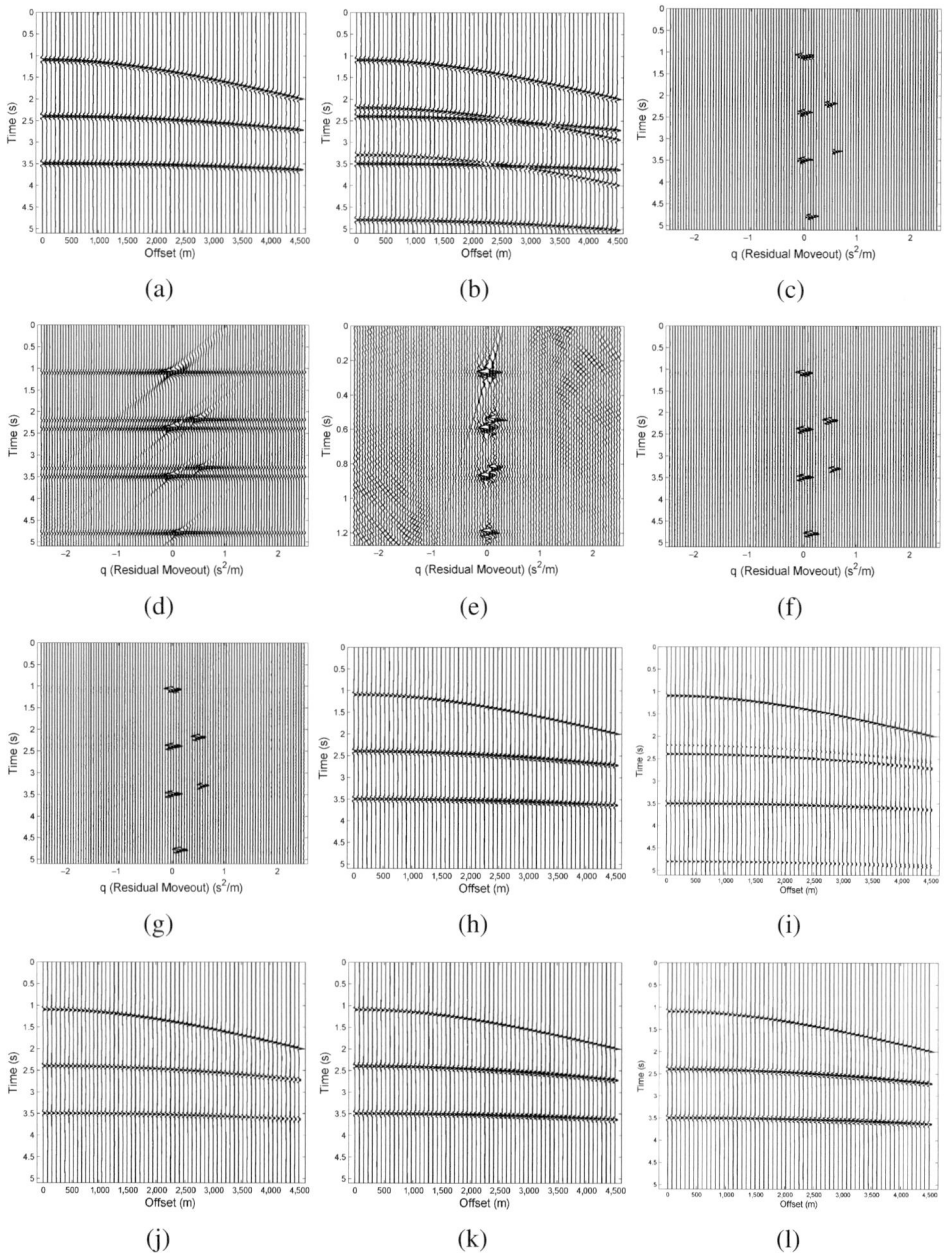

Figure 4.12 Multiple reflection attenuation of seismic data using different Radon transform. (a) and (b) Synthetic data without and with multiple reflections, respectively. (c)–(g) presents the $\tau - q$ transform of (b), by using different versions of Radon transforms: Proposed method, Adjoint, Least squares, Chloskey, and CG-FFT, respectively. Similarly, (h)–(l) presents the data after the multiple attenuation and inverse Radon transform transform of (c)–(g), respectively. The corresponding computation method for Radon transforms are as follows: Proposed method, Adjoint, Least squares, Chloskey, and CG-FFT, respectively.

Figure 4.12f and g show the $(\tau-q)$ domain for the Chloskey and CG-FFT, respectively. Clearly, from Figures 4.12c–g it is evident that the under-sampled Radon transform produces a high resolution Radon transform, with far fewer samples needed for the existing techniques. The results after muting the reflections ($q \neq 0$) are presented in Figure 4.12h–l, for the proposed method, least squares, Chloskey, CG-FFT, respectively. As expected, the high resolution as well as under-sampled Radon transform outperforms the low resolution Radon transform.

4.4 Summary

Proper sampling in time and space is extremely important when acquiring seismic data in order to avoid aliasing. Once aliasing exists, it can harm the seismic data, since the seismic data frequency-wavenumber components will not only be mixed-up but also the accompanying noise frequency-wavenumber components will. Compressive sensing (CS) provides an alternative means of acquiring seismic data. The theory provides promising results once the concept of CS can be realized at the seismic data acquisition stage.

Exercises

4.1 Consider the following discrete-time signal:

$$g[n] = \cos\left[\frac{n\pi}{8}\right].$$

If this $g[n]$ was sampled with $f_s = 10$ Hz, then obtain two different continuous-time signals that would produce $g[n]$.

4.2 Consider $g_a(t) = \cos(650\pi t) + 2\sin(720\pi t)$. Find f_N. Also, if $g_a(t)$ is sampled at twice the Nyquist rate, what are the frequencies of the sinusoids in $g[n]$?

4.3 Determine the Nyquist frequency and wavenumber values for $g(t,x) = 5\sin(175\pi x - 110\pi t)$.

4.4 If a seismic signal propagates with a frequency equal 60 Hz, in a dipping layer with $25°$ in an area of velocities of about 2,200 m/s, then what is the necessary spatial sampling interval to avoid aliasing?

4.5 Obtain the alias frequency for $g(t) = 10\sin(110\pi t)$ when it is sampled at $\Delta t = 4$ ms.

4.6 Based on the given seismic shot gathers and their corresponding $f - \grave{k}$ magnitude spectra in Figure 4.13, obtain the 2-D sampling interval parameters and the corresponding Nyquist rates. Discuss the impact of seismic signals aliasing based on Figure 4.13b and d.

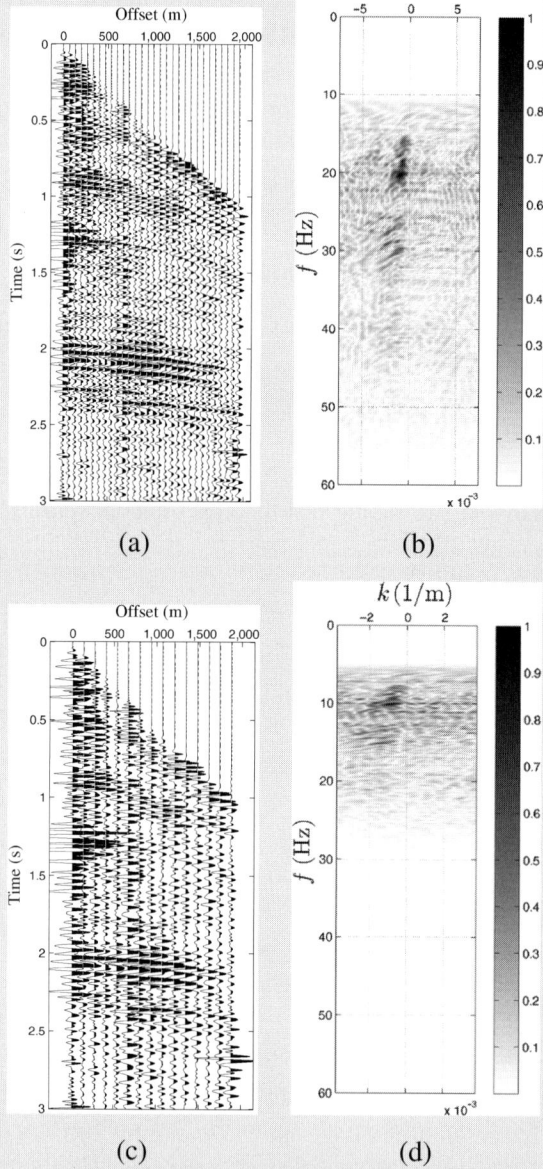

Figure 4.13 Problem 4.6. (a) A seismic shot gather from Mousa and Al-Shuhail (2011) and (b) its magnitude spectrum. (c) A down-sampled version of (a), and (d) its magnitude spectrum.

4.7 Using MATLAB, repeat Example 4.6 using the linear Radon transform. Comment on the differences in the recovered signal $g(t)$.

5

Seismic Applications of Digital Filtering Theory

5.1 Introduction

Filtering can be defined as the attenuation of signal components based on some measurable property like frequency, wavenumber, velocity, or even wave polarization. The process of filtering can be done either using analogue filters, where analogue signals are filtered (processed), or using digital filters, where the signal(s) of interest is (are) already digital (sampled and quantized). Digital filters are widely used in processing digital signals of various applications, including seismic signals, whether one-dimensional (1-D), two-dimensional (2-D), or three-dimensional (3-D). They can be classified into three many categories: Linear Shift Invariant (LSI) Filters, Adaptive Filters, and Non-linear Filters. The most commonly used class of digital filters are the LSI ones due to their simplicity in terms of analysis, design, and realization when compared to other digital filters. In most of cases, the subsurface structures, as a system with reflectivity signal (sequence) acting as the Earth's system impulse response, are usually assumed to represent an LSI system. Hence, the corresponding generated seismic signals can be digitally filtered using LSI filters. Sometimes, the recorded seismic signals show some time-varying behavior and, therefore, time-varying (adaptive) filters or filtering in the Time-Frequency domain can be utilized in such a case. The interested reader can read references, such as Sayed (2008), or Haykin (2002) for information on adaptive filters. LSI digital filters can uniquely be identified in the time and/or spatial domains by their impulse response sequences. Another way to characterize them is through their frequency and/or wavenumber responses. They are further classified into two main types:

1. Finite Impulse Response (FIR): where the impulse responses of such LSI digital filters are nonzero for only a finite number of samples.
2. Infinite Impulse Response (IIR): where their impulse responses come with an infinite number of nonzero samples.

The theory of designing FIR/IIR digital filters has been extensively addressed for more than 40 years, and it involves a blend of theory, applications, and technologies. The main objective of designing LSI digital filters is to determine the set of coefficients of either the FIR or IIR digital filters that closely approximate the desired prescribed frequency and/or wavenumber responses (in the form of ideal or template based). 1-D FIR and/or IIR digital filters, such as low-pass or band-pass, are used heavily to enhance the signal-to-noise (SNR) ratio of acquired seismic data. Furthermore, 2-D digital filters like fan filters have become standard in removal of surface waves accompanying seismic data records. Solving the wave equation numerically, for the purpose of seismic migration and imaging, may also require using FIR or IIR 1-D or 2-D digital filters such as the explicit depth wavefield extrapolation filters. This chapter will, therefore, deal with the analysis and design of many commonly and recently used filter types in seismic data processing. This includes interesting applications like seismic migration filters.

5.2 Filter Design

Ideal filters are non-causal and, hence, are physically unrealizable for real-time signal processing. Dealing with causal filters results in very important implications in the design of LSI digital filters. The first is that their frequency/wavenumber responses cannot be zero except at a finite set of points in frequency (or wavenumber). The second is that their magnitude responses cannot have an infinitely sharp cutoff from passband to stopband. Lastly, the real and imaginary parts of the frequency responses are independent. Hence, the magnitude and phase responses cannot be chosen arbitrarily.

Assume that one is interested in designing a 1-D digital filter that is LSI. Thus, the following discussions are going to be limited to LSI filters that are specified by the linear constant coefficients difference equation:

$$g[n] = -\sum_{k=1}^{M} a_k g[n-k] + \sum_{k=0}^{N} b_k w[n-k], \tag{5.1}$$

which is causal and physically realizable. Note that $w[n]$ is a 1-D seismic input signal and, $g[n]$ is the output of filtering $w[n]$ using the set of coefficients a_k and b_k. Without loss of generality, assume that $w[n]$ represents a seismic trace, where $w[n]$, in this case, is a function of a time index. Then, assuming zero initial conditions, by taking the z-transform of Equation (5.1) and dividing the z-transform of the output $G(z)$ by the z-transform of the input $W(z)$, one obtains:

$$\frac{G(z)}{W(z)} = H(z) = \frac{\sum_{k=0}^{N} b_k z^{-k}}{1 + \sum_{k=1}^{M} a_k z^{-k}}, \tag{5.2}$$

where $H(z)$ represents the LSI transfer or system function. Now, evaluating the complex variable z in $H(z)$ represented by Equation (5.2) on the unit circle, i.e., by substituting $z = e^{-j\omega}$, the frequency response of the LSI digital filter is:

$$H(e^{j\omega}) = \frac{\sum_{k=0}^{N} b_k e^{-j\omega k}}{1 + \sum_{k=1}^{M} a_k e^{-j\omega k}}. \tag{5.3}$$

The general filter design problem can then be stated as follows. Given an ideal (desired) frequency (wavenumber) response $H_d(\omega)$, find an implementable FIR or IIR digital filter whose frequency response $H(\omega)$ in Equation (5.3) approximates $H_d(\omega)$. That is, the set of coefficients a_k and b_k must be selected such that $H(\omega)$ in Equation (5.3) approximates $H_d(\omega)$ under certain criteria.

The careful selection of the filter coefficients is obtained in many ways, depending on the application of the digital filter. For example, one measure of performance for a designed digital filter can be minimizing the FIR filter length of IIR filter order, minimizing the width of the transition bands, reducing either the passband error or the stopband error or both. All of those characteristics depends on the filter type specifications.

Many common ideal filter types that are used when filtering seismic signals exist. These are listed as follows:

1. The ideal lowpass filter (LPF):

$$H_d(\omega) = \begin{cases} 1, & |\omega| \leq \omega_c \\ 0, & \omega_c < |\omega| \leq \pi, \end{cases} \tag{5.4}$$

   where ω_c is the cutoff frequency (wavenumber) and corresponds to the location where the frequency (wavenumber) components of the signal $w[n]$ will pass or become attenuated.

2. The ideal highpass filter (HPF):

$$H_d(\omega) = \begin{cases} 1, & \omega_c < |\omega| \leq \pi \\ 0, & |\omega| \leq \omega_c, \end{cases} \tag{5.5}$$

   where high frequency (wavenumber) component beyond ω_c will pass.

3. The ideal bandpass filter (BPF):

$$H_d(\omega) = \begin{cases} 1, & \omega_l \leq |\omega| \leq \omega_u \\ 0, & \text{othrewise.} \end{cases} \tag{5.6}$$

   Note that ω_u and ω_l represent, respectively, the upper and lower cutoff frequencies (wavenumbers) of the ideal BPF.

4. The ideal notch filter (NF):

$$H_d(\omega) = \begin{cases} 0, & \omega = \omega_c \\ 1, & \text{othrewise}, \end{cases} \tag{5.7}$$

where, ideally, spectrum will pas except for a single component.

The aforementioned ideal filters can be seen in Figure 5.1. In seismic data processing, both the LPF and BPF are used to increase the SNR. Recall from Chapter 2 that the bandwidth of seismic reflection exploration belongs to the range between 10–60 Hz. Thus, noise of high frequencies (such as scattering noise or ambient noise) can be attenuated via LPF or BPF digital filters. In land seismic data, ground-roll, a very troublesome coherent noise type (see Chapter 2), is of low frequency and low velocity with high amplitudes. One way to attenuate ground-roll is by applying a BPF on each affected seismic trace. In addition, whenever seismic signals are

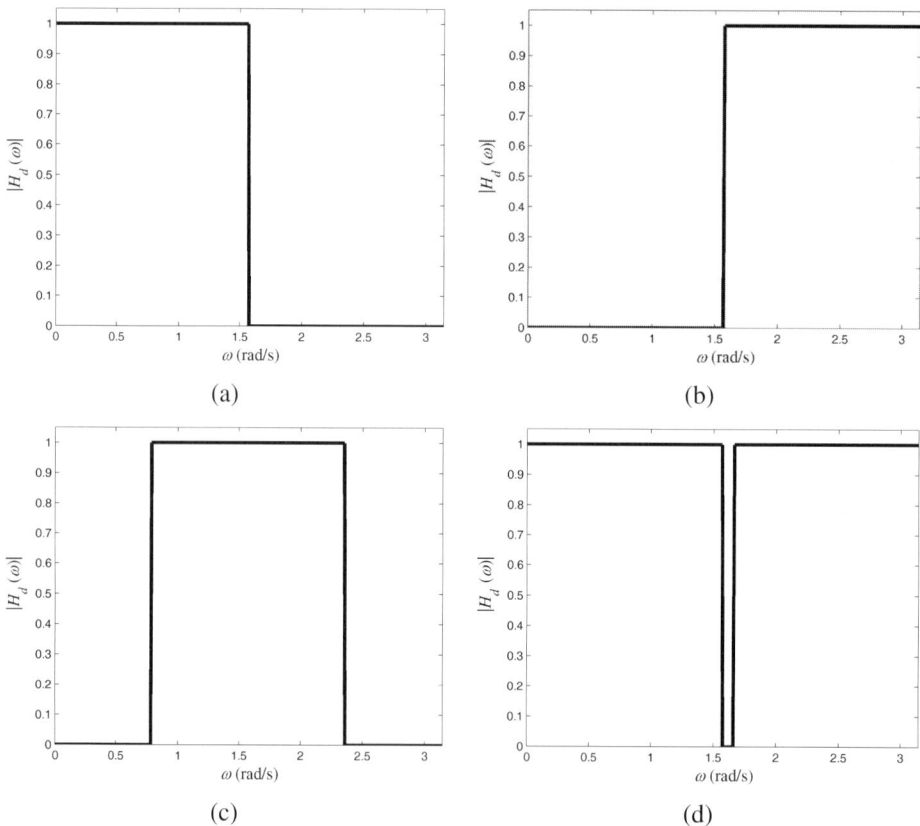

Figure 5.1 Common ideal filters: (a) lowpass at $\omega_c = \pi/2$ rad/s, (b) highpass at $\omega_c = \pi/2$ rad/s, (c) bandpass at $\omega_l = \pi/4$ rad/s and $\omega_u = 3\pi/4$ rad/s, and (d) notch filter at $\omega_c = \pi/2$ rad/s.

recorded near transmission power lines, the 50 Hz or 60 Hz sinusoid signals will be recorded on top of the seismic data. Therefore, notch filters are a very convenient digital filter to attenuate those sinusoidal harmonics.

As stated earlier, ideal filters, such as those seen in Equations (5.1, 5.4, 5.6, and 5.7), are not practical and, hence, some of their design conditions must be relaxed. Therefore, design templates for each ideal filter can be used instead of $H_d(\omega)$. Particularly, it is not necessary to insist that the magnitude response of Equation (5.3) be a constant in the entire passband of the filter. A small amount of ripple in the passband, say δ_p, of the filter magnitude response can be tolerable. Likewise, a small amount of stopband ripple, say δ_s, can also be tolerable. Sharp edges are replaced with the so-called *Transition Band*, in which the designed filter's frequency response would change smoothly, when moving from one band to another (passband to the stopband and vice-versa). The band edge frequency ω_p will define the edge of the passband, while the frequency ω_s denotes the beginning of the stopband. The width of the introduced transition band is, therefore, $\Delta\omega_t = \omega_s - \omega_p$. Note that the width of the passband is usually called the digital filter's bandwidth.

In general, in any LSI digital filter design problem, the following design template parameters must be specified:

1. The passband cutoff value ω_p (or cutoffs ω_{p_i} for $i = 1, 2, \ldots, c$, where c is the number of cutoff frequencies).
2. The stopband cutoff value ω_s (or cutoffs ω_{s_i} for $i = 1, 2, \ldots, c$).
3. The maximum allowable error in the passband δ_p.
4. The maximum allowable error in the stopband δ_s.

Note that, in general, the ideal filter's cutoff value ω_{c_i}, for $i = 1, 2, \ldots, c$, can be set to be equal to $\frac{\omega_{p_i} + \omega_{s_i}}{2}$. Based on these specifications, the filter coefficients a_k and b_k in Equation (5.3) can be determined using a certain design method. Figure 5.2 shows a corresponding design template for the ideal (desired) LPF. The degree to which the frequency response $H(\omega)$ given by Equation (5.3) approximates the specifications depends in part on the criterion used in selecting the filter coefficients a_k and b_k as well as on the FIR/IIR filter order. That is, the approximation will also depend on the number of sets of coefficients used to obtain $H(\omega)$ that approximates $H_d(\omega)$.

5.3 Design of FIR Digital Filters

Various ways to design FIR filters exist. That is, to find the values of b_k's, whereas $a_k = 0$ except $a_0 = 1$, in Equation (5.2 or 5.3) such that the desired frequency (wavenumber) response $H(\Omega)$ is close to the ideal one. A few such techniques are going to be discussed in the following subsections.

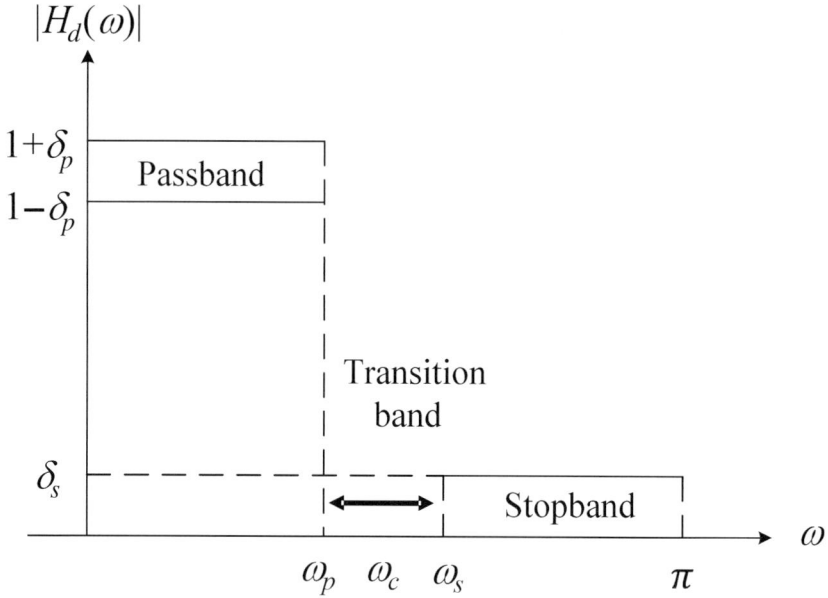

Figure 5.2 A design template for a LPF.

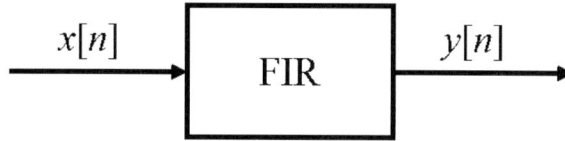

Figure 5.3 A LSI FIR filter input-output system.

5.3.1 Design of FIR Digital Filter Based on Windowing Techniques

Consider Figure 5.3, where an M-length FIR filter is described by:

$$g[n] = \sum_{k=0}^{M-1} b_k w[n-k], \tag{5.8}$$

$$= \sum_{k=0}^{M-1} h_k w[n-k]. \tag{5.9}$$

Such an FIR filter can also be characterized by its systems function:

$$H(z) = \sum_{k=0}^{M-1} h[k] z^{-k}. \tag{5.10}$$

On the other hand, the desired frequency response of the FIR filter is:

$$H_d(e^{j\omega}) = \sum_{n=0}^{\infty} h_d[n]e^{-j\omega n}, \tag{5.11}$$

where the ideal impulse response is:

$$h_d[n] = \frac{1}{2\pi} \int_{\pi}^{-\pi} H_d(e^{j\omega})e^{j\omega n} d\omega.$$

In general, $h_d[n]$ is infinite in duration and must be truncated at some point, say at $n = M - 1$, to yield an FIR filter of length M. The problem turns out to be determining the M coefficients $h[n]$, $n = 0, 1, M - 1$, from the specifications of the desired frequency response $H_d(\omega)$ of the FIR filter.

Truncation of $h_d[n]$ to a length M is equivalent to multiplying $h_d[n]$ by a rectangular window, defined as:

$$w[n] = \begin{cases} 1, & n = 0, 1, \ldots, M - 1 \\ 0, & otherwise. \end{cases} \tag{5.12}$$

Therefore,

$$h[n] = h_n[n]w[n] \tag{5.13}$$

$$= \begin{cases} h_d[n], & n = 0, 1, \ldots, M - 1 \\ 0, & otherwise. \end{cases} \tag{5.14}$$

In the frequency domain this is equivalent to the convolution of $H_d(\omega)$ with $W(\omega)$. That is:

$$h[n] = h_d[n]w[n] \longleftrightarrow H(e^{j\omega}) = \frac{1}{2\pi} \int_{-\pi}^{\pi} H_d(e^{j\lambda})W(e^{j\omega} - e^{j\lambda})d\lambda, \tag{5.15}$$

where,

$$W(e^{j\omega}) = \sum_{n=0}^{M-1} w[n]e^{-j\omega n}, \tag{5.16}$$

$$= \sum_{n=0}^{M-1} e^{-j\omega n}, \tag{5.17}$$

$$= \frac{1 - e^{-j\omega M}}{1 - e^{-j\omega}}, \tag{5.18}$$

$$= e^{-j\omega(M-1)/2} \frac{\sin(\omega M/2)}{\sin(\omega/2)}. \tag{5.19}$$

Thus, the convolution yields the following frequency response $H(\omega)$ of the truncated FIR filter

$$H(e^{j\omega}) = \frac{1}{2\pi} \int_{-\pi}^{\pi} H_d(e^{j\lambda}) W(e^{j\omega} - e^{j\lambda}) d\lambda,$$

where this convolution smoothes $H_d(\omega)$ since:

$$|W(e^{j\omega})| = \frac{|\sin(\omega M/2)|}{|\sin(\omega/2)|} \quad for \ |\omega| \leq \pi,$$

and

$$\angle W(e^{j\omega}) = \begin{cases} -\omega \left(\frac{M-1}{2} \right), & when \quad \sin(\omega M/2) \geq 0 \\ -\omega \left(\frac{M-1}{2} \right) + \pi, & when \quad \sin(\omega M/2) < 0. \end{cases}$$

Consider Figure 5.4. It can be seen that the sidelobes of $|W(e^{j\omega})|$ are relatively high and remain unaffected by increasing M. The width of each sidelobe decreases as M increases but its height increases. Hence, the area under each sidelobe remains invariant to changes in M. Also as M increases, $W(e^{j\omega})$ becomes narrower, and the smoothing provided by $W(e^{j\omega})$ is reduced and vice versa. On the other hand, the large sidelobes of $W(e^{j\omega})$ will result in some undesirable ringing effects in the FIR frequency response $H(e^{j\omega})$. Figure 5.5 shows the impulse response $h[n]$ and frequency response $H(e^{j\omega})$ of a LPF with a cutoff $\omega_c = \frac{\pi}{2}$ rad/s designed using a

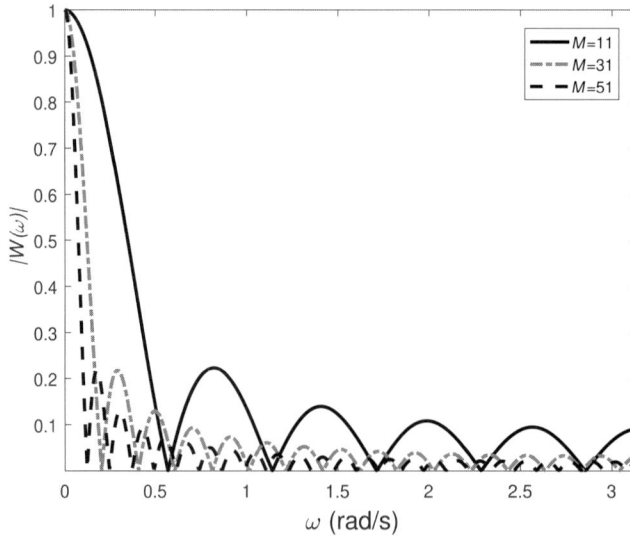

Figure 5.4 Rectangular window frequency response for various M values. The sidelobe amplitudes are relatively high and are unchanged as M increases.

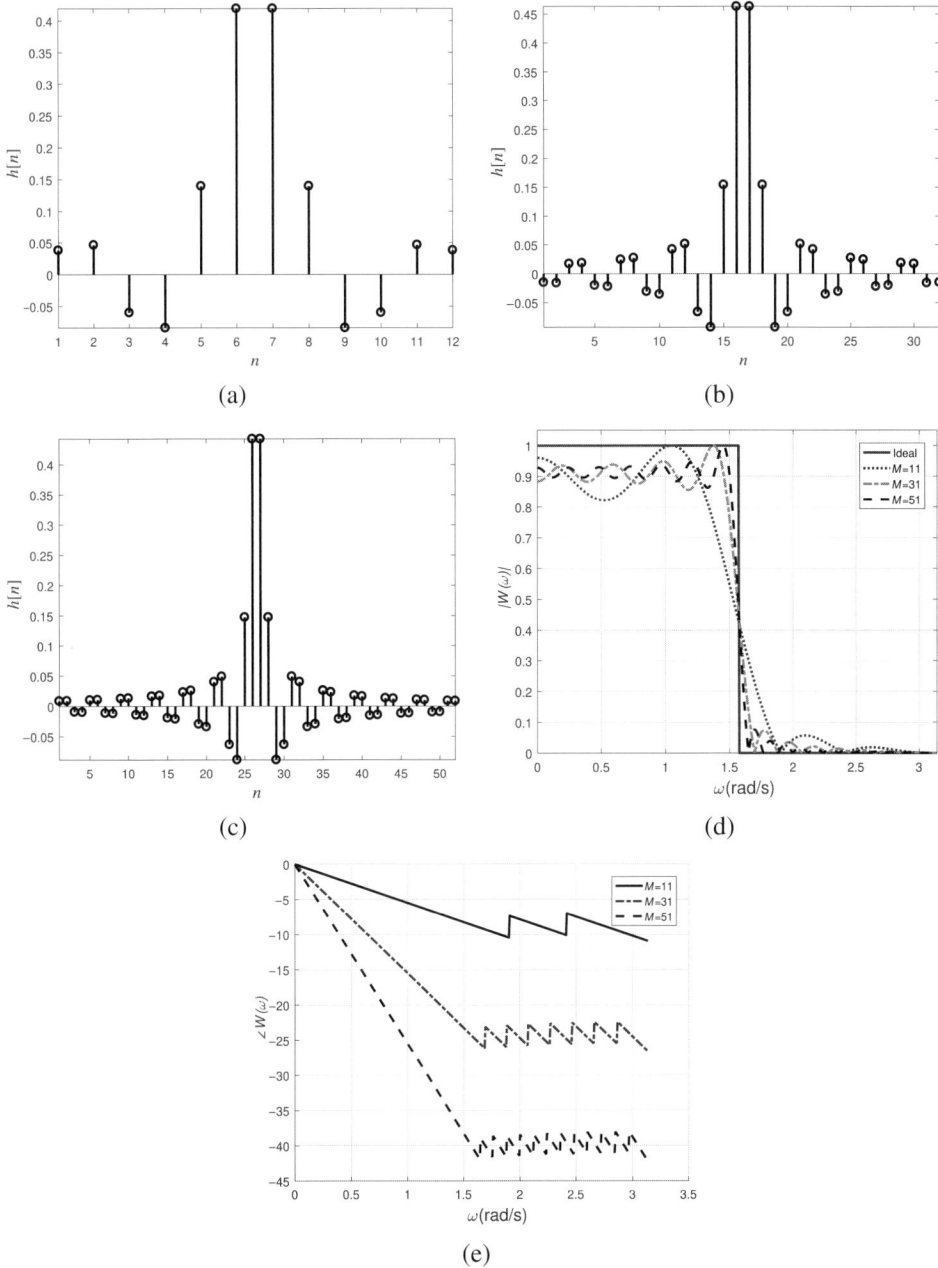

Figure 5.5 The design of a LPF linear-phase FIR filter with a cutoff $\omega_c = \frac{\pi}{2}$ rad/s using the rectangular window. (a)–(c) The impulse response $h[n]$ with $M = 11$, $M = 31$, and $M = 51$, respectively. The frequency response $H(e^{j\omega})$: (d) magnitude and (e) phase spectrum.

rectangular window with $M = 11$, $M = 31$, and $M = 51$, respectively. Clearly, near the cutoff, the magnitude spectrum becomes higher than 1 and oscillations start to become more visible as M increases. At the same time, the higher M is, the more frequency components will pass, when using this LPF to pass frequencies up to $\frac{\pi}{2}$ rad/s. Likewise, one can see that the phase is linear, and as M increases, so the slope of the phase response will become steeper.

Also, a real seismic trace from a shot gather (Yilmaz, 2001) is shown in Figure 5.6a and its magnitude spectrum in Figure 5.6b. Using the windowing method with a rectangular window, a BPF linear-phase FIR filter with $\omega_l = 0.08\pi$ rad/s and $\omega_u = 0.24\pi$ rad/s, and for $M = 35$ was designed (see Figure 5.6c) to attenuate low and high frequency noise. The amount of ripples are high in both the passband and stopband. Figure 5.6d displays the filtered seismic trace and Figure 5.6e its magnitude spectrum. Finally, Figure 5.7 shows the seismic shot gather from (Yilmaz, 2001) and its $f - x$ magnitude spectrum before and after applying on each seismic trace the BPF FIR digital filter. The ground-roll noise was not attenuated totally due to the sidelobe amplitude of the filter. Therefore, other seismic events did not became as visible as expected. Additionally, the harmonic noise seen near offset -1000 became more visible and required filtering.

Instead of the rectangular window, one can use other windows that do not contain abrupt discontinuities in their time-domain characteristics and have correspondingly low sidelobes in their frequency-domain characteristics. That is, it is desirable to use other window functions that are tapered at the ends and, therefore, produce smaller ripple in the filter's frequency response. This, course, comes at the expense of having wider transition bands. Examples of other windows include:

1. The Hanning window: corresponds to a simple raised cosine function and can be given by:

$$w[n] = \frac{1}{2}\left(1 - \cos\left[\frac{2\pi n}{M}\right]\right), \quad for\ n = 0, 1, \ldots, M - 1, \quad (5.20)$$

   where its actual length is equal to $M - 1$, since the two end values are zero.
2. The Hamming window: also corresponds to a raised cosine function, which is:

$$w[n] = 0.54 - 0.46\cos\left[\frac{2\pi n}{M}\right], \quad for\ n = 0, 1, \ldots, M - 1. \quad (5.21)$$

   However, unlike the Hanning window, it tapers to an end of 0.08 and not zero. Hanning and Hamming windows perform almost the same when designing FIR filters, where they preserve low sidelobe amplitudes.

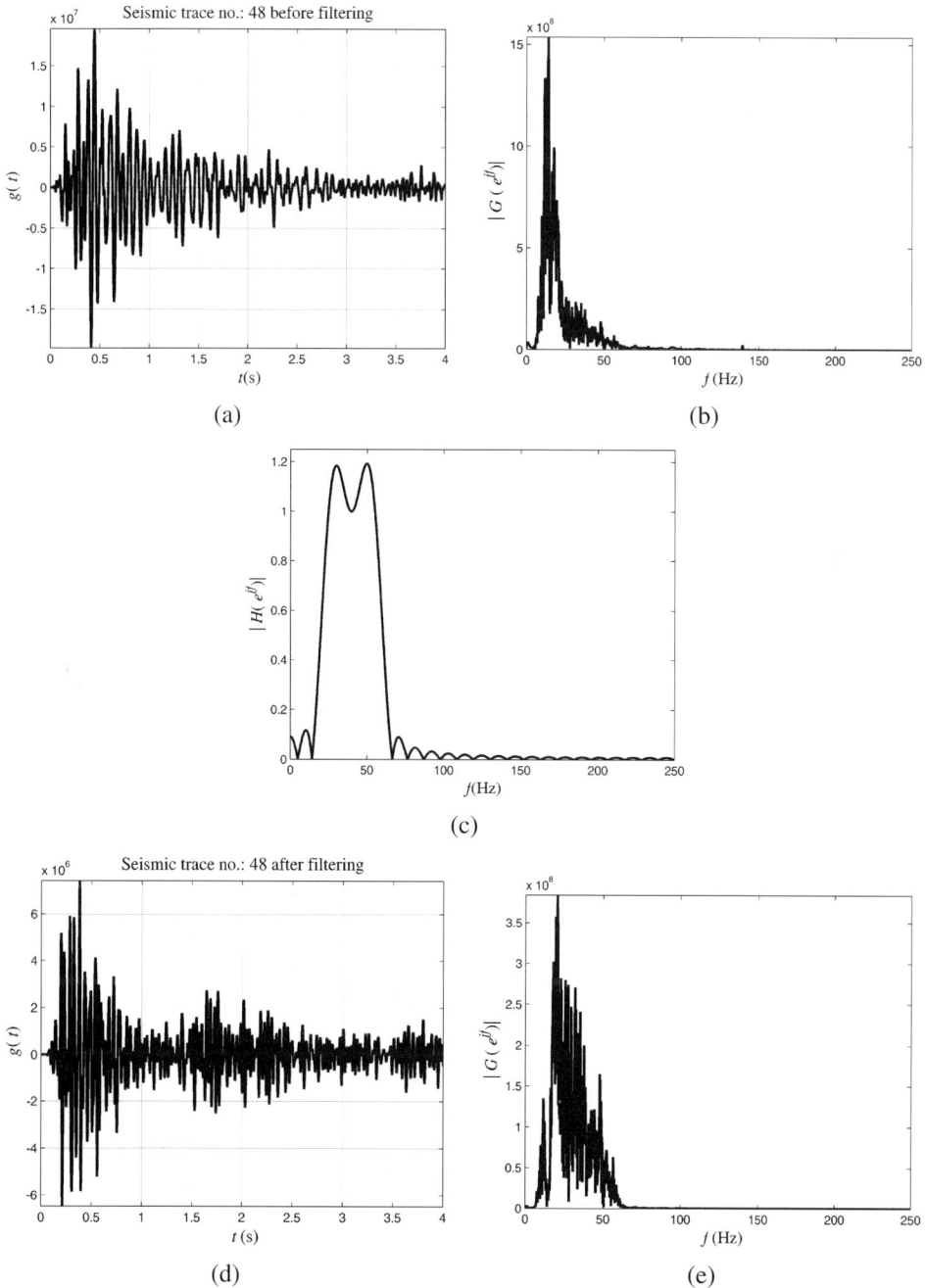

(a)

(b)

(c)

(d)

(e)

Figure 5.6 (a) A real seismic trace from a shot gather (modified from Yilmaz, 2001. (b) Its magnitude spectrum. (c) The design, using the windowing method with a rectangular window, of a BPF linear-phase FIR filter with $\omega_l = 0.08\pi$ rad/s and $\omega_u = 0.24\pi$ rad/s, and for $M = 45$. (d) The filtered seismic trace and its (e) magnitude spectrum. Note that $\Delta t = 2$ ms.

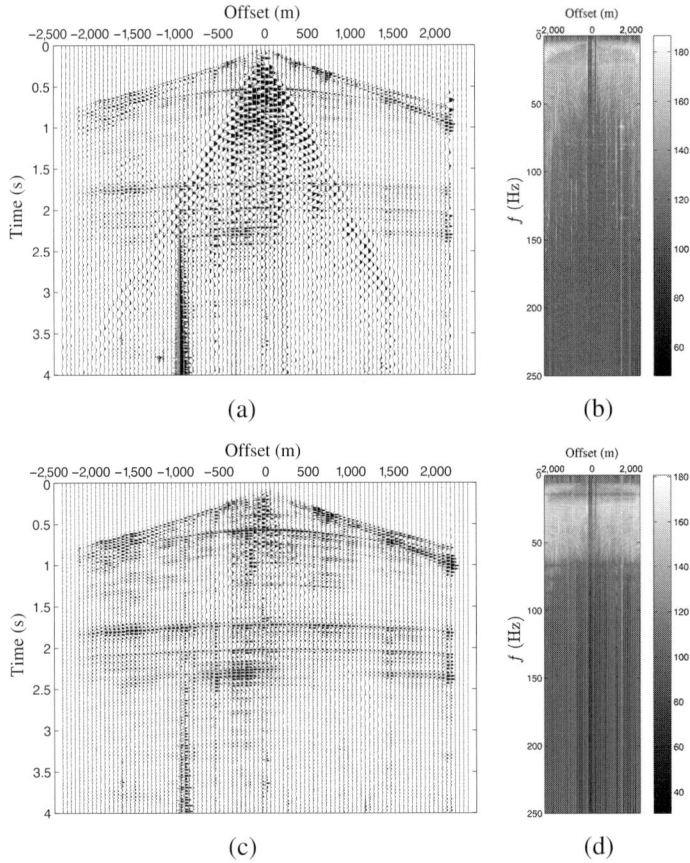

Figure 5.7 (a) A real seismic shot gather (modified from Yilmaz, 2001). (b) Its $f - x$ magnitude spectrum. (c) The data after applying for each trace in (a) the BPF used in Figure 5.6. (d) The $f - x$ magnitude spectrum of (c).

3. The Blackman window: comes with lower sidelobe amplitudes than Hanning or Hamming windows and can be given by:

$$w[n] = 0.42 - 0.5 \cos\left[\frac{2\pi n}{M}\right] + 0.08 \cos\left[\frac{4\pi n}{M}\right], \quad for\ n = 0, 1, \ldots, M - 1.$$

$$(5.22)$$

4. The Kaiser window: is a very flexible and useful window. It is optimum in the sense of having the largest energy in the main lobe for a given peak sidelobe level δ_2. The Kaiser window can be given by:

$$w_k[n] = \frac{I_0[\beta\sqrt{1 - (1 - 2n/M)^2}]}{I_0[\beta]}, \quad for\ n = 0, 1, \ldots, M - 1, \quad (5.23)$$

where $I_0[.]$ is the modified zeroth-order Bessel function of the first kind. β is the shape parameter determining the tradeoff between the main lobe width and the peak side-lobe level. Typical values for β are between 4 and 9. Note that one can compute $I_0[.]$ using the power series expression:

$$I_0[x] = 1 + \sum_{m \geq 0} \left[\frac{(x/2)^m}{m!} \right]^2, \tag{5.24}$$

with the first 15 terms being sufficient for many applications. Note that to design an FIR filter using a Kaiser window, one can perform the following. First, compute the normalized transition bandwidth:

$$\Delta\omega = \frac{\omega_s - \omega_p}{2\pi}. \tag{5.25}$$

Also, find the stop-band attenuation:

$$A = -20\log_{10}\delta_2. \tag{5.26}$$

Then, one can determine the FIR filter order assuming that $\delta_1 \approx \delta_2$ using:

$$M \approx \frac{A - 7.95}{14.36\Delta\omega}. \tag{5.27}$$

Finally, β can be computed using:

$$\beta = \begin{cases} 0.1102(A - 8.7) & A \geq 50 \\ 0.5842(A - 21)^{0.4} + 0.07886(A - 21) & 21 < A < 50. \end{cases} \tag{5.28}$$

Figure 5.8 shows various windows in the time-index domain and the corresponding frequency responses. All the windows, except for the rectangular, have low sidelobe amplitudes. Figure 5.9 shows the design of a LPF linear-phase FIR filter with a cutoff $\omega_c = \frac{\pi}{2}$ rad/s using various windows with $M - 21$. The Kaiser FIR filter outperforms all the designs, where the Hanning and Hamming filters come next with similar performance, while the Blackman requires a larger transition bandwidth compared to all designs. Finally, the rectangular window design comes with the largest sidelobe amplitudes and low passband performance compared to the rest.

The same real seismic trace from a shot gather (Yilmaz, 2001) a shown in Figure 5.6 is shown in Figure 5.10a. Its magnitude spectrum is shown in Figure 5.10b. Using the windowing method but with a Kaiser window ($\beta = 4$), a BPF linear-phase FIR filter with $\omega_l = 0.08\pi$ rad/s and $\omega_u = 0.24\pi$ rad/s, and for $M = 35$ was designed (see Figure 5.10c) to attenuate low and high frequency

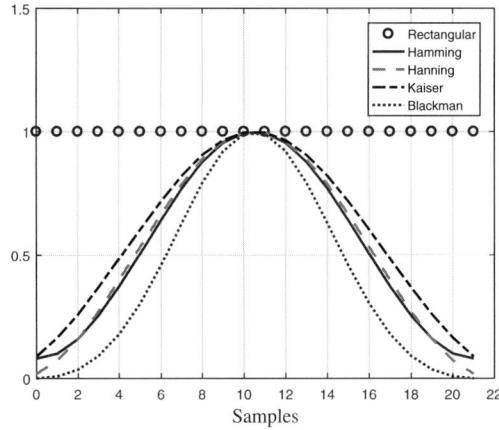

(a)

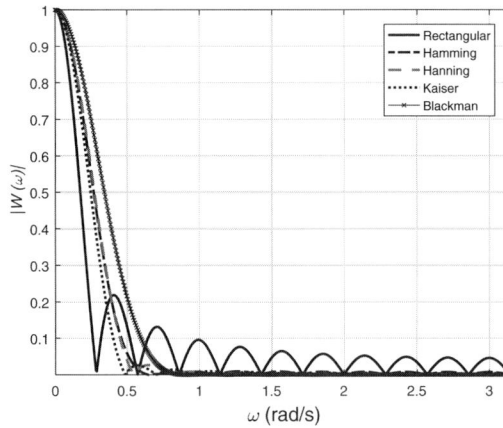

(b)

Figure 5.8 Various windows for $M = 21$ in the: (a) time-index domain, and (b) their frequency responses.

noise. The amount of ripples is reduced (compared to that of the filter-designed rectangular window) in both the passband and stopband. Figure 5.10d displays the filtered seismic trace and Figure 5.10e its magnitude spectrum. Finally, Figure 5.11 shows the seismic shot gather from (Yilmaz, 2001) and its $f - x$ magnitude spectrum before and after applying on each seismic trace the BPF FIR digital filter. The ground-roll noise was better attenuated compared to that of the rectangular window FIR filter due to the reduction in sidelobe amplitudes. While other seismic events became more visible, some ground-roll noise is present near the zero-offset. Also, the harmonic noise seen near offset -1000 became more visible and required filtering.

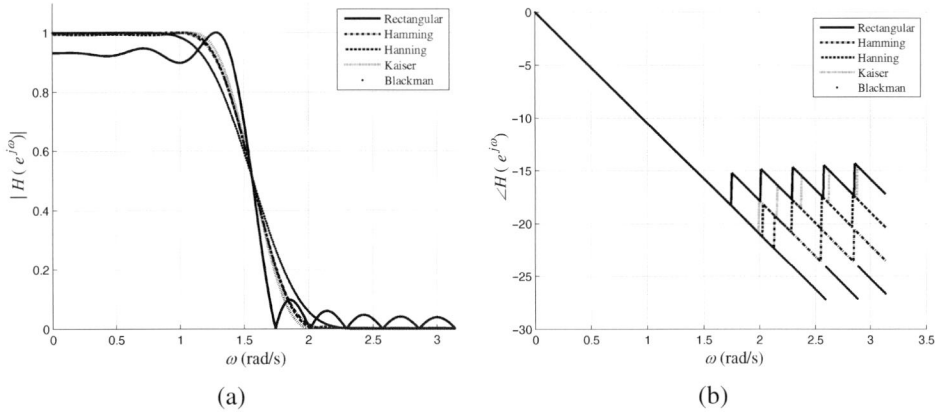

Figure 5.9 The design of a LPF linear-phase FIR filter with a cutoff $\omega_c = \frac{\pi}{2}$ rad/s using various windows with $M - 21$. The frequency response $H(e^{j\omega})$: (a) magnitude and (b) phase spectrum.

5.3.2 Design of FIR Filters Using Frequency Sampling

Sometimes it may be preferable to employ a design technique that utilizes speci- fied values of the desired frequency response $H_d(\omega)$ directly. This is without the necessity of determining $h_d[n]$, as is the case in the Windowing techniques. In this approach, $H_d(e^{j\omega})$ is uniformly sampled at M equidistant points between $[-\pi, \pi]$. That is:

$$H(l) = H_d(e^{j\omega})|_{\omega = \frac{2\pi l}{M}}, \text{ for } l = 0, 1, \ldots, M - 1.$$

These frequency samples constitute an M-point DFT whose inverse will be an FIR filter of length M (order $M - 1$).

$$h[n] = \frac{1}{M} \sum_{k=0}^{M-1} H[l]e^{j\frac{2\pi l}{M}}, \text{ for } n = 0, 1, \ldots, M - 1.$$

In general, the designed filters using this approach come with a response very simi- lar to the ones obtained via rectangular windows. Hence, the method is modified by introducing one or more frequency samples within the transition band, i.e., expanding the transition band. The frequency sampling FIR filter design approach is, hence, optimized in an iterative manner to maximize the stopband attenuation or minimize the passband ripple (Proakis and Manolakis, 2006). Figure 5.12 shows the frequency response of various LPF linear-phase FIR filters designed using the frequency sampling method. The filter cutoff is $\omega_c = \frac{\pi}{2}$ rad/s, and lengths are $M = 11$, $M = 31$, and $M = 51$, respectively. For $M = 11$, the filter will attenuate more frequencies than the other two, since its transition band is wider

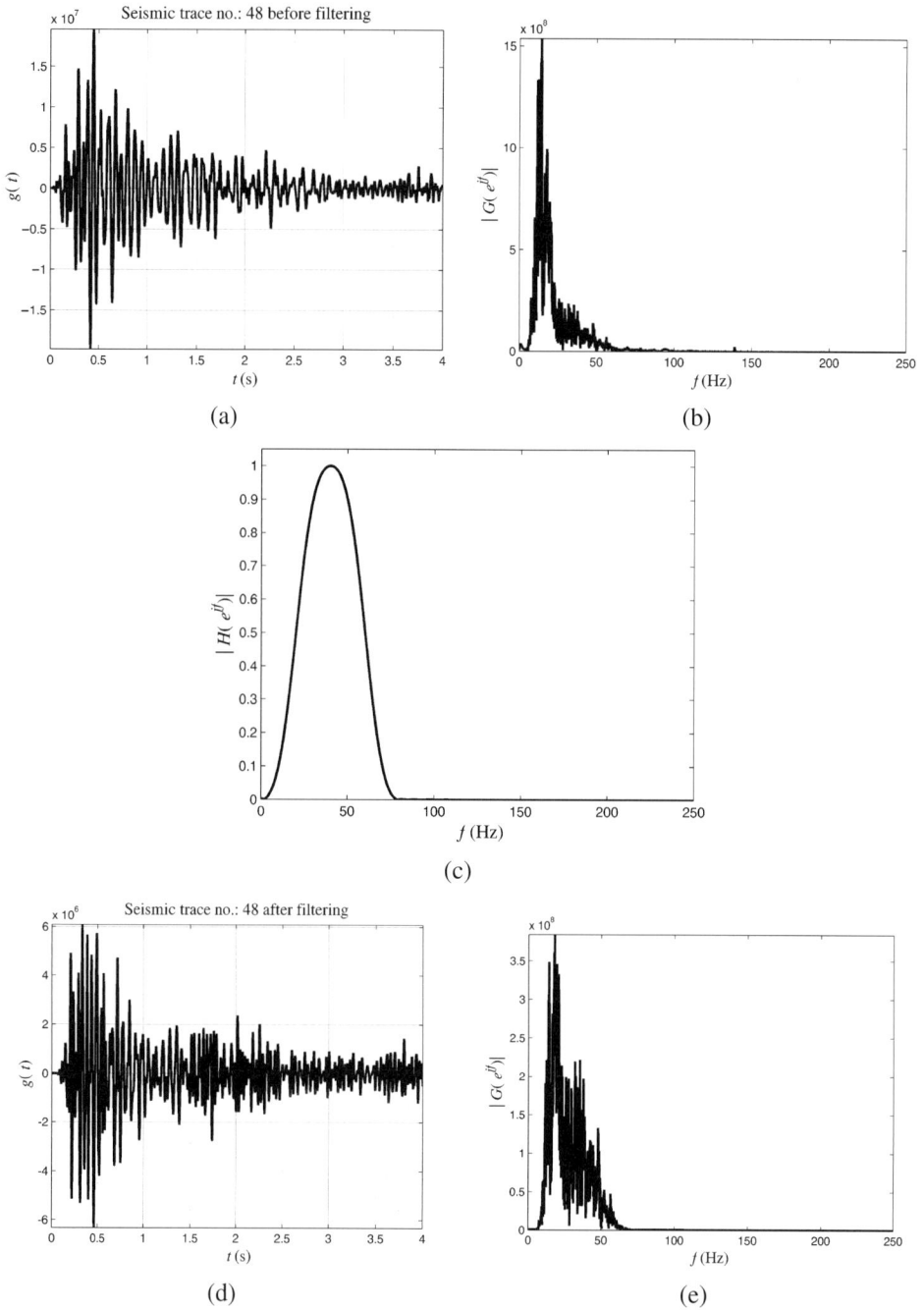

Figure 5.10 (a) A real seismic trace from a shot gather (modified from Yilmaz, 2001). (b) Its magnitude spectrum. (c) The design, using the windowing method with a Kaiser window ($\beta = 4$), of a BPF linear-phase FIR filter with $\omega_l = 0.08\pi$ rad/s and $\omega_u = 0.24\pi$ rad/s, and for $M = 45$. (d) The filtered seismic trace and its (e) magnitude spectrum. Note that $\Delta t = 2$ ms.

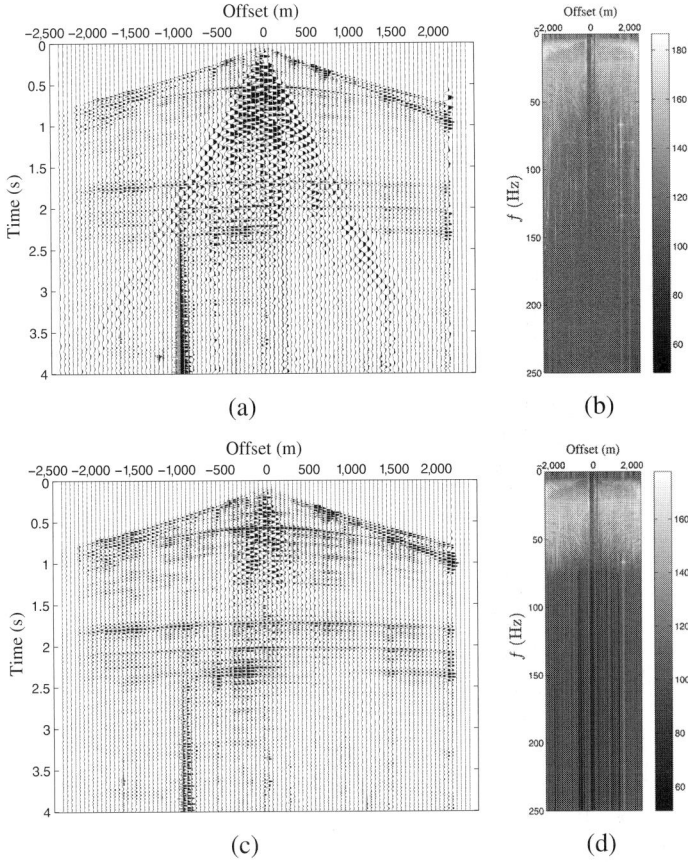

Figure 5.11 (a) A real seismic shot gather (modified from Yilmaz, 2001). (b) Its $f - x$ magnitude spectrum. (c) The data after applying for each trace in (a) the BPF used in Figure 5.10. (d) The $f - x$ magnitude spectrum of (c).

than the other two filters with $M = 31$ and 51. At the same time, all of them have very small ripples in both the passband ($\delta_p \approx 0.002$) and stopband ($\delta_s \approx 0.0025$). Furthermore, a real seismic trace from a shot gather (Yilmaz, 2001) is shown in Figure 5.13a and its magnitude spectrum in Figure 5.13b. Using the frequency sampling method, a BPF linear-phase FIR filter with $\omega_l = 0.08\pi$ rad/s and $\omega_u = 0.24\pi$ rad/s, and for $M = 35$ was designed (see Figure 5.13c) to attenuate low and high frequency noise. Figure 5.13d displays the filtered seismic trace and Figure 5.13e its magnitude spectrum. Finally, Figure 5.14 shows the seismic shot gather from (Yilmaz, 2001) and its $f - x$ magnitude spectrum before and after applying on each seismic trace the BPF FIR digital filter. Clearly, the ground-roll noise was attenuated to a great degree except near the zero-offset region, while the other seismic events became more visible. At the same time, some of the harmonic

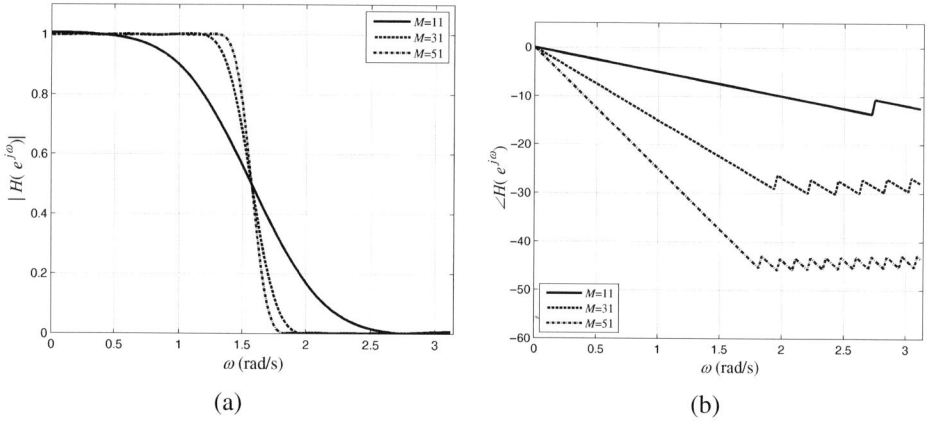

(a)   (b)

Figure 5.12 The design, using the frequency sampling method, of a LPF linear-phase FIR filter with a cutoff $\omega_c = \frac{\pi}{2}$ rad/s, and for $M = 11$, $M = 31$, and $M = 51$, respectively. The frequency response $H(e^{j\omega})$: (a) magnitude and (b) phase spectrum.

noise seen near offset -1000 became more visible and required filtering. This result is better than the result of filtering with the windowing method based on a rectangular window; however, it is similar to that of filtering using the windowing method with a Kaiser window shown earlier.

5.3.3 *FIR Projections onto Convex Sets Based Digital Filters*

A novel iterative FIR filter design algorithm was introduced (based on one forward and one inverse Fast Fourier Transform (FFT)) to design zero-phase FIR filters (Çetin et al., 1997). The algorithm alternately satisfies the frequency domain constraints on the magnitude response bounds, as well as time domain constraints on the impulse response support (Çetin et al., 1997; Özbek et al., 2004). The main advantages of this method are based on its implementation simplicity and versatility. This algorithm is known as *Projection onto Convex Sets (POCS)*. The idea was similarly used in Hermanowicz and Blok (2000) for the design of an arbitrary complex frequency response. However, both algorithms were derived heuristically without explicitly defining the constraint sets properly and deriving their associated projections. In addition, the heuristic nature of such approaches does not obviously lend itself to the design of filters with other constraints. Afterward, a more rigouros mathematical way to derive the design algorithm for 1-D real-valued linear-phase FIR filters using POCS was shown in Haddad et al. (2000), but for the 2-D real-valued linear-phase FIR filters' case in Stark and Yang (1998). The POCS theory leads to a feasible solution, which satisfies all predefined constraint sets. The

(a)

(b)

(c)

(d)

(e)

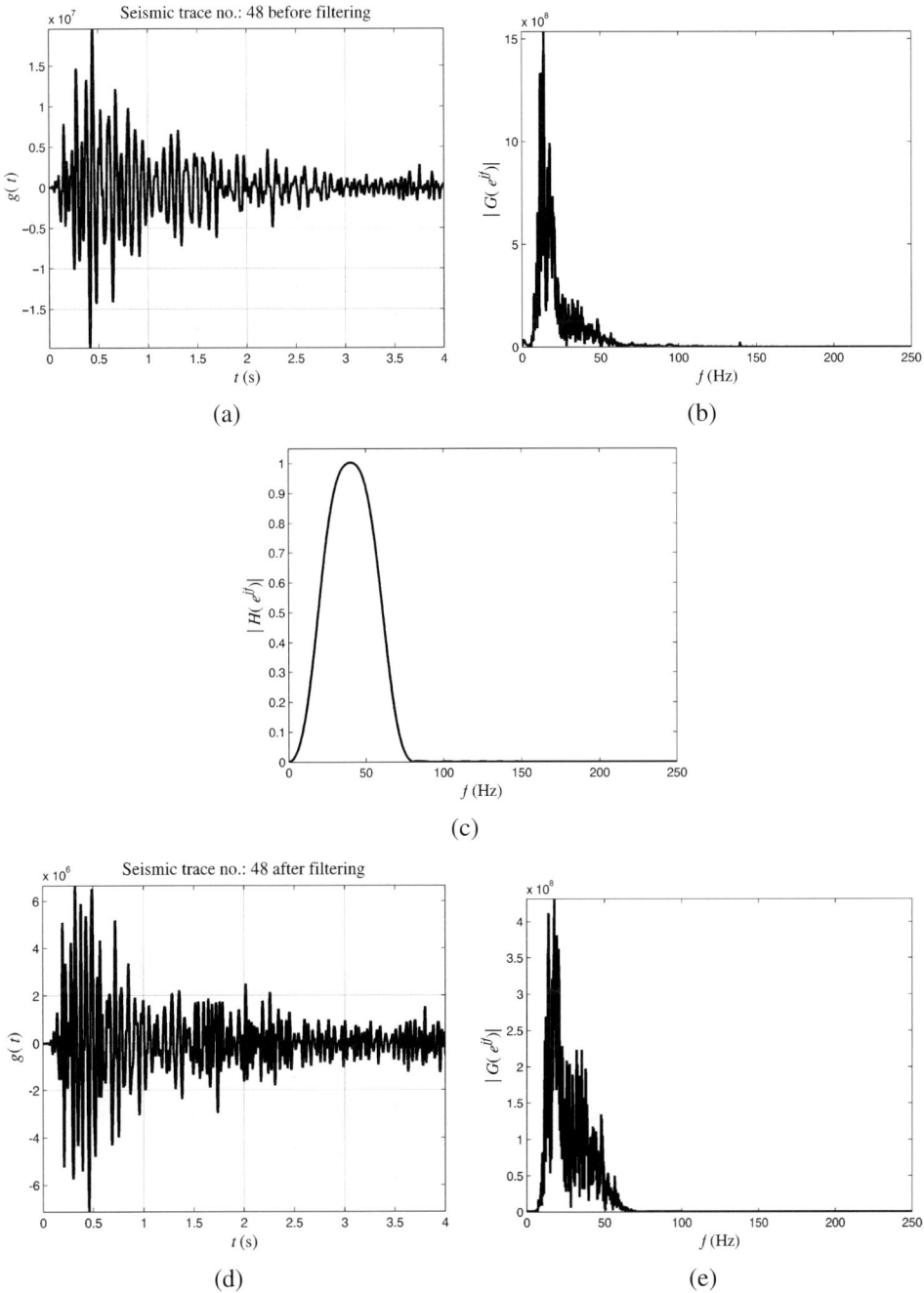

Figure 5.13 (a) A real seismic trace from a shot gather (modified from Yilmaz, 2001). (b) Its magnitude spectrum. (c) The design, using the frequency sampling method, of a BPF linear-phase FIR filter with $\omega_l = 0.08\pi$ rad/s and $\omega_u = 0.24\pi$ rad/s, and for $M = 45$. (d) The filtered seismic trace and its (e) magnitude spectrum. Note that $\Delta t = 2$ ms.

(a)

(b)

(c)

(d)

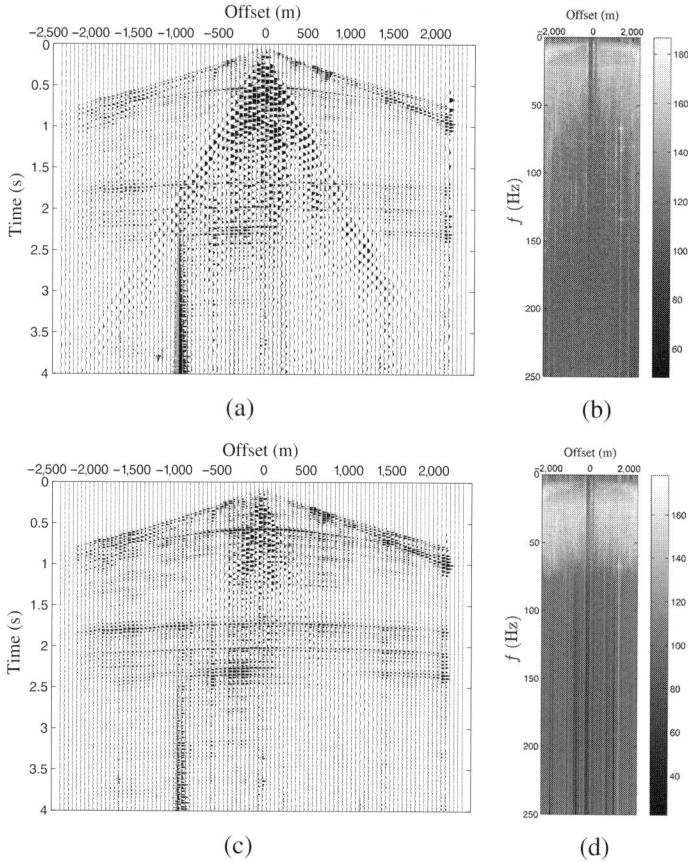

Figure 5.14 (a) A real seismic shot gather (modified from Yilmaz, 2001). (b) Its $f - x$ magnitude spectrum. (c) The data after applying for each trace in (a) the BPF used in Figure 5.13. (d) The $f - x$ magnitude spectrum of (c).

constraint sets are formulated as sets that are closed and convex within a suitable Hilbert space (Haddad et al., 2000). In Haddad et al. (2000) and Stark and Yang (1998), the Hilbert space was the M-dimensional real (Euclidean) space. Moreover, unlike many FIR filter design techniques, the POCS approach for designing FIR filters can easily be extended to the design of m-D real-valued filters. This particular approach is encouraged, due to the existence and advances of m-D FFT algorithms (Boussakta et al., 2001a,b). Furthermore, when designing FIR digital filters, the POCS will require only two FFT computations per iteration (Çetin et al., 1997).

Projections Onto Convex Sets Overview

POCS has been used in a number of applications (Levi and Stark, 1983; Sezan, 1992; Haddad et al., 1999; Oskoui-Fard and Stark, 1988; Oh et al., 1994; Bjarnason and Menke, 1993). However, it is necessary to briefly review the theory of POCS to help facilitate a better understanding of this paper. To begin, let all the FIR filters of interest be elements of a Hilbert space $\mathbf{H}$, and consider a closed convex set C, which is a subset of $\mathbf{H}$. Then, for any vector (FIR filter) $\mathbf{h} \in \mathbf{H}$, the projection $P_C\mathbf{h}$ of $\mathbf{h}$ onto C (where P_C is an operator) is the nearest neighbor element in C to $\mathbf{h}$ (i.e., $\mathbf{y}$) and is determined by

$$\|\mathbf{h} - P_C\mathbf{h}\| = \min_{\mathbf{y} \in C}\|\mathbf{h} - \mathbf{y}\|, \tag{5.29}$$

where $\| \cdot \|$ is the Euclidean norm. Without ambiguity, the operator P_C is in general a nonlinear projection operator that maps any vector $\mathbf{h} \in \mathbf{H}$ to a vector that belongs to C. The basic idea of POCS follows: every known property (e.g., say the ith property) of the unknown $\mathbf{h} \in \mathbf{H}$ will restrict $\mathbf{h}$ to lie in a closed convex set, say $C_i \in \mathbf{H}$. Assume that $C_1, C_2, \ldots, C_m$ denote m (for m known properties) closed convex sets in a Hilbert space $\mathbf{H}$, and C_o denotes their intersection set given by,

$$C_o = \bigcap_{i=1}^{m} C_i. \tag{5.30}$$

The set C_o, which is considered the solution set, will contain elements that satisfy all the constraint sets and will therefore represent feasible solutions. For each $i = 1, 2, \ldots, m$, let P_{C_i} denote the projection operator onto the set C_i. Then, the Fundamental Theorem of POCS is given as follows (Stark and Yang, 1998):

> **Theory 1** *Assume that C_o is non-empty. Then, for every $\mathbf{h} \in \mathbf{H}$ and $i = 1, 2, \ldots, m$, the sequence $\{P_n\mathbf{h}\}$ converges weakly to a point of C_o.*

In other words, theorem 1 states that the vector iterates $\{\mathbf{h}_k\}$ generated by

$$\mathbf{h}_{k+1} = P_{C_m} P_{C_{m-1}} \cdots P_{C_1}\mathbf{h}_k, \tag{5.31}$$

with an arbitrary starting point $\mathbf{h}_0$, will converge weakly to a point of C_o, and since the Hilbert space is of finite dimension, the algorithm will strongly converge to a point within C_o (Haddad et al., 2000; Kreyszig, 1978).

Complex-Valued FIR Filter Design Using POCS

To properly design N-length complex-valued FIR digital filters using POCS, one needs to first define the required filter properties in constraint sets that are closed convex belonging to $\mathbf{H} = \mathbb{C}^M$. That is, since the filter coefficients are complex-valued, then the Hilbert space must be the set of M-dimensional complex vectors where the dimension M is much greater than the filter length N. More specifically, to design a complex-valued N-length FIR filter $h[n]$ using POCS theory, the FIR filter properties from both the time and frequency domains are required to be placed in proper closed and convex sets. The magnitude spectrum of the Discrete Time

Fourier Transform (DTFT) of $h[n]$ must be upper and lower bounded by $1 + \delta_p$ and $1 - \delta_p$, respectively, in the passband. In addition, the stopband magnitude spectrum must be bounded by δ_s. Finally, the phase spectrum must be equal to a predefined phase, say $\phi(\omega)$. If these constraint sets are closed convex sets, and happen to intersect, then one can guarantee strong convergence of the algorithm, since $\mathbf{H} = \mathbb{C}^M$ space is of finite dimension. Therefore, the following represent the constraint sets:

The Constraint Set C_1

Let $C_1 = \{\mathbf{h} \in \mathbb{C}^M : h[n] = 0 \text{ for } n \notin S\}$, where S is the set of points on which the filter coefficients of length N are not equal to zero. Basically speaking, C_1 is the set of all complex-valued vectors of length M with, at most, N nonzero filter coefficients. By inspection, C_1 is closed and convex. Therefore, the projection of an arbitrary vector, which belongs to the Hilbert space onto C_1, can be derived as shown in the following equations. According to the POCS theorem, the projection will be unique. Let $\mathbf{x} \in \mathbb{C}^M$ be an arbitrary vector in the Hilbert space $\mathbb{C}^M$ and let $\mathbf{x} \notin C_1$. Then, for $\mathbf{x} \notin C_1$, and $\mathbf{h} \in C_1$:

$$\|\mathbf{x} - \mathbf{h}\|^2 = \sum_{i \in S} |x[i] - h[i]|^2 + \sum_{i \notin S} |x[i] - h[i]|^2$$
$$= \sum_{i \in S} |x[i] - h[i]|^2 + \sum_{i \notin S} |x[i]|^2. \tag{5.32}$$

However, $\|\mathbf{x} - \mathbf{h}\|^2$ is minimized with respect to $\mathbf{h}$, if:

$$h[n] = x[n] \text{ for } n \in S. \tag{5.33}$$

Therefore, the projection onto C_1, i.e., P_{C_1}, can be given by the following relationship:

$$P_{C_1}\mathbf{x} = \begin{cases} x[n], \text{ if } n \in S \\ 0, \text{ if } n \notin S. \end{cases} \tag{5.34}$$

The Constraint Set C_2

$C_2 = \{\mathbf{h} \in \mathbb{C}^M \text{ with } h[n] \leftrightarrow H(e^{j\omega}) : \angle H(e^{j\omega}) = \phi(\omega)\}$. That is, C_2 is basically the set of all sequences that are complex-valued and whose phase response is constrained to be equal to a predefined phase response $\phi(\omega)$.

Convexity of C_2: Let $\mathbf{h}_1, \mathbf{h}_2 \in C_2$, where $H_1(e^{j\omega})$ and $H_2(e^{j\omega})$ are the DTFT of $\mathbf{h}_1$ and $\mathbf{h}_2$, respectively, and so $H_1(e^{j\omega})$ and $H_2(e^{j\omega})$ must have the same phase (since $\mathbf{h}_1, \mathbf{h}_2 \in C_2$), say $\phi(\omega)$. Using the definition of convexity, let $\mathbf{h}_3 = \mu\mathbf{h}_1 + (1 - \mu)\mathbf{h}_2$, where $0 \leq \mu \leq 1$. In the DTFT domain, this is equivalent to $H_3(e^{j\omega}) = \mu H_1(e^{j\omega}) + (1 - \mu)H_2(e^{j\omega})$, by linearity. However, $H_3(e^{j\omega})$ can be written as:

$$H_3(e^{j\omega}) = \mu|H_1(e^{j\omega})|\exp(j\phi(\omega)) + (1-\mu)|H_2(e^{j\omega})|\exp(j\phi(\omega))$$
$$= [\mu|H_1(e^{j\omega})| + (1-\mu)|H_2(e^{j\omega})|]\exp(j\phi(\omega))$$
$$= |H_3(e^{j\omega})|\exp(j\phi(\omega)). \tag{5.35}$$

Hence, $\mathbf{h}_3 \in C_2$. Therefore, C_2 is convex.

Closure of C_2: Let $\{\mathbf{h}_n\}$ be a convergent sequence in C_2, where it converges to $\hat{\mathbf{h}}$, and let $H_n(e^{j\omega})$ and $\hat{H}(e^{j\omega})$ be the DTFT's of $\mathbf{h}_n$ and $\hat{\mathbf{h}}$, respectively. It is intended to show that $\hat{\mathbf{h}} \in C_2$, also. If the phase of $H_n(e^{j\omega})$ is denoted by $\phi(\omega)$, and the phase of $\hat{H}(e^{j\omega})$ is equal to $\psi(\omega)$, then:

$$\|\mathbf{h}_n - \hat{\mathbf{h}}\|^2 = \frac{1}{2\pi}\int_{-\pi}^{\pi}|H_n(e^{j\omega}) - \hat{H}(e^{j\omega})|^2 d\omega$$
$$= \frac{1}{2\pi}\int_{-\pi}^{\pi}||H_n(e^{j\omega})|\exp(j\phi(\omega))$$
$$- |\hat{H}(e^{j\omega})|\exp(j\psi(\omega))|^2 d\omega. \tag{5.36}$$

Now, by using the fact that (Stark and Yang, 1998):

$$||H_n(e^{j\omega})|\exp(j\phi(\omega)) - |\hat{H}(e^{j\omega})|\exp(j\psi(\omega))|^2 \geq$$
$$|\hat{H}(e^{j\omega})|^2 \sin^2(\phi(\omega) - \psi(\omega)), \tag{5.37}$$

then

$$\|\mathbf{h}_n - \hat{\mathbf{h}}\|^2 \geq \frac{1}{2\pi}\int_{-\pi}^{\pi}|\hat{H}(e^{j\omega})|^2$$
$$\times \sin^2(\phi(\omega) - \psi(\omega))d\omega. \tag{5.38}$$

As $n \to \infty$, then for the right-hand side of Equation (5.38), one can see that:

$$\frac{1}{2\pi}\int_{-\pi}^{\pi}|\hat{H}(e^{j\omega})|^2 \sin^2(\phi(\omega) - \psi(\omega))d\omega = 0. \tag{5.39}$$

In this case, either $\hat{H}(e^{j\omega}) = 0$, which contradicts the fact that it can be shown as a member of C_2, or $\sin^2(\phi(\omega) - \psi(\omega)) = 0$. In this case, $\phi(\omega) - \psi(\omega) = m\pi$ or $\phi(\omega) = \psi(\omega) + m\pi$, where $m \in \mathbb{Z}$, and $\mathbb{Z}$ is the set of all integers. However, m must be even since $\mathbf{h}_n \to \hat{\mathbf{h}}$. Therefore, $\hat{\mathbf{h}} \in C_2$, and C_2 is closed.

Projection onto C_2: Since C_2 is a closed and convex set, one is left with the derivation of the projection of an arbitrary vector $\mathbf{x} \in \mathbb{C}^M$ ($\mathbf{x} \notin C_2$) onto C_2. Based on the proposed constraint set C_2, it is required to minimize

$$\min_{\mathbf{h}\in C_2} J(H(e^{j\omega})) = \|X(e^{j\omega}) - H(e^{j\omega})\|^2 \tag{5.40}$$

subject to $\angle H(e^{j\omega}) = \phi(\omega)$, where $X(e^{j\omega})$ is the DTFT of $\mathbf{x}$, and $\| \cdot \|$ is now the L_2 norm with respect to continuous functions. Construct the following Lagrangian (Sydsaeter and Hammond, 2001) equation in the DTFT domain:

$$J(H(e^{j\omega})) = \|X(e^{j\omega}) - H(e^{j\omega})\|^2 - \lambda \left[\arctan \frac{\Im H(e^{j\omega})}{\Re H(e^{j\omega})} - \phi(\omega) \right], \quad (5.41)$$

where $\Re\{.\}$ and $\Im\{.\}$ denote, respectively, the real and imaginary parts for a given complex function. Now, let $X_r = \Re X(e^{j\omega})$, $X_i = \Im X(e^{j\omega})$, $H_r = \Re H(e^{j\omega})$, and $H_i = \Im H(e^{j\omega})$. Then, Equation (5.41) can be rewritten as

$$J(H_r, H_i) = (X_r - H_r)^2 + (X_i - H_i)^2 - \lambda \left[\arctan \frac{H_i}{H_r} - \phi(\omega) \right]. \quad (5.42)$$

The first-order conditions are:

$$\frac{\partial J(H_r, H_i)}{\partial H_r} = -2X_r + 2H_r + \lambda \frac{H_i}{H_r^2 + H_i^2} = 0, \quad (5.43)$$

and

$$\frac{\partial J(H_r, H_i)}{\partial H_i} = -2X_i + 2H_i - \lambda \frac{H_r}{H_r^2 + H_i^2} = 0, \quad (5.44)$$

and the constraint is

$$\frac{H_i}{H_r} = \tan(\phi(\omega)). \quad (5.45)$$

From Equation (5.43), solve for λ to obtain:

$$\lambda = \frac{2(X_r - H_r)(H_r^2 + H_i^2)}{H_i}. \quad (5.46)$$

Similarly, by using Equation (5.44), solve for λ to get:

$$\lambda = \frac{-2(X_i - H_i)(H_r^2 + H_i^2)}{H_r}. \quad (5.47)$$

Now, by equating Equations (5.46 and 5.47), and performing some mathematical simplifications, one can get:

$$\frac{(X_r - H_r)}{(X_i - H_i)} = -\tan \phi(\omega). \quad (5.48)$$

Recall that $\mathbf{x} \notin C_2$, then by substituting Equation (5.45) into Equation (5.48)

$$\frac{(X_r - H_r)}{(X_i - H_i)} = -\frac{H_i}{H_r}, \quad (5.49)$$

which is equivalent to

$$(H_r^2 + H_i^2) = X_r H_r + X_i H_i. \tag{5.50}$$

Now, since $X_r = |X(e^{j\omega})| \cos \theta_x$, and $X_i = |X(e^{j\omega})| \sin \theta_x$ (where $\theta_x = \arctan \frac{X_i}{X_r}$), and since one knows that $H_r = |H(e^{j\omega})| \cos \phi(\omega)$, $H_r = |H(e^{j\omega})| \sin \phi(\omega)$, and $|H(e^{j\omega})|^2 = H_r^2 + H_i^2$, then Equation 5.50 can be rewritten as:

$$|H(e^{j\omega})| = |X(e^{j\omega})|(\cos \theta_x \cos \phi(\omega) + \sin \theta_x \sin \phi(\omega))$$
$$= |X(e^{j\omega})| \cos (\theta_x - \phi(\omega)). \tag{5.51}$$

However, Equation (5.51) can only be true if $\theta_x - \phi(\omega) \in [-\pi/2 + n\pi, \pi/2 + n\pi]$ for an even integer n, and so reduces to the following two cases:

- If $\theta_x - \phi(\omega) \in [-\pi/2 + n\pi, \pi/2 + n\pi]$, where n is an even integer, then $\cos (\theta_x - \phi(\omega)) \geq 0$, and hence, $H(e^{j\omega}) = |X(e^{j\omega})| \cos (\theta_x - \phi(\omega)) \exp(j\phi(\omega))$.
- If $\theta_x - \phi(\omega) \in (-\pi/2 + l\pi, \pi/2 + l\pi)$, where l is an odd integer, then $\cos (\theta_x - \phi(\omega)) < 0$, and hence, $H(e^{j\omega}) = -|X(e^{j\omega})| \cos (\theta_x - \phi(\omega)) \exp(j\phi(\omega))$, since its required that $|H(e^{j\omega})| \geq 0$.

Therefore, the projection onto C_2, i.e., P_{C_2} can be given by the following equation:

$$P_{C_2}\mathbf{x} \leftrightarrow \begin{cases} |X(e^{j\omega})| \cos (\theta_x - \phi(\omega)) \exp(j\phi(\omega)) \\ \quad , \text{if } \cos (\theta_x - \phi(\omega)) \geq 0 \\ -|X(e^{j\omega})| \cos (\theta_x - \phi(\omega)) \exp(j\phi(\omega)) \\ \quad , \text{if } \cos (\theta_x - \phi(\omega)) < 0. \end{cases} \tag{5.52}$$

The Constraint Sets C_3 and C_4

Define C_3 as the set of complex-valued sequences whose DTFT passband magnitude spectrum minus one should not exceed the limit δ_p. That is, $C_3 = \{\mathbf{h} \in \mathbb{C}^M$ with $h[n] \leftrightarrow H(e^{j\omega}) : (|H(e^{j\omega})| - 1) \leq \delta_p$ for $\omega \in \Omega_p\}$, where Ω_p is the passband interval equal to $[-\omega_p, \omega_p]$, ω_p is the cut-off frequency, and δ_p is the maximum passband allowable tolerance. Finally, let C_4 be the set of all sequences, which are complex-valued, and whose DTFT magnitude are bounded by δ_s in the stopband Ω_s where $\Omega_s = [-\pi, -\omega_s) \cap (\omega_s, \pi]$, ω_s is the stopband cut-off frequency, and δ_s is the maximum allowable stopband tolerance. So, mathematically, $C_4 = \{\mathbf{h} \in \mathbb{C}^M$ with $h[n] \leftrightarrow H(e^{j\omega}) : |H(e^{j\omega})| \leq \delta_s$ for $\omega \in \Omega_s\}$. By following the same methodology as developed for the constraint set C_2, one can easily show that C_3 and C_4 are closed convex sets in $\mathbb{C}^M$. The projection P_{C_3} onto C_3 of an arbitrary vector $\mathbf{x} \in \mathbb{C}^M$, where $\mathbf{x} \notin C_3$, can be written as

$$P_{C_3}\mathbf{x} \leftrightarrow \begin{cases} -\delta_p \frac{X(e^{j\omega})-1}{|X(e^{j\omega})-1|}, \text{if } |X(e^{j\omega})| - 1| \geq \delta_p \text{ for } \omega \in \Omega_p \\ X(e^{j\omega}) - 1, \text{ otherwise.} \end{cases} \tag{5.53}$$

Finally, the projection of an arbitrary vector $\mathbf{x} \in \mathbb{C}^M$, where $\mathbf{x} \notin C_4$, can be shown as:

$$P_{C_4}\mathbf{x} \leftrightarrow \begin{cases} -\delta_s \frac{X(e^{j\omega})}{|X(e^{j\omega})|}, \text{ if } |X(e^{j\omega})| \geq \delta_s \text{ for } \omega \in \Omega_s \\ X(e^{j\omega}), \text{ otherwise.} \end{cases} \quad (5.54)$$

The POCS Design Algorithm for m-D Complex-Valued FIR Filters

Based on Equation (5.31), and also on the above derived POCS operators (given in Equations (5.34, 5.52, 5.53, and 5.54), respectively) for designing 1-D complex-valued FIR digital filters, the m-D POCS algorithm for designing complex-valued FIR digital filters can be stated as follows:

1. Project $\mathbf{h}_k$ onto C_4, that is

$$\mathbf{g}_{1,k} = P_{C_4}\mathbf{h}_k \leftrightarrow \begin{cases} -\delta_s \frac{H_k(e^{j\Omega})}{|H_k(e^{j\Omega})|}, \text{ if } |H_k(e^{j\Omega})| \geq \delta_s \text{ for } \Omega \in \Omega_s \\ H_k(e^{j\Omega}), \text{ otherwise.} \end{cases} \quad (5.55)$$

2. Project $\mathbf{g}_{1,k}$ onto C_3 using

$$\mathbf{g}_{2,k} = P_{C_3}\mathbf{g}_{1,k} \leftrightarrow \begin{cases} -\delta_p \frac{G_{1,k}(e^{j\Omega})-1}{|G_{1,k}(e^{j\Omega})|-1} \\ \text{, if } |G_{1,k}(e^{j\Omega})| - 1 \geq \delta_p \text{ for } \Omega \in \Omega_p \\ G_{1,k}(e^{j\Omega}) - 1, \text{ otherwise.} \end{cases} \quad (5.56)$$

3. Project $\mathbf{g}_{2,k}$ onto C_2 using

$$\mathbf{g}_{3,k} = P_{C_2}\mathbf{g}_{2,k} \leftrightarrow \begin{cases} |G_{2,k}(e^{j\Omega})| \cos(\theta_{G_{2,k}} - \phi(\Omega)) \\ \times \exp(j\phi(\Omega)) \\ \text{if } \cos(\theta_{G_{2,k}} - \phi(\Omega)) \geq 0 \\ -|G_{2,k}(e^{j\Omega})| \cos(\theta_{G_{2,k}} - \phi(\Omega)) \\ \times \exp(j\phi(\Omega)) \\ \text{if } \cos(\theta_{G_{2,k}} - \phi(\Omega)) < 0. \end{cases} \quad (5.57)$$

4. Finally, project $\mathbf{g}_{3,k}$ onto C_1 by

$$\mathbf{h}_{k+1} = P_{C_1}\mathbf{g}_{3,k} = \begin{cases} \mathbf{g}_{3,k}[\mathbf{n}], \text{ for } \mathbf{n} \in \mathbf{S} \\ 0, \text{ otherwise.} \end{cases} \quad (5.58)$$

Note that $\mathbf{h}_0$ is an arbitrary complex-valued vector (where its size depends on the number of dimensions and the dimension M of the Hilbert space), $\Omega = (\omega_1, \omega_2, \dots, \omega_m)$ and $\mathbf{n} = (n_1, n_2, \dots, n_m)$, $\mathbf{S}$ is the m-D finite extent support, Ω_p is

(a)

(b)

(c)

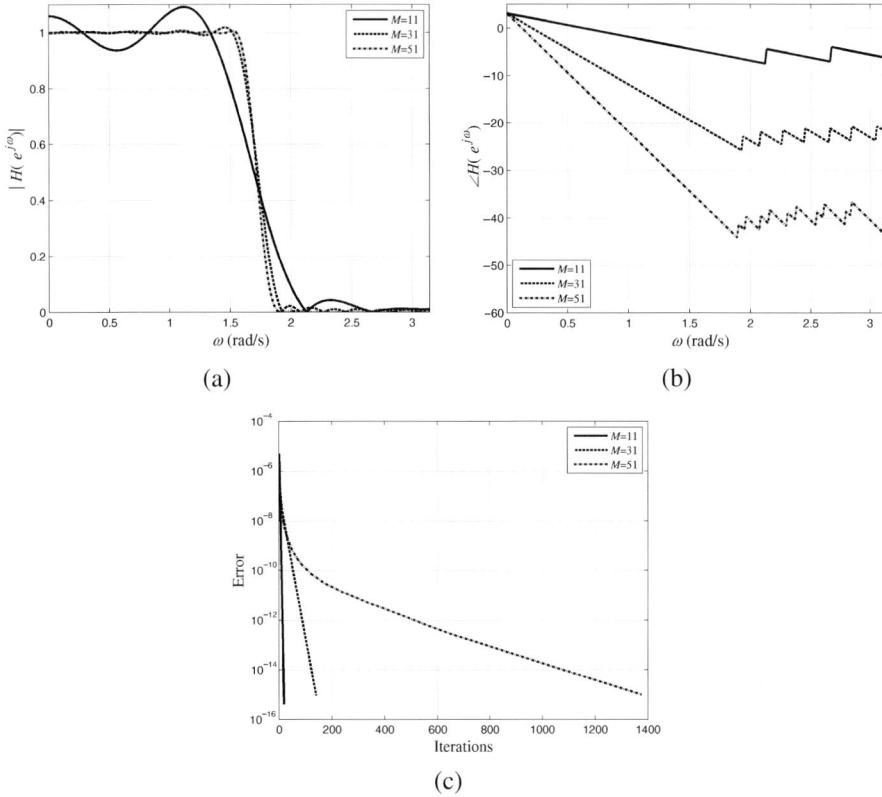

Figure 5.15 The design, using the POCS method, of a LPF linear-phase FIR filter with a cutoff $\omega_c = \frac{\pi}{2}$ rad/s, and for $M = 11$, $M = 31$, and $M = 51$, respectively. All are designed with a stopping threshold $\epsilon = 10^{-15}$. The frequency response $H(e^{j\omega})$: (a) magnitude and (b) phase spectrum. (c) Shows the absolute error between $\|\mathbf{h}_{k+1} - \mathbf{h}_k\|$ for the three FIR filters.

m-D passband region, and $\mathbf{\Omega}_s$ is m-D stopband region. The same stopping criterion reported in Haddad et al. (2000) is going to be used in this paper. That is, if the error distance is less than or equal to a predefined threshold ϵ, i.e., if $\|\mathbf{h}_{k+1} - \mathbf{h}_k\| \leq \epsilon$, then stop the algorithm. Otherwise, repeat steps 1–4. Figure 5.15 shows the frequency response of various LPF linear-phase FIR filters designed using the POCS method. The filters cutoff is $\omega_c = \frac{\pi}{2}$ rad/s, and lengths are $M = 11$, $M = 31$, and $M = 51$, respectively. The magnitude response in Figure 5.15a indicates that filters approximate equiripple behavior (Stark and Yang, 1998). Also, the absolute error curve for the three filters is given in Figure 5.15c, where, as expected, as M increases, more iterations are required to converge with an error less than or equal to $\epsilon = 10^{-15}$.

5.4 Design of IIR Digital Filters

Analog filter design is a mature and well-developed field. The design of IIR digital filters will start in the analog domain and then convert the design into the digital domain. An analog filter can be described by (Madisetti and Williams, 1998):

$$H_a(s) = \frac{G(s)}{W(s)} = \frac{\sum_{k=0}^{M} \beta_k s^k}{1 + \sum_{k=1}^{N} \alpha_k s^k}, \tag{5.59}$$

where α_k and β_k are the analog filter coefficients. Also, we can describe such a filter by its impulse response, which is related to $H_a(s)$ by the Laplace transform:

$$H_a(s) = \int_{-\infty}^{\infty} h_a(t) e^{-st} dt. \tag{5.60}$$

Note that $w(t)$ is the input to the filter and $g(t)$ is the output of the filter. Equations (5.59 and 5.60) are equivalent characterizations of an analog filter. Each of them will lead to alternative methods for converting the filter into the digital domain. Many transformation techniques exist to convert analog IIR filters into digital IIR filters. However, the focus will be on the design of IIR filters by Bilinear Transformation, for many reasons. The bilinear transformation design method can be used to design LPF, BPF, and HPF, unlike other techniques. Also, the bilinear transformation is a one-to-one mapping, unlike other techniques.

The design of IIR filters by bilinear transformation is a conformal mapping that transforms the $j\Omega$-axis (Ω is the analog angular frequency $\Omega = 2\pi F$) into the unit circle in the z-plane only once. In this way, one basically avoids aliasing of frequency components. In addition, all points in the left-hand s-plane are mapped inside the unit circle in the z-plane, while all points in the right-hand s-plane are mapped into the corresponding points outside the unit circle in the z-plane (see Figure 5.16). The mapping from the s-plane into the z-plane is performed through the following equation:

$$s = \frac{2}{\Delta t} \left(\frac{1 - z^{-1}}{1 + z^{-1}} \right), \tag{5.61}$$

where Δt is the time sampling interval. To investigate the characteristics of this transformation, rewrite Equation (5.61) to be:

$$s = \frac{2}{\Delta t} \left(\frac{z - 1}{z + 1} \right). \tag{5.62}$$

Let $z = re^{j\omega}$, and substituting this into Equation (5.61) results in:

$$s = \frac{2}{\Delta t} \left(\frac{re^{j\omega} - 1}{re^{j\omega} + 1} \right). \tag{5.63}$$

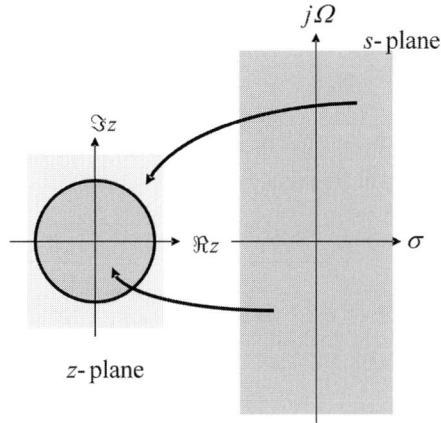

Figure 5.16 The bilinear transformation method is a conformal mapping that transforms the $j\Omega$-axis in the s-plane into the unit circle in the z-plane only once.

By using Eulers identity, and after some mathematical manipulation, one can obtain the following:

$$s = \frac{2}{\Delta t}\left(\frac{r^2 - 1}{1 + r^2 + 2r\cos(\omega)}\right) + j\frac{2}{T}\left(\frac{2r\sin(\omega)}{1 + r^2 + 2r\cos(\omega)}\right). \qquad (5.64)$$

However, $s = \sigma + j\Omega$ and, hence, from Equation (5.64):

$$\sigma = \frac{2}{\Delta t}\left(\frac{r^2 - 1}{1 + r^2 + 2r\cos(\omega)}\right), \qquad (5.65)$$

and,

$$\Omega = \frac{2}{\Delta t}\left(\frac{2r\sin(\omega)}{1 + r^2 + 2r\cos(\omega)}\right). \qquad (5.66)$$

Now, if $r < 1$, then $\sigma < 0$ and the left-hand s-plane maps into the inside of the unit circle in the z-plane, while when $r > 1$, then $\sigma > 0$ and the right-hand s-plane maps into the inside of the unit circle in the z-plane. If $r = 0$, then $\sigma = 0$, and, hence, one can show, based on Equation (5.66), that:

$$\omega = 2\tan^{-1}\left(\frac{\Omega\Delta t}{2}\right). \qquad (5.67)$$

This mapping is highly nonlinear where it is lower and upper bounded by $-\pi$ and π, respectively.

Example 5.1 Convert the analog filter with the transfer function:

$$H_a(s) = \frac{s + 0.1}{(s + 0.1)^2 + 9},\tag{5.68}$$

into a digital IIR filter by means of the bilinear transformation method. Let $\Delta t = 0.1$ s. What is the resonant frequency of the designed IIR digital filter?

Solution: Using Equation (5.61) and the given $\Delta t = 0.1$ s yields a substitution for s, which is:

$$s = \frac{2}{0.1}\left(\frac{1 - z^{-1}}{1 + z^{-1}}\right) = 20\left(\frac{1 - z^{-1}}{1 + z^{-1}}\right),$$

into the given analog filter:

$$H_a(s) = \frac{s + 0.1}{s^2 + 0.2s + 9.01}.$$

That is,

$$H(z) = \frac{20\left(\frac{1-z^{-1}}{1+z^{-1}}\right) + 0.1}{\left[20\left(\frac{1-z^{-1}}{1+z^{-1}}\right)\right]^2 + 0.2\left[20\left(\frac{1-z^{-1}}{1+z^{-1}}\right)\right] + 9.01}.$$

Now, after mathematical manipulation, the IIR digital filter is given by:

$$H(z) = \frac{4.8667 \times 10^{-2} + 4.8425 \times 10^{-4}z^{-1} - 4.8183 \times 10^{-2}z^{-2}}{1 - 1.8934z^{-1} + 9.8063 \times 10^{-1}z^{-2}}.$$

Note that the resonant frequency is $\omega = 2\tan^{-1}(\frac{3 \times 0.1}{2}) = 0.095\pi$ rad/s, a normalized frequency of 0.0475 (at 3 Hz). Figure 5.17 shows the magnitude spectrum of the designed IIR analog and digital filters.

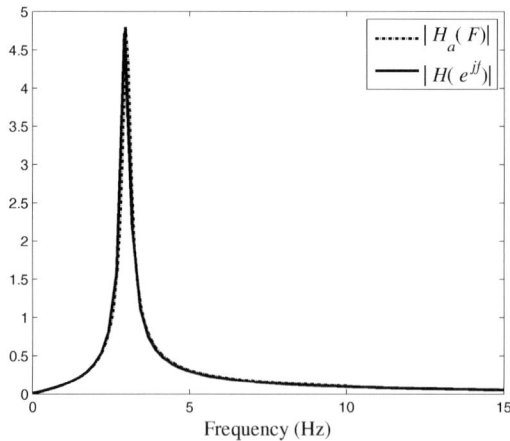

Figure 5.17 The frequency response magnitude spectrum of the designed IIR analog and digital filters in Example 5.1.

As mentioned earlier, the d IIR digital filters from analog ones requires knowing certain models in the analog form. Many standard LPF, BPF, HPF, analog prototypes exist such as the Butterworth, Chebyshave types I and II, and elliptic (Oppenheim and Schafer, 1989). Through the following examples, a LPF and a BPF IIR digital filter will be designed using the bilinear transformation technique based on the Butterworth prototypes.

Example 5.2 Consider the second-order Butterworth LPF IIR analog filter prototype having a 3-dB bandwidth of a cutoff Ω_c rad/s (Ziemer et al., 1998b):

$$H_a(s) = \frac{\Omega_c^2}{s^2 + \sqrt{2}\Omega_c s + \Omega_c^2}. \tag{5.69}$$

Design the IIR digital filter using the bilinear transformation method for a frequency cutoff $F_c = 60$ Hz and $\Delta t = 4$ ms.

Solution: Using Equation 5.61, one can show that the IIR digital filter transfer function $H(z)$ is:

$$H(z) = \frac{\Omega_c^2}{\left[\frac{2}{\Delta t}\left(\frac{1-z^{-1}}{1+z^{-1}}\right)\right]^2 + \sqrt{2}\Omega_c\left[\frac{2}{\Delta t}\left(\frac{1-z^{-1}}{1+z^{-1}}\right)\right] + \Omega_c^2}. \tag{5.70}$$

After mathematical manipulation, $H(z)$ will be equal to:

$$H(z) = \frac{1 + 2z^{-1} + z^{-2}}{(\psi^2 + \sqrt{2}\psi + 1) + 2(1 - \psi^2)z^{-1} + (\psi^2 - \sqrt{2}\psi + 1)z^{-2}}, \tag{5.71}$$

where $\psi = \frac{2}{\Delta t \Omega_c}$, which can be shown, based on Equation 5.67, to be equal to:

$$\psi = \cot\left(\frac{\Omega_c}{2}\right) = \cot\left(\frac{\pi F_c}{f_s}\right), \tag{5.72}$$

where F_c is the required frequency cutoff and f_s is the sampling frequency. So to attenuate frequency components of a seismic signal above 50 Hz, which are sampled at $\Delta t = 4$ ms, $F_c = 50$ Hz and $f_s = 250$ Hz. Hence, $F_c/f_s = 0.24$ and $\psi = 1.0649$. Therefore, the IIR digital LPF transfer function will be:

$$H(z) = \frac{1 + 2z^{-1} + z^{-2}}{3.64 - 0.268z^{-1} + 0.628z^{-2}}. \tag{5.73}$$

The frequency response (magnitude and phase) is presented in Figure 5.18.

To design other filter types, i.e., HPF, BPF, and BSF, one can apply transformation for converting a prototype digital LPF into other types (Proakis and Manolakis, 2006; Ziemer et al., 1998b). For example, frequency transformation from a LPF into

(a)

(b)

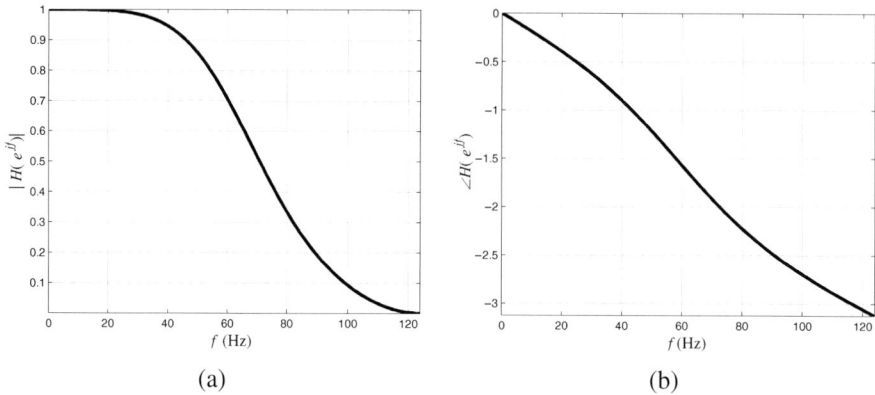

Figure 5.18 The LPF IIR digital filter in Example 5.2: (a) magnitude spectrum, and (b) phase spectrum.

a bandpass filter can by done by replacing every s in the filter's transfer function $H_a(s)$ by:

$$s = A \frac{1 - Bz^{-1} + z^{-2}}{1 - z^{-2}}, \tag{5.74}$$

where,

$$A = \cot\left[(\Omega_u - \Omega_l)\frac{\Delta t}{2}\right] = \cot\left[\pi\left(\frac{F_u}{f_s} - \frac{F_l}{f_s}\right)\right], \tag{5.75}$$

$$B = 2\frac{\cos\left[(\Omega_u + \Omega_l)\frac{\Delta t}{2}\right]}{\cos\left[(\Omega_u - \Omega_l)\frac{\Delta t}{2}\right]}, \tag{5.76}$$

$$= \frac{\cos\left[\pi\left(\frac{F_u}{f_s} + \frac{F_l}{f_s}\right)\right]}{\cos\left[\pi\left(\frac{F_u}{f_s} - \frac{F_l}{f_s}\right)\right]},$$

where Ω_u (or F_u) is the upper band cutoff frequency, and Ω_l (or F_l) is the lower band cutoff frequency.

Example 5.3 Design a BPF based on the LPF transfer function $H(z)$ shown in Example 5.2 using digital transformation with $F_l = 10$ Hz and $F_u = 60$ Hz. Plot the frequency response of the resultant digital filter.

Solution: Using the transformation Equation (5.74), one can show that the IIR PBF digital filter can be given by the following transfer function:

$$H(z) = \frac{1 + 2z^{-2} + z^{-4}}{C + Dz^{-1} + Ez^{-2} + Fz^{-3} + Gz^{-4}}, \tag{5.77}$$

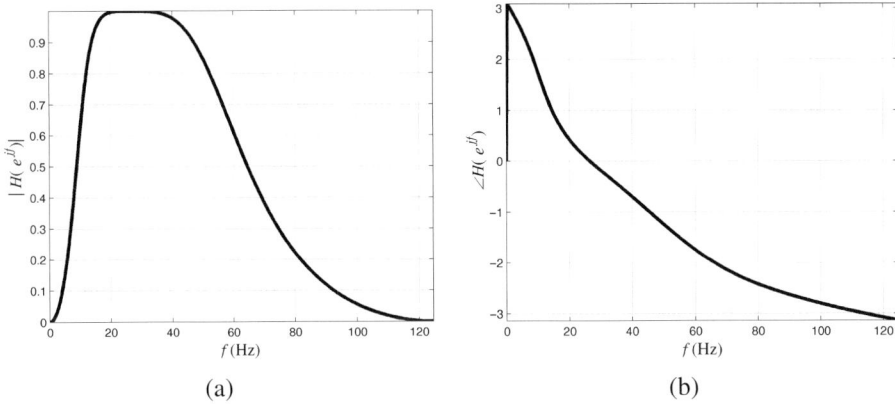

Figure 5.19 The BPF IIR digital filter in Example 5.3: (a) magnitude spectrum, and (b) phase spectrum.

where $C = (\psi^2 + \sqrt{2}\psi + 1)$, $D = -B(2\psi^2 + \sqrt{2}\psi)$, $E = ((2 + B^2)\psi^2 - 2)$, $F = -B(2\psi^2 - \sqrt{2}\psi)$, $G = (\psi^2 - \sqrt{2}\psi + 1)$, and $\psi = A/(\Omega_c\Delta t)$. Hence, for $F_l = 10$ Hz and $F_u = 60$ Hz with $\Delta t = 4$ ms, the filter will be:

$$H(z) = \frac{1 + 2z^{-2} + z^{-4}}{5.661 - 11.203z^{-1} + 8.976z^{-2} - 4.229z^{-3} + 1.235z^{-4}}. \tag{5.78}$$

Figure 5.19 shows both the magnitude and phase response of the designed BPF IIR digital filter.

The same real seismic trace used in the FIR case (from a shot gather (Yilmaz, 2001)) is shown in Figure 5.20a and its magnitude spectrum in Figure 5.20b. Using the bilinear transformation method, a BPF IIR filter with $\omega_l = 0.08\pi$ rad/s and $\omega_u = 0.24\pi$ rad/s, and for $M = N = 9$ was designed to attenuate low and high frequency noise. The filter is designed via a Butterworth prototype. From its magnitude response, seen in Figure 5.20c, the filter has very short transition bands and is well-designed, compared to those shown earlier in the case of FIR filters. Figure 5.20d displays the filtered seismic trace and Figure 5.20e its magnitude spectrum. Finally, Figure 5.21 shows the seismic shot gather from Yilmaz (2001) and its $f - x$ magnitude spectrum before and after applying on each seismic trace the BPF IIR digital filter. It is evident that the ground-roll noise was attenuated to a great degree and other seismic events, such as reflections, became very visible. Still some of the harmonic noise seen near offset -1000 became more visible and required filtering. This result is better than the result of FIR filtering seen previously in this chapter.

(a)

(b)

(c)

(d)

(e)

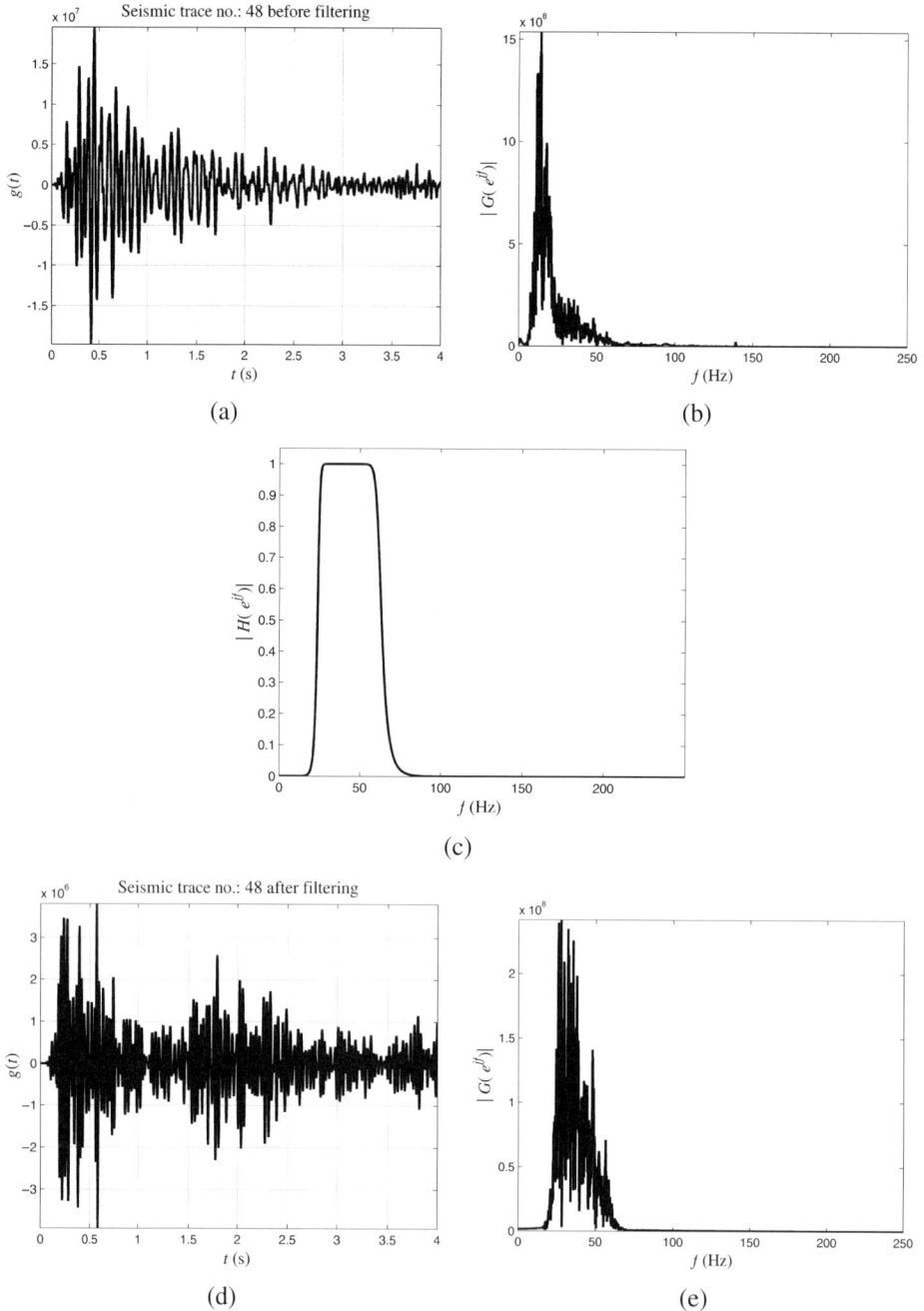

Figure 5.20 (a) A real seismic trace from a shot gather (modified from Yilmaz, 2001). (b) Its magnitude spectrum. (c) The design, using the bilinear transformation method, of a BPF IIR filter with $\omega_l = 0.08\pi$ rad/s and $\omega_u = 0.24\pi$ rad/s, and for $M = N = 9$. (d) The filtered seismic trace and its (e) magnitude spectrum. Note that $\Delta t = 2$ ms.

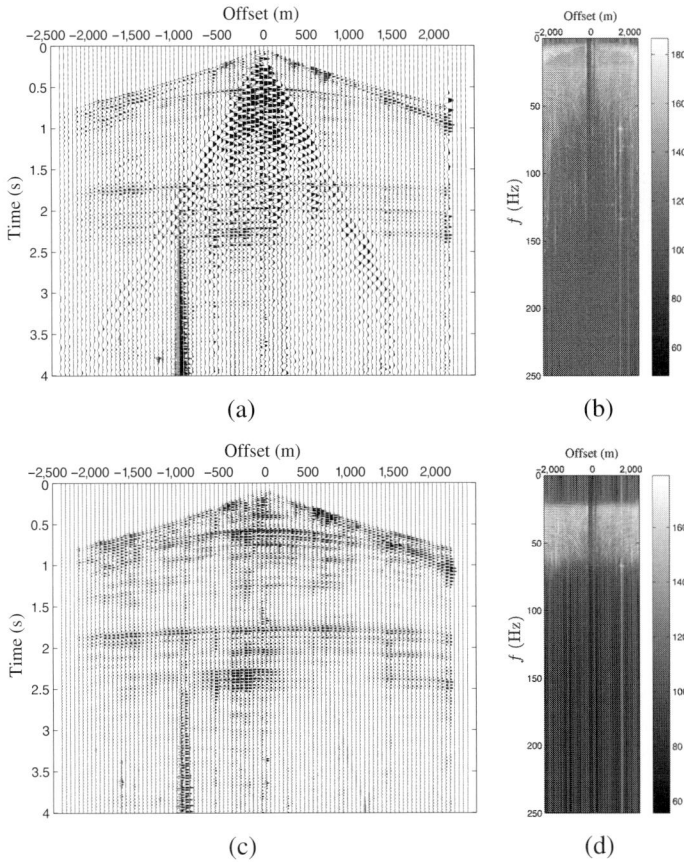

Figure 5.21 (a) A real seismic shot gather (modified from Yilmaz, 2001). (b) Its $f - x$ magnitude spectrum. (c) The data after applying for each trace in (a) the BPF used in Figure 5.20. (d) The $f - x$ magnitude spectrum of (c).

5.5 Seismic Wavefield Extrapolation 1-D FIR and IIR Filters

This section describes the migration as a filtering process and shows the complex-valued seismic migration filter's origins. Here, the migration is based on the concept that a wavefield (acquired seismic data) that is measured at the Earth's surface, $u(x, t, z = 0)$, represents the boundary values of the wavefield $u(x, t, z)$ reflected in the sub-Earth. The main objective of this migration is to determine the true position of the reflectors by downward continuation of the wavefield that is measured at the surface. This can mathematically be explained through the wave equation and its solution, where we deal with the 2-D case.

Let us assume that an acoustic seismic wave propagates upward through the Earth. For simplicity we assume that the wave propagates in a homogeneous medium with constant velocity c. This wave can be described by the following 2-D hyperbolic wave equation (Yilmaz, 2001):

$$\frac{\partial^2 u}{\partial x^2} + \frac{\partial^2 u}{\partial z^2} = \frac{1}{c^2}\frac{\partial^2 u}{\partial t^2}. \tag{5.79}$$

where $u(x,t,z)$ is the propagating wavefield (displacement), t stands for the time variable, x represents the lateral spatial axis, and z denotes the depth axis.

Now, define $j = \sqrt{-1}$ and let $U(K_x,\Omega_t,z)$ be the 2-D Fourier transform of $u(x,t,z)$ with respect to the variables x and t as given by:

$$U(K_x,\Omega_t,z) = \int_{-\infty}^{\infty}\int_{-\infty}^{\infty} u(x,t,z)e^{j(K_x x - \Omega_t t)}dxdt. \tag{5.80}$$

Note that K_x and Ω_t are the analog wavenumber and analog angular frequency. Taking the Fourier transform of both sides of Equation (5.79) and carrying out the necessary rearrangement gives:

$$\frac{\partial^2 U(K_x,\Omega_t,z)}{\partial z^2} = \left(\frac{\Omega_t^2}{c^2} - K_x^2\right)U(K_x,\Omega_t,z). \tag{5.81}$$

Equation (5.81) is a second-order ordinary differential equation with respect to the variable z. Its well-known solution is given by:

$$U(K_x,\Omega_t,z) = A\exp\left(jz\sqrt{\frac{\Omega_t^2}{c^2} - K_x^2}\right) + B\exp\left(-jz\sqrt{\frac{\Omega_t^2}{c^2} - K_x^2}\right), \tag{5.82}$$

where A and B are constants that can be found using the initial conditions related to the differential equation. Following the sign convention of Dudgeon and Mersereau (1984), the positive exponent refers to an upgoing propagation, while the negative solution refers to a downgoing propagation. Since it is assumed that the wave is propagating upward, one can set $B = 0$. To find A, consider the 2-D inverse Fourier transform of $u(x,t,z)$ given as:

$$u(x,t,z) = \int_{-\infty}^{\infty}\int_{-\infty}^{\infty} U(K_x,\Omega_t,z)e^{-j(K_x x - \Omega_t t)}dK_x d\Omega_t. \tag{5.83}$$

Substituting Equation (5.82) (with $B = 0$) into Equation (5.83) yields:

$$u(x,t,z) = \int_{-\infty}^{\infty}\int_{-\infty}^{\infty} A\exp\left(jz\sqrt{\frac{\Omega_t^2}{c^2} - K_x^2}\right)e^{-j(K_x x - \Omega_t t)}dK_x d\Omega_t. \tag{5.84}$$

Now, by putting $z = 0$, it can be clearly seen from Equations (5.83 and 5.84) that the boundary condition $A = U(K_x,\Omega_t,0)$ and, therefore Equation (5.82) becomes:

$$U(K_x, \Omega_t, z) = \begin{cases} U(K_x, \Omega_t, 0) \exp(jz\sqrt{\frac{\Omega_t^2}{c^2} - K_x^2}) & \text{if } |\Omega_t| > c|K_x| \\[2ex] U(K_x, \Omega_t, 0) \exp(-z\sqrt{K_x^2 - \frac{\Omega_t^2}{c^2}}) & \text{if } |\Omega_t| \le c|K_x|, \end{cases} \tag{5.85}$$

where $U(K_x, \Omega_t, 0) = \mathcal{F}\{u(x,t,0)\}$ represents the boundary condition for the differential equation given in Equation (5.81). Then the record $u(x,t,z)$ will be $u(x,t,z) = \mathcal{F}^{-1}\{U(K_x, \Omega_t, z)\}$. Note that $\mathcal{F}$ and $\mathcal{F}^{-1}$ are the forward and inverse Fourier transform operators, respectively. In general, at $z = z_o + \Delta z$ and given the initial condition at $z = z_o$, the general solution in terms of boundaries at a width equal to Δz is given by:

$$U(K_x, \Omega_t, z_o + \Delta z) = \begin{cases} U(K_x, \Omega_t, z_o) \exp\left(j\Delta z\sqrt{\frac{\Omega_t^2}{c^2} - K_x^2}\right) & \text{if } |\Omega_t| > c|K_x| \\[2ex] U(K_x, \Omega_t, z_o) \exp\left(-\Delta z\sqrt{K_x^2 - \frac{\Omega_t^2}{c^2}}\right) & \text{if } |\Omega_t| \le c|K_x|. \end{cases} \tag{5.86}$$

Clearly, Equation (5.86) can be rewritten as:

$$U(K_x, \Omega_t, z_o + \Delta z) = U(K_x, \Omega_t, z_o) H_d(K_x, \Omega_t), \tag{5.87}$$

where

$$H_d(K_x, \Omega_t) = \begin{cases} \exp\left(j\Delta z\sqrt{\frac{\Omega_t^2}{c^2} - K_x^2}\right) & \text{if } |\Omega_t| > c|K_x| \\[2ex] \exp\left(-\Delta z\sqrt{K_x^2 - \frac{\Omega_t^2}{c^2}}\right) & \text{if } |\Omega_t| \le c|K_x|. \end{cases} \tag{5.88}$$

Equation (5.88) can be seen as the wavenumber-angular frequency response of a space-time linear-shift invariant (LSI) analog filter (Dudgeon and Mersereau, 1984; Karam and McClellan, 1997) and it is known as a *Seismic Migration Filter*. In the geophysics community, this filter is known as the extrapolation operator, since one extrapolates the previous depth seismic section to the next one (Thorbecke, 1997; Yilmaz, 2001).

To carry on, the wavefield in practice is presented in sampled form, so let us define Δt to be the temporal sampling interval, Δx to be the horizontal spatial (trace) sampling interval, and Δz as the depth sampling interval. Then, the discrete version of the acoustic wavefield $u(x,t,z)$ is $u(x_i, t_l, z_k)$, where $x_i = i\Delta x$, $t_l = l\Delta t$, and $z_k = k\Delta z$ for all $i,l,k \in \mathbb{Z}$ with $\mathbb{Z}$ the set of integers. Also, define k_x as the digital wavenumber counterpart of K_x and ω as the digital angular frequency counterpart of Ω_t. The migration operation is performed using a digital filter (derived from its analogue counterpart in Equation (5.88) (Garibotto, 1979; Hale, 1991; Yilmaz, 2001)) whose frequency-wavenumber $(\omega - k_x)$ response is given by:

$$H_d(e^{jk_x}, e^{j\omega}) = \exp\left(j\frac{\Delta z}{\Delta x}\sqrt{\frac{\Delta x^2}{\Delta t^2}\frac{\omega^2}{c^2} - k_x^2}\right). \tag{5.89}$$

This is the ideal frequency-wavenumber response of an all-pass filter with nonlinear phase. For a single angular frequency ω_o, Equation (5.89) becomes a 1-D digital filter given by:

$$H_d(e^{jk_x}, e^{j\omega_o}) := H_d(e^{jk_x}) = \exp\left(j\frac{\Delta z}{\Delta x}\sqrt{\frac{\Delta x^2}{\Delta t^2}\frac{\omega_o^2}{c^2} - k_x^2}\right). \qquad (5.90)$$

From the signal processing point of view, Equation (5.90) shows that $H_d(e^{jk_x})$ is a complex-valued even function with $H_d(e^{jk_x}) = H_d(e^{-jk_x})$. Also, it was shown in Garibotto (1979) that the migration (extrapolation) process cannot be obtained by a causal filter since its response is defined for negative values of the variable x. Therefore, Equation (5.90) can be approximated by a non-causal even symmetric FIR digital filter of length N (N is odd) (Hale, 1991; Karam and McClellan, 1997; Yilmaz, 2001). That is,

$$H_d(e^{jk_x}) = h[0] + 2\sum_{n=1}^{\frac{N+1}{2}-1} h[n]\cos(nk_x), \qquad (5.91)$$

where $h[n] \in \mathbb{C}$ ($\mathbb{C}$ is the set of complex numbers), i.e., the FIR filter coefficients are complex-valued. Now, the filter response cutoff is given by the wavenumber k_c, where:

$$k_x = k_c = \frac{\Delta x}{\Delta t}\frac{\omega_o}{c}, \qquad (5.92)$$

where the approximation needs only to be accurate for $|k_x| < |k_c|$, since this corresponds to the wavenumbers k_x for which the waves are propagating (Hale, 1991; Yilmaz, 2001). Finally, let $b = \Delta z/\Delta x$, and then the 1-D migration filter simply becomes:

$$H_d(e^{jk_x}) = \exp\left(jb\sqrt{k_c^2 - k_x^2}\right). \qquad (5.93)$$

Then, for each frequency sample we can extrapolate a spatially sampled seismic wavefield $u(x_i, e^{j\omega_o}, z_k)$ from, say, depth z_k to $z_{k+1} = z_k + \Delta z$ using:

$$U(e^{jk_x}, e^{j\omega_o}, z_o + \Delta z) = U(e^{jk_x}, e^{j\omega_o}, z_o)H_d(e^{jk_x}, e^{j\omega_o}), \qquad (5.94)$$

where $U(e^{jk_x}, e^{j\omega_o}, z_o + \Delta z)$ is equal to the discrete-space Fourier transform (DSFT) of $u(x_i, e^{j\omega_o}, z_k)$ with respect to x_i.

Equation (5.94) informs us that the propagating wavefield on the z_kth layer to the next layer at z_{k+1} is easily accomplished in the $\omega - k_x$ domain. This type of extrapolation is called the phase shift method and was first introduced by Gazdag (1978) for performing post-stack migration. His migration algorithm starts with Fourier transforming the input traces from their $t - x$ domain to the $\omega - k_x$ domain and then using the extrapolation form given in Equation (5.94). This method is accurate up to dip angles of 90° (Yilmaz, 2001). In addition, this method is accurate when a constant velocity is used for each extrapolation step Δz. However, the velocity may vary for each depth step size. In this case, an accurate migration is allowed when the velocities vary only with depth, i.e., when the velocities are vertically varying. The main objective is to design a finite impulse response (FIR) or an infinite impulse response (IIR) frequency-space ($f - x$) wavefield extrapolation filter that approximates the following wavenumber response (as described, for example, by Mousa et al. (2006)):

$$H_d(k_x) = \exp\left(jb\sqrt{k_c^2 - k_x^2}\right),\tag{5.95}$$

where k_c is the normalized wavenumber cutoff, k_x is the normalized horizontal wavenumber, and b is a constant that is equal to the ratio of depth sampling interval to spatial sampling interval. Equation (5.95) is usually approximated via an Nth-order non-causal finite impulse response (FIR) filter with a transfer function with even symmetry, as described by Mousa (2006):

$$H_{FIR}(z) = h_0 + \sum_{n=1}^{N} h_n(z^n + z^{-n}),\tag{5.96}$$

where $h_n \in \mathbb{C}$. Such filter coefficients h_n are stored in a look-up table and are used to perform wavefield extrapolation via a spatially varying convolution operation (Holberg, 1988; Mousa, 2014; Naseer and Mousa, 2015). The transfer function for the symmetrical IIR $f - x$ extrapolation filter (for Equation (5.95)), on the other hand, can be given by the non-causal transfer function according to Nguyen and Castagna (2002), and Mousa (2012a, and b):

$$H_{IIR}(z) = \frac{b_0 + b_1(z + z^{-1}) + \cdots + b_M(z^M + z^{-M})}{1 + a_1(z + z^{-1}) + \cdots + a_N(z^N + z^{-N})},\tag{5.97}$$

where $M \leq N$. Due to the even symmetry in both the magnitude and phase responses in Equation (5.95), the IIR coefficients belong to the set of complex numbers, i.e., $a_n, b_m \in \mathbb{C}$ for $n = 1, \ldots, N$ and $m = 0, \ldots, M$. Evaluating the IIR $f - x$ extrapolation filter transfer function in Equation 5.97 on the unit circle and performing a few mathematical simplifications results in the following transfer function in the wavenumber k_x domain (Mousa, 2012a,b):

$$H(e^{jk_x}) = \frac{B(e^{jk_x})}{A(e^{jk_x})} = \frac{b_0 + 2\sum_{m=1}^{M} b_m \cos(k_x m)}{1 + 2\sum_{n=1}^{N} a_n \cos(k_x n)}.\tag{5.98}$$

Figure 5.22 shows the spectra of 1-D FIR $\omega - x$ seismic extrapolation digital filters ($M = 25$, $M = 35$ and $M = 55$) designed using the POCS algorithm,

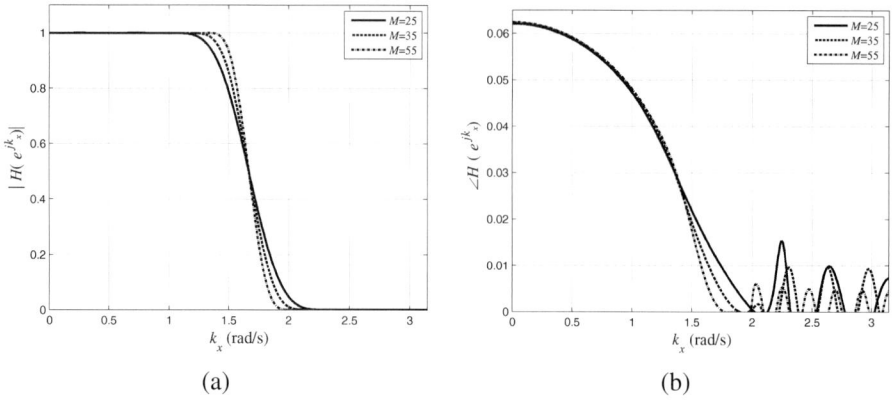

(a)   (b)

Figure 5.22 1-D FIR $\omega - x$ seismic extrapolation digital filters ($M = 25, M = 35$ and $M = 55$) designed using the POCS algorithm, for a normalized wavenumber cutoff $k_c = 0.25$ or $(kc = \pi/4$ rad/s), and a stopping algorithm threshold $\epsilon = 10^{-15}$. (a) Magnitude response and (b) phase response.

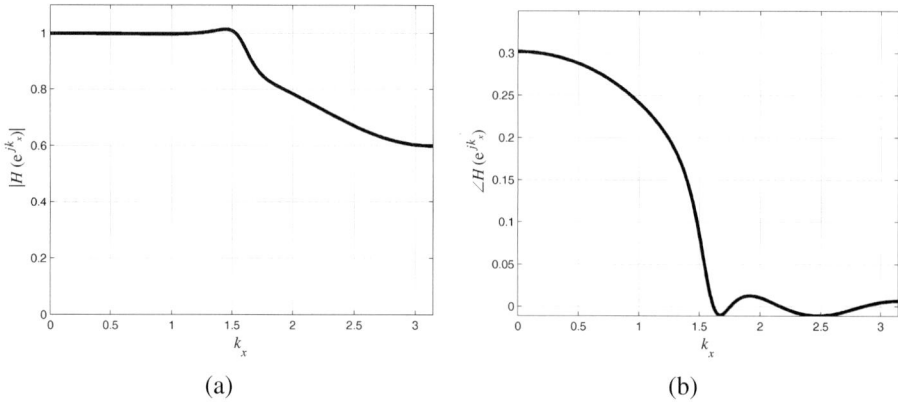

(a)   (b)

Figure 5.23 1-D IIR $\omega - x$ seismic extrapolation digital filters ($M = 2$ and $N = 2$) designed using the IRLS algorithm, for a normalized wavenumber cutoff $k_c = 0.25$ or $(kc = \pi/4$ rad/s). (a) Magnitude response and (b) phase response.

for a normalized wavenumber cutoff $k_c = 0.25$ or $(kc = \pi/4$ rad/s). The POCS stopping algorithm threshold $\epsilon = 10^{-15}$. The magnitude response in Figure 5.22a is of even symmetry and attenuates wavefields that are above the cutoff, while the phase response, in Figure 5.22b, is also even and is nonlinear. Finally, Figure 5.23 shows a 1-D IIR $\omega - x$ seismic extrapolation digital filter ($M = 2$ and $N = 2$) designed using the IRLS algorithm (Mousa, 2012b), for a normalized wavenumber cutoff $k_c = 0.25$ or $(kc = \pi/4$ rad/s).

5.6 Two-dimensional Filters for Seismic Data

The 1-D filters can be extended as 2-D filters via many techniques, assuming separability of the filter operating coefficients. Also, 2-D FIR or IIR filters can directly be designed using windowing techniques, POCS, Weighted Least Squares, etc., to name a few. Here, the focus will be on designing 2-D FIR fan filters. Fan, which is also known as phase velocity, velocity, or pie-slice filtering, has a wide range of applications in exploration seismology. It is used to remove coherent noise events from seismic records on the basis of particular angles at which the events dip. Basically, they can be used either to pass or reject seismic events that propagate as a function of time and space (Dudgeon and Mersereau, 1984; Yilmaz, 2001).

A seismic pulse traveling with velocity v at an angle θ to the vertical will propagate across the spread with an apparent velocity $v_a = v \sin(\theta)$, as illustrated in Figure 5.24. Along the spread direction, each individual sinusoidal component of the pulse will have an apparent wavenumber, say k, related to its individual frequency f or v_a, where;

$$\omega = 2\pi v_a k, \tag{5.99}$$

which basically forms a linear relation between the frequency and the apparent wavenumber for that pulse with a slope of v_a (see Figure 5.25). Any seismic event propagating across a surface spread will be characterized by an $\omega - k$ curve radiating from the origin at a particular slope determined by the apparent velocity with which the event passes across the spread. The overall set of curves for a typical shot gather containing reflected and surface propagating seismic events is shown in Figure 5.26.

As explained in Chapter 3, the fact that different types of seismic event fall within different zones of the $\omega - k$, plot (spectrum) provides a means of filtering to suppress unwanted events (noise regions in the $f - k$, spectrum) on the basis

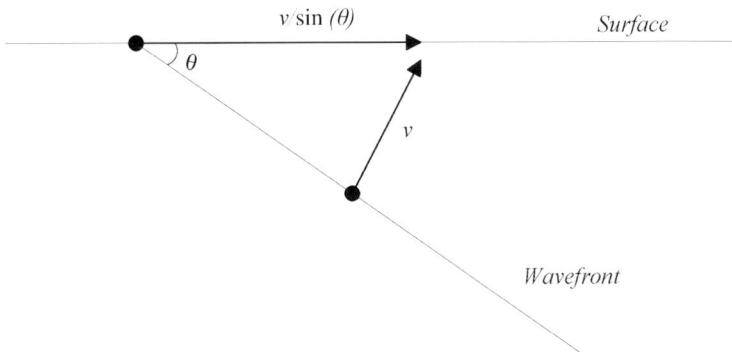

Figure 5.24 A wave traveling at angle θ to the vertical will pass across an inline spread of surface detectors at a velocity $v_a = v/\sin(\theta)$.

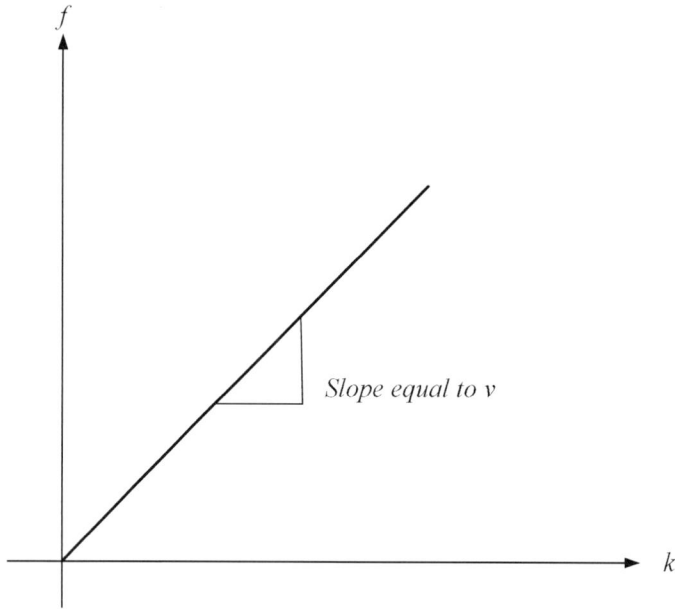

Figure 5.25 An $\omega - k_x$ plot for a seismic pulse passing across a surface spread of detectors.

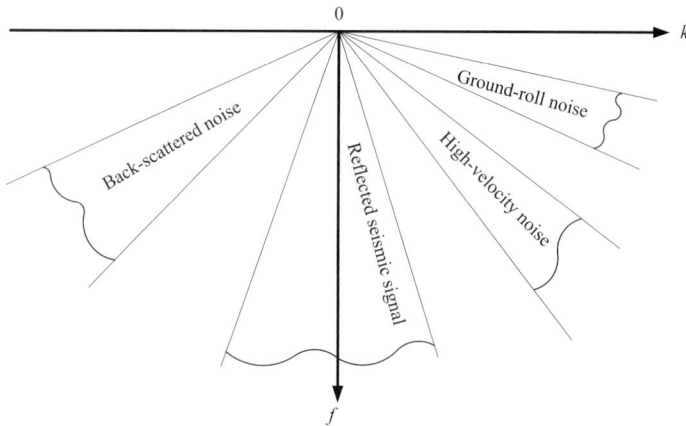

Figure 5.26 Typical band region of 2-D seismic data events.

of their apparent velocities. In 2-D seismic data, the normal means by which this is achieved, known as $f - k$ filtering, is to enact a 2-D Fourier transformation on the seismic data from the $t - x$ domain to the $\omega - k$ domain, then to filter that spectrum by removing wedge-shaped zone/zones containing the unwanted noise regions. Finally, transform back into the $t - x$ domain to have a filtered seismic section. An important application of velocity filtering, as mentioned earlier, is the removal of ground-roll noise (Lu and Antoniou, 1992; Yilmaz, 2001).

Mathematically, an ideal amplitude response of a pass velocity filter may be given by (Dudgeon and Mersereau, 1984; Lu and Antoniou, 1992):

$$H(e^{jk_x}, e^{j\omega}) = \begin{cases} 1 & \text{if} \quad |\arctan(\frac{\omega}{k_x})| \leq \theta \\ 0 & \text{if} \quad otherwise. \end{cases} \tag{5.100}$$

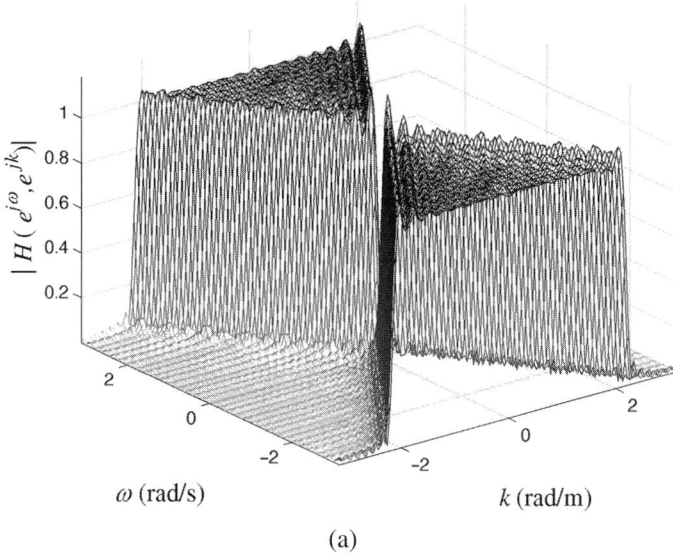

(a)

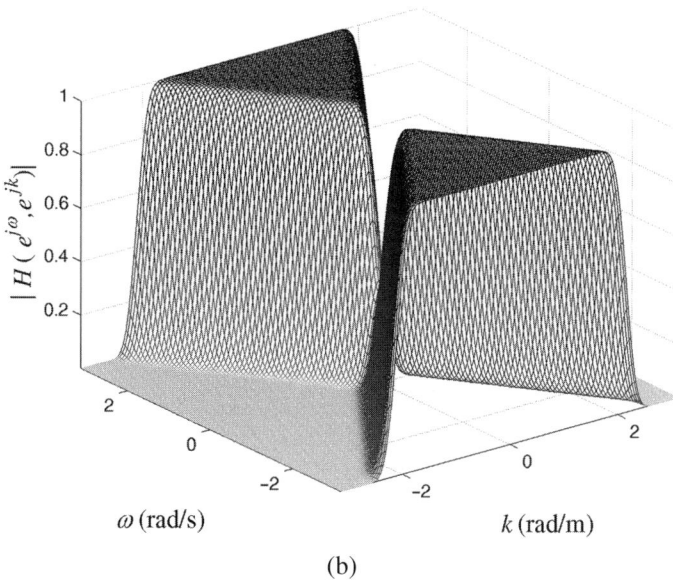

(b)

Figure 5.27 The magnitude response of a 2-D FIR fan digital filter designed at $\theta = 35°$ using: (a) a 2-D rectangular window, and (b) a 2-D Kaiser window with $\delta_p = \delta_s = 10^{-5}$.

Also, the ideal amplitude response of a reject velocity filter (Dudgeon and Mersereau, 1984; Lu and Antoniou, 1992) may be given by:

$$H(e^{jk_x}, e^{j\omega}) = \begin{cases} 0 & \text{if} \quad |\arctan(\frac{\omega}{k_x})| \leq \theta \\ 1 & \text{if} \quad otherwise. \end{cases} \qquad (5.101)$$

It has been a while since the problem of designing such fan filters was considered. The methodologies for designing 2-D FIR fan filters include the use of transformation of 1-D designs, the use of optimal Chebyshev design techniques, windowing, and so on. See, for example, the work by Dudgeon and Mersereau (1984), Lu and Antoniou (1992), and Yilmaz (2001). Figure 5.27 shows the magnitude response of a 2-D FIR fan digital filter designed using a 2-D rectangular and a 2-D Kaiser window (Lu and Antoniou, 1992). In order to apply fan filtering in the $f - \hat{k}$ domain, Figure 5.28 can be used as a workflow. This was used to filter the real seismic data seen in Figure 5.29. Interestingly, the data contain high amplitude, low

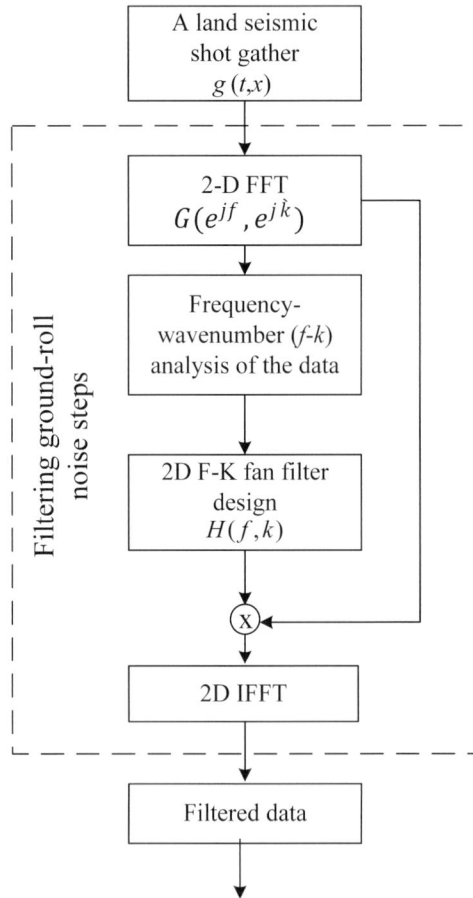

Figure 5.28 Workflow chart for ground-roll noise attenuation using 2-D frequency-wavenumber fan filters (modified from Yilmaz, 2001).

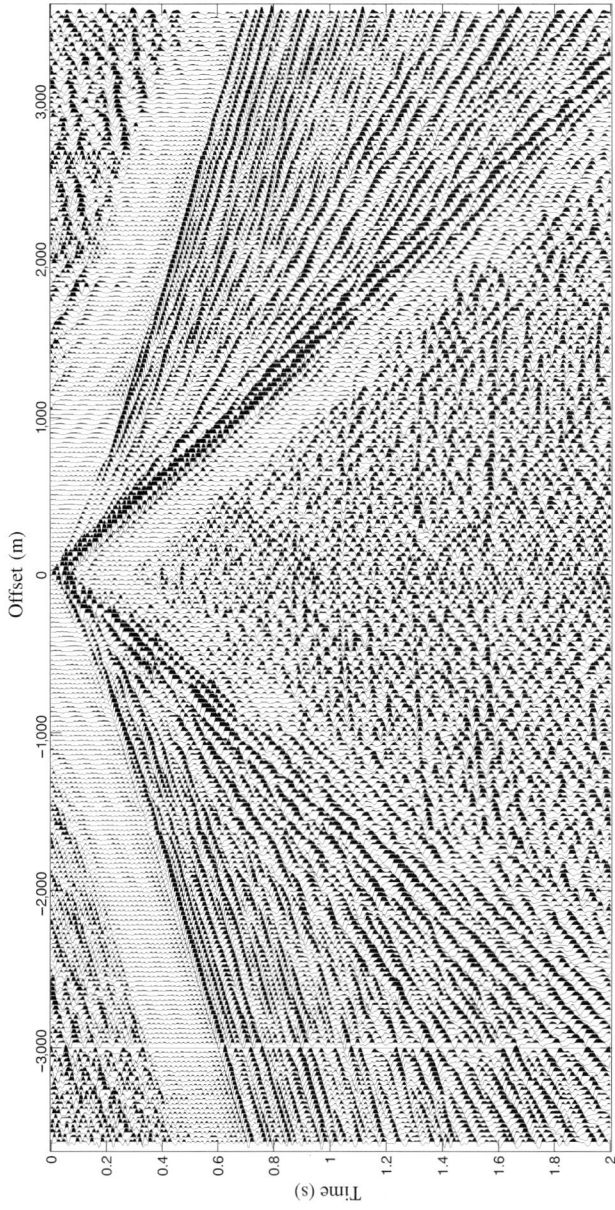

Figure 5.29 Land seismic data (see Figure 2.38 in Chapter 2) before applying fan filtering. The data contain high amplitude, low frequency dispersive ground-roll noise.

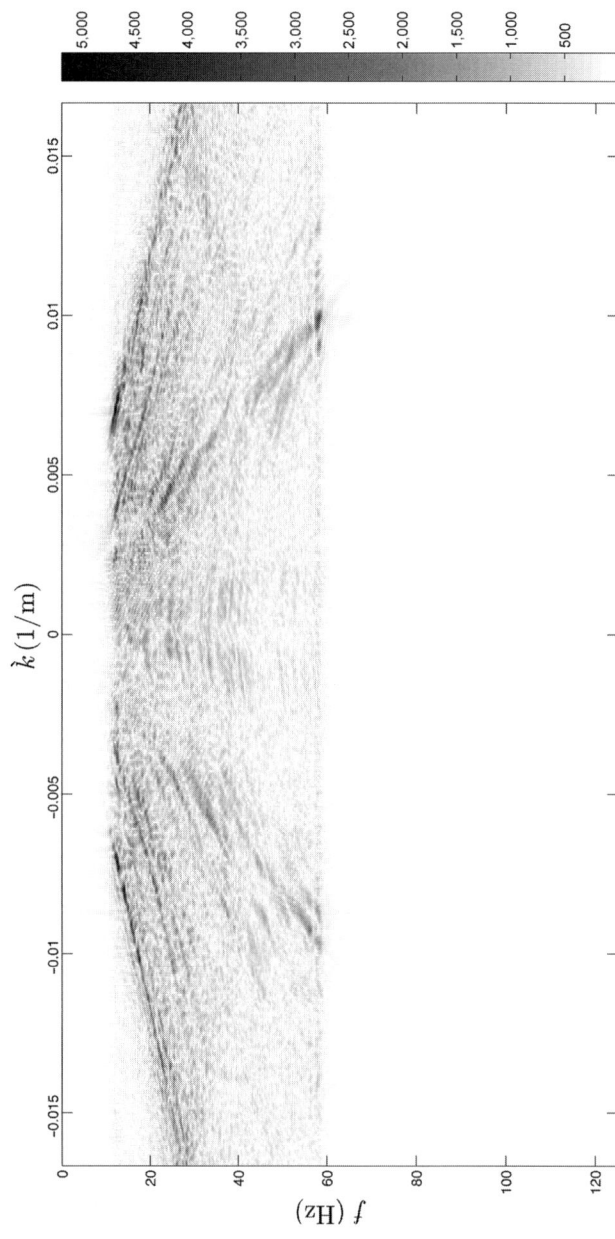

Figure 5.30 The $f - \hat{k}$ magnitude spectrum of the land seismic data seen in Figure 5.29. The low frequency/low wavenumber ground-roll noise can be filtered using fan filters.

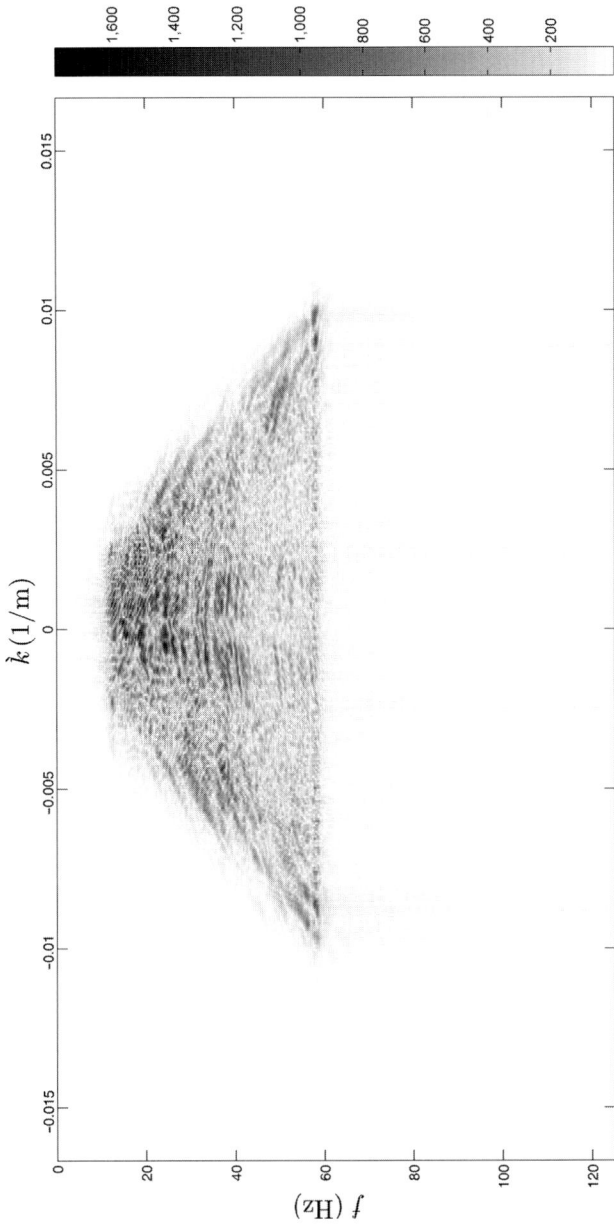

Figure 5.31 The $f - \grave{k}$ magnitude spectrum of the land seismic data seen in Figure 5.29 after applying fan filtering to attenuate the low frequency/low wavenumber ground-roll noise seen in Figure 5.30.

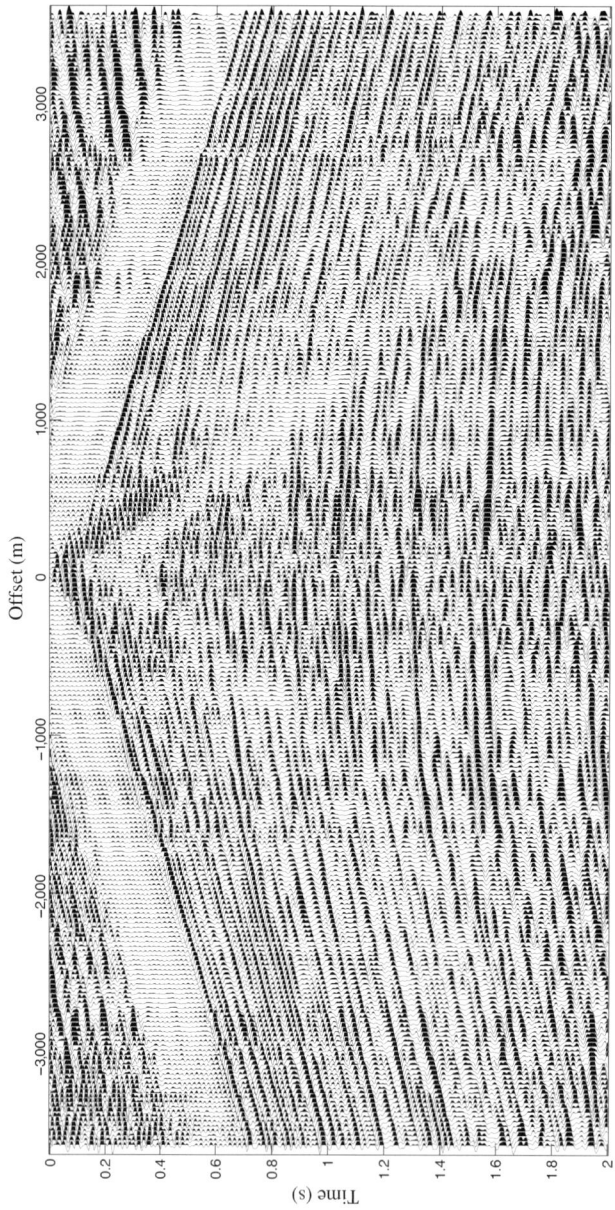

Figure 5.32 The land seismic data (seen in Figure 5.29) after applying fan filtering. The hidden primary reflections (hyperbolic events) became visible.

frequency dispersive ground-roll noise, seen in both the $t - x$ domain (Figure 5.29) and the $f - \hat{k}$ domain (Figure 5.30). A 2-D FIR fan digital filter was designed based on a Kaiser window and Equation (5.100) with $\theta = 50°$. The magnitude response of the data after fan filtering is shown in Figure 5.31, while its inverse 2-D Fourier transform ($t - x$ domain) is shown in Figure 5.32. It is very evident that the hidden primary reflections (seismic hyperbolic events) became visible after attenuating the ground-roll strong presence.

5.7 Summary

In this chapter, the design of LSI FIR and IIR filters was introduced. Various design methods with application to real seismic data were presented and compared. FIR filters tend to be more straightforward and require less effort than designing IIR ones. At the same time, IIR filters provide, basically, better designs than that of the FIR case. It is worth mentioning that filtering noise such as ground-roll, in the case of seismic reflection data, might be best performed using fan filters. Such filters are part of the standard workflow of processing seismic data. However, if the ground-roll noise is aliased, then filtering the data using linear filters such as 1-D BPF or 2-D fan filters will not be the best option to attenuate such a troublesome noise type. Finally, filtering seismic data can be done via convolution in the time (or time-space) domain or, alternatively, in the frequency (or $f - \hat{k}$) domain using one FFT and one inverse FFT, as explained in Chapter 3.

Exercises

5.1 Design a three-length lowpass FIR digital filter using the Hanning window:

$$w[n] = \frac{1}{2}\left(1 - \cos\left[\frac{2\pi n}{M - 1}\right]\right),$$

with a cutoff $\omega_c = \pi/2$. The impulse response of the ideal lowpass filter can be given by:

$$h[n] = \begin{cases} \frac{\omega_c}{\pi}, & n = 0 \\ \frac{\omega_c}{\pi}\frac{\sin[\omega_c n]}{\omega_c}, & n \neq 0. \end{cases}$$

Is it a causal or a non-causal filter?

5.2 Specify the FIR filter order, cutoff, and other parameters for a highpass filter with $\omega_p = 0.4\pi$ rad/s, $\omega_s = 0.2\pi$ rad/s and $\delta_s = 2.5 \times 10^{-4}$, using Kaiser windowing.

5.3 Design a first-order LPF IIR digital filter with a 3-dB bandwidth of a cutoff $\Omega_c = 0.2\pi$ rad/s, using the bilinear transformation, with $\Delta t = 2$ ms, applied to the following Butterworth analog filter:

$$H_a(s) = \frac{\Omega_c}{s + \Omega_c}. \tag{5.102}$$

Plot the ploe-zero map of $H(z)$ and compute the frequency response of the filter.

5.4   Convert the designed IIR digital filter (with transfer function $H(z)$) in Problem 5.3 into a a bandpass filter with a 3-dB upper cutoff $\omega_u = 3\pi/5$ rad/s and a lower cutoff $\omega_l = 2\pi/5$ rad/s. Compute and plot its frequency response.

5.5   Using the bilinear transformation, design a notch IIR digital filter to attenuate a 60 Hz signal with a signal sampled at 4 ms. Use the second-order Butterworth LPF digital filter prototype in Equation (5.71) and apply the bandstop transformation:

$$s = A \frac{1 - z^{-2}}{1 - Bz^{-1} + z^{-2}},$$

where,

$$A = \tan\left[\pi\left(\frac{F_u}{f_s} - \frac{F_l}{f_s}\right)\right],$$

$$B = \frac{\cos\left[\pi\left(\frac{F_u}{f_s} - \frac{F_l}{f_s}\right)\right]}{\cos\left[\pi\left(\frac{F_u}{f_s} + \frac{F_l}{f_s}\right)\right]}.$$

Plot its frequency response. Comment on your findings.

5.6   Write a MATLAB code to design 2-D fan filters (based on Equations (5.100 and 5.101)) using Hamming and Blackman windows. Compare your designs via those designed using rectangular windows. Use your codes to reject reflections and pass ground-roll noise of a seismic data of your choice. Comment on your results.

Part III
Statistical Digital Signal Processing for Seismic Data

6

Fundamentals of Digital Optimal Filtering

6.1 Introduction

In reflection seismology, the spectrum of noise may cover the same frequency range as the seismic signals, i.e., both the seismic signal and unwanted energy do not, in this case, have different frequency content. The use of conventional linear frequency filters like band-pass or low-pass filters may not be the best choice. Such filters may extract frequency components of large SNR when compared to other frequency ranges. On the other hand, the SNR may differ as well for different frequencies, when seismic and noise frequency ranges are similar. Furthermore, when using low-pass or band-pass filters to attenuate unwanted seismic energy from the seismic data records, the seismic vertical resolution will be affected, i.e., the high frequency components of seismic waves are attenuated. Therefore, one needs to obtain a filter that does the best job of converting an input into a desired output. Optimum filters, or Wiener filters, are considered to be that type of filters. They can be used to separate seismic signals and noise when, for example, both possess the same frequency ranges but under the following conditions:

1. The seismic signal and noise must differ in some way, such as differences in their frequency or wavenumber spectra, velocity, and so on.
2. Certain signal and noise parameters must be known.

Multiples or ghost noise types can greatly be attenuated when using predication error filters: an application of Wiener optimum filtering processes. The seismic trace (vertical) resolution, which might be lost due to previous linear filtering processes, can be improved in order to better view and recognize important subsurface structures such as thin layers and faults. This can be done via the so-called *spiking* deconvolution, which is another application of Wiener optimum filtering processes. In this chapter, the basic fundamentals of optimum filtering will be provided. Also, different applications of this important theory in seismic data processing will be highlighted.

$$d[n]$$

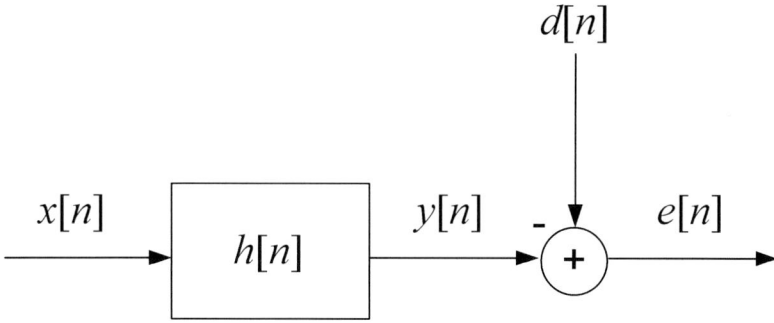

Figure 6.1 The Wiener optimum filter model.

6.2 The Wiener Optimum Filter

The most important approach to designing optimum filters is based on the work done by Wiener in 1949 (Robinson, 1967; Robinson and Treitel, 2000). Consider the block diagram in Figure 6.1, where $x[n]$ is a random process (signal) input, $h[n]$ is the required optimum filter set of coefficients that formulates a stationary random process, $d[n]$ is the desired (signal) random process, and $e[n]$ is the resulting error between the filter output $y[n]$ and $d[n]$. It is aimed to filter $x[n]$ to make $y[n]$ look as close as possible to $d[n]$ in a least square sense (Moon and Stirling, 2000; Robinson and Treitel, 2000). That is, a set of filter coefficients $h[n]$ will be found that minimizes the following objective function:

$$J = E\{e^2[n]\}, \tag{6.1}$$

where $E\{.\}$ is the expected value operator. Without loss of generality, assume that $h[n]$ is a causal Finite Impulse Response (FIR) filter, where:

$$y[n] = \sum_{k=0}^{N-1} h[k]x[n-k]. \tag{6.2}$$

Then,

$$e[n] = d[n] - y[n], \tag{6.3}$$

$$= d[n] - \sum_{k=0}^{N-1} h[k]x[n-k],$$

which implies from Equation (6.1) that the objective function J is:

$$J = E\left\{ \left(d[n] - \sum_{k=0}^{N-1} h[k]x[n-k] \right)^2 \right\}. \tag{6.4}$$

Hence, the optimum causal FIR filter coefficients can be obtained by minimizing Equation (6.4) in the least square sense. That is, by taking the partial derivative of J with respect to $h[n]$ as follows:

$$\frac{\partial J}{\partial h} = 2E\left\{e[n]\frac{\partial e[n]}{\partial h[m]}\right\} = 0, \text{ for } m = 0, \dots, N-1. \tag{6.5}$$

However,

$$\frac{\partial e[n]}{\partial h[m]} = \frac{\partial}{\partial h[m]}(d[m] - h[m]x[n-m]) = -x[n-m]. \tag{6.6}$$

Substituting Equation (6.6) into Equation (6.5) and, since the expected value operator is linear, should yield:

$$E\{d[n]x[n-m]\} - E\left\{x[n-m]\sum_{k=0}^{N-1}h[k]x[n-k]\right\} = 0. \tag{6.7}$$

The first term of Equation (6.7) is, by definition, the crosscorrelation between $d[n]$ and $x[n]$ and is denoted by:

$$R_{xd}[m] = E\{d[n]x[n-m]\}]. \tag{6.8}$$

The second term of Equation (6.7) can be rewritten as:

$$E\{x[n-m]\sum_{k=0}^{N-1}h[k]x[n-k]\} = E\left\{\sum_{k=0}^{N-1}h[k]x[n-m]x[n-k]\right\}, \tag{6.9}$$

$$= \sum_{k=0}^{N-1}h[k]E\{x[n-m]x[n-k]\},$$

$$= \sum_{k=0}^{N-1}h[k]R_{xx}[m-k],$$

where $R_{xx}[.]$ is the autocorrelation function of $x[n]$. Hence, substituting Equations (6.8 and 6.9) into Equation (6.7) results in the following relation:

$$\sum_{k=0}^{N-1}h[k]R_{xx}[m-k] = R_{xd}[m], \text{ for } m = 0, 1, \dots, N-1. \tag{6.10}$$

Note that Equation (6.10) formulates a system of N linear equations. Therefore, a more attractive form of Equation (6.10) is in a matrix form. That is:

$$
\begin{pmatrix}
R_{xx}[0] & R_{xx}[-1] & \cdots & R_{xx}[1-N] \\
R_{xx}[1] & R_{xx}[0] & \cdots & R_{xx}[2-N] \\
\vdots & \vdots & \ddots & \vdots \\
R_{xx}[N-1] & R_{xx}[N-2] & \cdots & R_{xx}[0]
\end{pmatrix}
\begin{pmatrix}
h[0] \\
h[1] \\
\vdots \\
h[N-1]
\end{pmatrix}
=
\begin{pmatrix}
R_{xd}[0] \\
R_{xd}[1] \\
\vdots \\
R_{xd}[N-1]
\end{pmatrix}.
$$

$$(6.11)$$

However, the autocorrelation function is of even symmetry, i.e., $R_{xx}[-k] = R_{xx}[k]$, since all the considered signals here are real random processes. Therefore, Equation (6.11) can be rewritten as:

$$
\begin{pmatrix}
R_{xx}[0] & R_{xx}[1] & \cdots & R_{xx}[N-1] \\
R_{xx}[1] & R_{xx}[0] & \cdots & R_{xx}[N-2] \\
\vdots & \vdots & \ddots & \vdots \\
R_{xx}[N-1] & R_{xx}[N-2] & \cdots & R_{xx}[0]
\end{pmatrix}
\begin{pmatrix}
h[0] \\
h[1] \\
\vdots \\
h[N-1]
\end{pmatrix}
=
\begin{pmatrix}
R_{xd}[0] \\
R_{xd}[1] \\
\vdots \\
R_{xd}[N-1]
\end{pmatrix}.
$$

$$(6.12)$$

The previous set of equations are called the *Wiener-Hopf* normal equations. So in order to obtain the optimal FIR filter coefficients, it is required to solve N simultaneous equations. In an algebraic form, Equation (6.12) can be written as:

$$\mathbf{R}_{xx}\mathbf{h}_{opt} = \mathbf{R}_{xd}, \tag{6.13}$$

where $\mathbf{R}_{xx}$ is the autocorrelation matrix of $x[n]$ with size $N \times N$, $\mathbf{R}_{xd}$ is the crosscorrelation vector between $x[n]$ and $d[n]$ of size $N \times 1$, and $\mathbf{h}_{opt}$ is the optimal filter vector of size $N \times 1$. In this case, $\mathbf{h}_{opt}$ can be obtained by first computing the inverse of $\mathbf{R}_{xx}$ and then multiplying the result by $\mathbf{R}_{xd}$, that is:

$$\mathbf{h}_{opt} = \mathbf{R}_{xx}^{-1}\mathbf{R}_{xd}, \tag{6.14}$$

under the condition that $\mathbf{R}_{xx}$ is non-singular. A few remarks to be mentioned for the optimum filters designed using Equation (6.14) are as follows:

1. To obtain the optimum filters based on Figure 6.1 where $h_{opt}[n]$ is assumed to be a causal FIR filter, one then wants to:

   (a) Compute $\mathbf{R}_{xx}$ and $\mathbf{R}_{xd}$ given in Equation (6.12).
   (b) Compute the inverse of $\mathbf{R}_{xx}$.
   (c) Solve for $\mathbf{h}_{opt}$ using Equation (6.14).

2. The set of obtained optimal N-length filter coefficients in Equation (6.14) is optimal in the sense where no other causal N-length FIR filter exists for which the error energy is smaller. To further explain this, reconsider Equation (6.1), which can be expressed in the following manner for a causal N-length FIR filter:

$$J = E\{e^2[n]\}, \qquad (6.15)$$

$$= R_{dd}(0) - 2\sum_{k=0}^{N-1} h[k]R_{xd}[k] + \sum_{k=0}^{N-1}\sum_{m=0}^{N-1} h[k]h[m]R_{xx}(k-m).$$

This is a quadratic function of $h[n]$, for $k = 0, 1, \dots, N-1$, with a unique minimum. See Figure 6.2 for the case when $N = 2$.

3. The autocorrelation matrix $\mathbf{R}_{xx}$ in Equation (6.12) is symmetric and has the same elements along the diagonal from the upper left to the lower right. This type of symmetric matrix is known as a *Toeplitz* matrix.

4. Many efficient algorithms exist to solve Equation (6.12) recursively, such as the Levinson algorithm (Levinson, 1947), iteratively using gradient algorithms and/or in the batch mode using the inverse of $\mathbf{R}_{xx}$ and then Equation (6.14), since $\mathbf{R}_{xx}$ is a Toeplitz matrix.

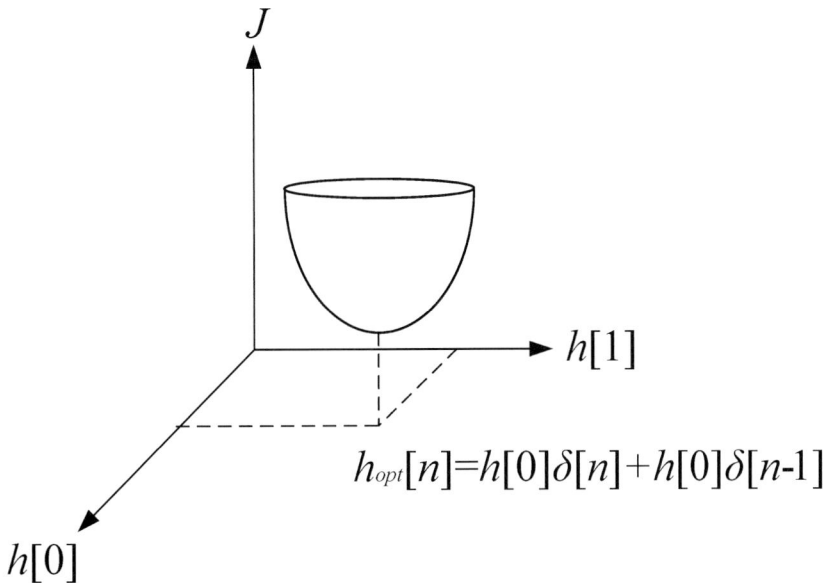

Figure 6.2 The Wiener filter solution $h_{opt}[n]$ is optimal in the least square sense. Here is a paraboloid showing the solution for $N = 2$ based on the quadratic Equation (6.15).

5. The Wiener-Hopf system of equations in Equation (6.12) is called the normal system of equations since the error signal $e[n]$ in Figure 6.1 when expressed in the following vector form:

$$\mathbf{e} = \begin{pmatrix} e[0] \\ e[1] \\ \vdots \\ e[N-1] \end{pmatrix} = \begin{pmatrix} d[0] - y[0] \\ d[1] - y[1] \\ \vdots \\ d[N-1] - y[N-1] \end{pmatrix} = \mathbf{d} - \mathbf{y}, \qquad (6.16)$$

will be perpendicular or normal to the vector containing the values of the filter output $y[n]$.

6. Sometimes $\mathbf{R}_{xx}^{-1}$ may not exist because $\mathbf{R}_{xx}$ is singular. This is due to the fact that some $\mathbf{R}_{xx}$ matrix eigenvalues are zeros. Hence, in practice, one needs to add a small amount of white Gaussian noise to $\mathbf{R}_{xx}$, which ultimately makes the eigenvalues nonzero. Then, $\mathbf{R}_{xx}^{-1}$ will exist and can then be computed.

Example 6.1 Consider Figure 6.1, where the input signal $x[n] = 2\delta[n] + \delta[n-1]$ and the desired signal $d[n] = \delta[n]$. Assuming that the optimum filter to be designed is a causal FIR filter, then:

(a) Obtain the set of optimum filter coefficients $h[n]$ for $N = 2$.
(b) Compute the filter output $y[n]$.
(c) Compute the error signal $e[n]$.
(d) Show that the output signal and the error signal are normal.

Solution: (a) In order to obtain $h_{opt}[n]$ for $N = 2$ and the given values of $x[n]$ and $d[n]$, one needs first to compute $\mathbf{R}_{xx}$ and $\mathbf{R}_{xd}$ based on Equation (6.12). That is:

$$\begin{pmatrix} R_{xx}[0] & R_{xx}[1] \\ R_{xx}[1] & R_{xx}[0] \end{pmatrix} \begin{pmatrix} h[0] \\ h[1] \end{pmatrix} = \begin{pmatrix} R_{xd}[0] \\ R_{xd}[1] \end{pmatrix}. \qquad (6.17)$$

To find $R_{xx}[n]$, one can use:

$$R_{xx}[n] = E\{x[x]x[n+m]\}, \qquad (6.18)$$

$$= \sum_{n=0}^{N_x-1} x[x]x[n+m], \text{ for } m = 0, 1, \ldots, N,$$

where N_x is the length of the sequence $x[n]$. Similarly, $R_{xd}[n]$ can be computed using

$$R_{xd}[n] = E\{x[x]d[n+m]\}, \qquad (6.19)$$

$$= \sum_{n=0}^{N_{x,d}-1} x[x]d[n+m], \text{ for } m = 0, 1, \ldots, N,$$

where $N_{x,d}$ is equal to the largest length between $x[n]$ and $d[n]$. Hence, by using Equations (6.19 and 6.20) to compute R_{xx} and R_{xd}, respectively, the filter design problem is given by:

$$\begin{pmatrix} 5 & 2 \\ 2 & 5 \end{pmatrix} \begin{pmatrix} h[0] \\ h[1] \end{pmatrix} = \begin{pmatrix} 2 \\ 0 \end{pmatrix}. \tag{6.20}$$

Therefore, by solving Equation (6.20), the vector $\mathbf{h}_{opt}$ is equal to:

$$\mathbf{h}_{opt} = \begin{pmatrix} 10/21 \\ -4/21 \end{pmatrix}, \tag{6.21}$$

or $h_{opt}[n] = \frac{10}{21}\delta[n] - \frac{4}{21}\delta[n-1]$.

(b) The output $y[n]$ is simply the convolution between $x[n]$ and the causal FIR filter $h_{opt}[n]$, where:

$$y[n] = \left(\frac{10}{21}\delta[n] - \frac{4}{21}\delta[n-1] \right) * \left(\frac{10}{21}\delta[n] - \frac{4}{21}\delta[n-1] \right), \tag{6.22}$$

$$= \frac{1}{21}(20\delta[n] + 2\delta[n-1] - 4\delta[n-2]).$$

(c) The error $e[n]$ is equal to:

$$e[n] = d[n] - y[n] = \frac{1}{21}(\delta[n] - 2\delta[n-1] + 4\delta[n-2]). \tag{6.23}$$

(d) To show that $y[n]$ is normal to $e[n]$, one can use the dot product between the vectors representing them, i.e.,

$$\mathbf{e}^T \cdot \mathbf{y} = \frac{1}{21} \begin{pmatrix} 1 & -2 & 4 \end{pmatrix} \begin{pmatrix} 20 \\ 2 \\ -4 \end{pmatrix} = 0. \tag{6.24}$$

Hence, $\mathbf{e} \perp \mathbf{y}$ and, hence, $y[n]$ and $e[n]$ are normal to each other.

For seismic applications, when practically designing optimum filters, it is required to specify the desired signal, the filter length, the filter type, and the amount of white Gaussian noise to be added during the filtering process. More details will be given in the remaining chapters of this book.

6.3 Application of Optimum Filters to Reflection Seismology

The Earth works as a filter to seismic energy signals, where, in reality, one will not only record primary seismic reflections but also will record other types of energy signals. Examples include multiples, diffractions, surface waves, scattered waves from near-surface irregularities, reflected refractions, and many others (refer to Chapter 2). These are modified by filtering because of absorption and other causes. Random noise such as ambient or wind noise are also superimposed on primary seismic reflection signals. In addition to the seismic source effect $s_s[n]$, the used recorder impulse response $r_g[n]$, and the A/D convertor impulse response $g_{A/D}[n]$ (refer to Equation (1.7), Chapter 1), in general, some major zones affecting the recorded seismic signals exist. These are listed as follows:

1. The zone near the source, where the stresses and absorption of energy, especially at higher frequencies, are often extreme. The impulse response of this zone can be denoted $s[n]$.
2. The sequence of reflectors with impulse response $r[n]$, which is the signal where seismic data processing is intended to ultimately obtain.
3. The near-surface zone, which has a disproportionate effect in modifying the recorded seismic signals. Its impulse response can be denoted $n_s[n]$.
4. Additional modifying effects due to absorption, wave mode conversion, multiples, diffractions, etc. can all be combined by the impulse response $p[n]$.

Combining these effects, the obtained recorded seismic signal can be given as the following convolutional model:

$$g[n] = r[n] * s_s[n] * s[n] * n_s[n] * p[n] * r_g[n] * g_{A/D}[n]. \tag{6.25}$$

In other words, a seismic trace signal $g[n]$ can be thought of as a series of convolutions. The convolution model is central to most seismic data processing. In a more compact form, Equation (6.25) can be written as:

$$g[n] = r[n] * w[n], \tag{6.26}$$

where,

$$w[n] = s_s[n] * s[n] * n_s[n] * p[n] * r_g[n] * g_{A/D}[n]. \tag{6.27}$$

$w[n]$ is called the embedded or equivalent wavelet, which would be affected from a single isolated interface. Wavelets will be discussed in more detail in Chapter 8. The convolution model in Equation (6.26) often includes additive noise $n_r[n]$, which is usually but not necessarily assumed to be random. Therefore, the convolution model of a noisy seismic trace is modeled as follows:

$$g[n] = r[n] * w[n] + n_r[n]. \tag{6.28}$$

Figure 6.3 shows a block diagram for the seismic convolution model of Equation (6.28).

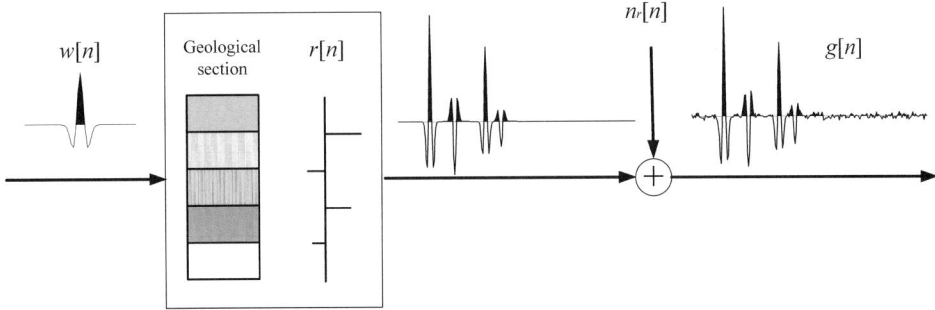

Figure 6.3 Convolution seismic data model with noise. In principle, a seismic wavelet $w[n]$, representing the major zones affecting a recorded seismic signal (Equation (6.27)), which is convolved with the reflectivity function $r[n]$, to yield a seismic trace. This result is added with noise $n_r[n]$.

Many useful applications of Wiener optimum filters exist in seismic reflection surveying. Those applications may be classified into three main categories:

1. Seismic deconvolution, inverse filtering or sometimes called equalization. This is a process that is designed to restore a waveshape to the form it had before it underwent a linear filtering action, as in Equation (6.28). The main objective of this process is to ultimately improve the recognizability and resolution of seismic reflections. It may have the following forms:

   (a) System deconvolution by which the filtering effects of the recording system are removed.
   (b) Whitening, sometimes called equalization, by which all the frequency components within a pass-band range become of equal amplitude. This is also known as spiking.
   (c) Wavelet processing as an attempt to determine an estimate of the embedded seismic wavelet shape $w[n]$.
   (d) Reverberation for marine seismic data by which the filtering action of a water layer is removed.

   So in the context of Figure 6.1, the Wiener optimum filtering problem for seismic deconvolution can, theoretically, be described by Figure 6.4a.
2. M-step linear predication. This application of Wiener optimum filters is used in seismic reflection, mainly to attenuate multiples that involve the surface or near-surface reflectors. In this case, Figure 6.1 becomes the Wiener filtering problem described in Figure 6.4b.
3. Noise cancelation, which is mainly used in seismic reflection, to attenuate, for example, ambient noise, power transmission line harmonics noise, wind noise,

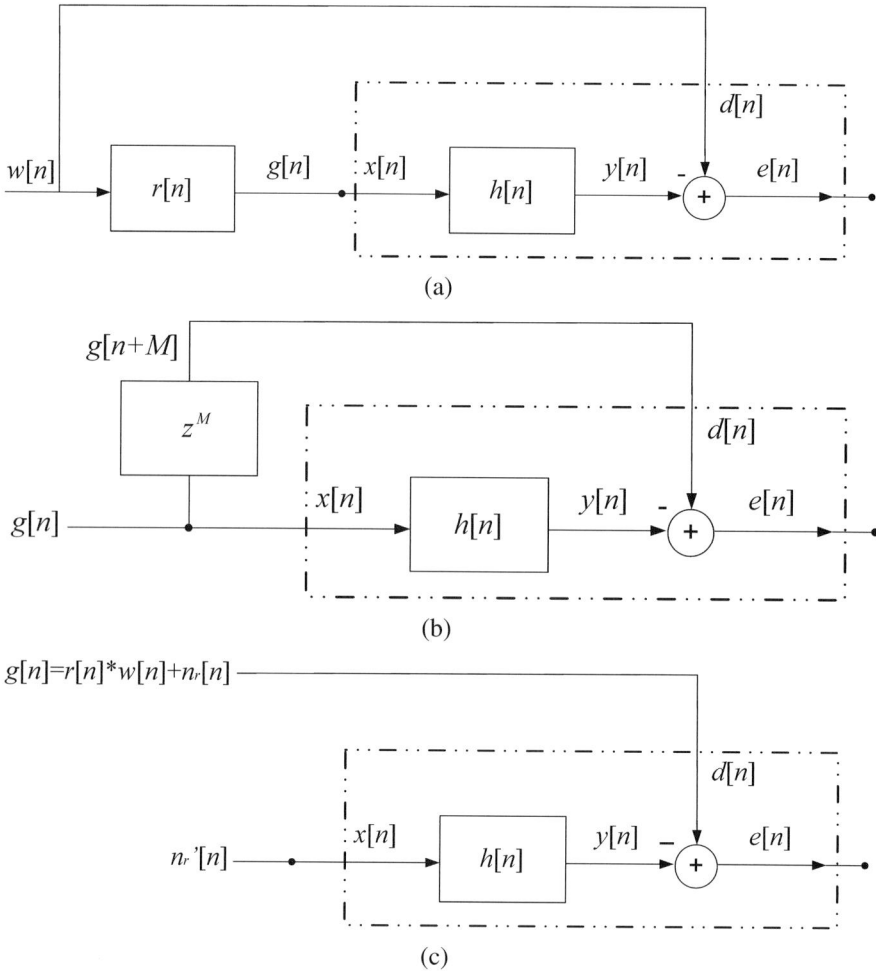

(a)

(b)

(c)

Figure 6.4 Block diagrams of various seismic data processing applications of the Wiener optimal filtering model in Figure 6.1 (inside the dash double-dotted box): (a) deconvolution, (b) linear predication, and (c) noise cancellation.

or even sometimes ground-roll noise attenuation. The Wiener filtering noise cancellation problem turns to be that described in Figure 6.4c, where it is assumed that the noise $n_r[n]$ is statistically correlated with $n'_r[n]$.

Example 6.2 Consider the block diagram in Figure 6.4c, where $x[n] = \sin[n\pi/4] + \sin[n\pi/8]$ and $d[n] = 2\sin[n\pi/4]$. Determine the following:

(a) The autocorrelation function $R_{xx}[m]$.
(b) The crosscorrelation function $R_{xd}[m]$.
(c) The optimum filter coefficients $h_{opt}[n]$ for $N = 25$, where $h[n]$ is an FIR filter. Comment on the frequency response of the designed optimum filter.

Solution: (a) The autocorrelation function of $x[n]$ $R_{xx}[n]$ can be calculated as follows:

$$R_{xx}[m] = E\{x[n]x[n+m]\}, \tag{6.29}$$

$$= E\left\{\left(\sin\left[\frac{n\pi}{4}\right] + \sin\left[\frac{n\pi}{8}\right]\right)\left(\sin\left[\frac{(n+m)\pi}{4}\right] + \sin\left[\frac{(n+m)\pi}{8}\right]\right)\right\},$$

$$= \frac{1}{2}\left(\cos\left[\frac{m\pi}{4}\right] + \cos\left[\frac{m\pi}{4}\right]\right).$$

(b) The crosscorrelation function between $x[n]$ and $d[n]$ $R_{xd}[n]$ can be calculated as follows:

$$R_{xd}[m] = E\{x[n]d[n+m]\}, \tag{6.30}$$

$$= E\left\{\left(\sin\left[\frac{n\pi}{4}\right] + \sin\left[\frac{n\pi}{8}\right]\right)\left(2\sin\left[\frac{(n+m)\pi}{4}\right]\right)\right\},$$

$$= \cos\left[\frac{m\pi}{4}\right].$$

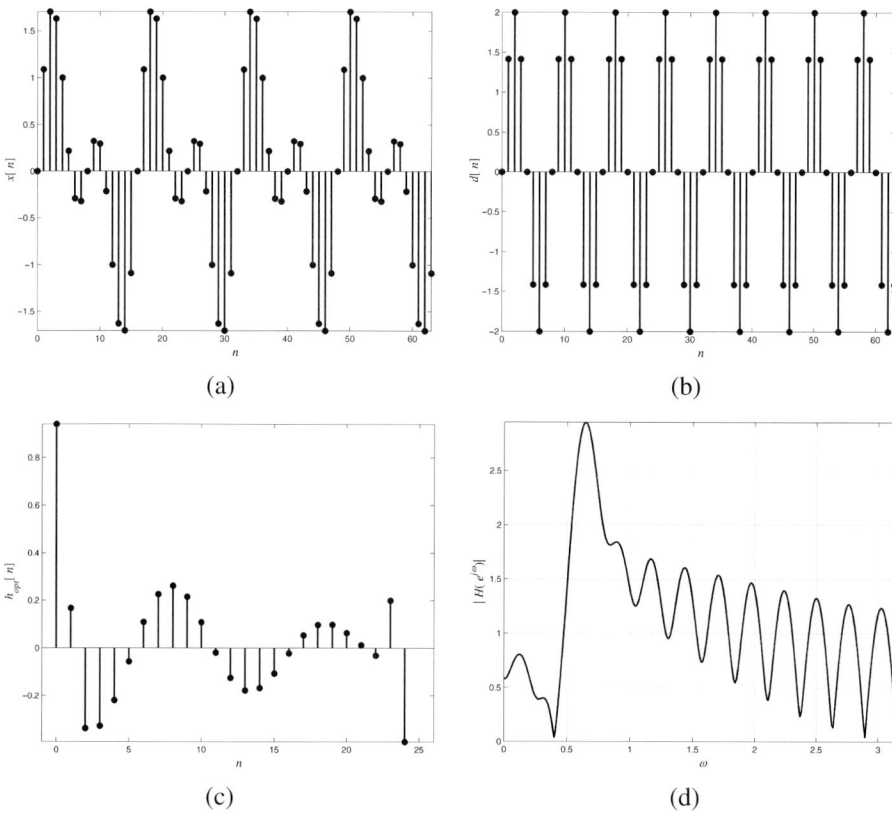

Figure 6.5 Example 6.2 for 64 discrete time samples with $N = 25$. (a) $x[n]$, (b) $d[n]$, (c) $h_{opt}[n]$, and (d) the magnitude response $H(e^{j\omega})$ of $h_{opt}[n]$.

(c) Figure 6.5a–c show (for 64 discrete time samples) $x[n]$, $d[n]$, $h_{opt}[n]$, respectively. Also, Figure 6.5d shows $h_{opt}[n]$ magnitude response $H(e^{j\omega})$, where the filter will mainly retain the frequency component of $\omega = \frac{\pi}{4}$.

6.4 Summary

This chapter introduced Wiener optimal filters for causal FIR systems, where it was shown that such filters are optimal in the least square sense. It also highlighted many usages of the Wiener optimal filter model for seismic data processing. Applications include seismic deconvolution, linear prediction filtering, and noise cancellation.

Exercises

6.1 Based on the Wiener filter model in Figure 6.1, derive the optimal Wiener filters for non-causal FIR systems. How does the result differ from Equation (6.10)?

6.2 If the input to a causal FIR filter is $x[n] = 2\delta[n] + \delta[n-1]$, the desired signal is $d[n] = \delta[n-1] + 2\delta[n-2] + \delta[n-3]$, then determine the optimal filter coefficients $h[n]$ for $N = 3$. Calculate and plot the frequency response of $h[n]$.

6.3 Given the input signal $x[n] = \delta[n] - \frac{1}{2}\delta[n-1]$ and a desired signal $d[n] = \delta[n]$ to a the Wiener filter model in Figure 6.1.

(a) Sketch the input and desired signals.
(b) Design a causal FIR filter of first order and second order.
(c) Convolve the obtained filters with the input signal. Sketch the result of the convolution.
(d) Compute the energy of the error signal $e[n]$ for both cases.

6.4 Repeat Problem 6.3 but for $d[n] = \delta[n-1]$. Comment on the results as compared to Problem 6.3.

6.5 Write a MATLAB script for designing N-length Wiener optimal FIR filters based on Equation (6.14) for any input signal $x[n]$ and desired signal $d[n]$. Validate your written script by repeating Problems 6.2 and 6.3 using MATLAB.

7

Seismic Deconvolution

7.1 Introduction

Physically, the Earth is composed of layers of rocks with different lithology and physical properties, where, in seismic studies, rock layers are defined by the densities and velocities with which seismic waves propagate through them. The product of density and velocity, as explained in Section 2.8, yields the seismic acoustic impedance. The impedance contrast between adjacent rock layers causes the reflections that are recorded along a seismic surface profile, and the recorded seismic trace, are modeled using the convolution model of the Earth's impulse response $r[n]$ with the seismic wavelet $w[n]$. Recall that $w[n]$ has many components, including: the source signature, the recording filter, the receiver array response, etc. The Earth's impulse response $r[n]$ is what would be recorded if the wavelet was just a simple unit impulse $w[n] = \delta[n]$, known in the geophysics community as a spike. One can basically define deconvlution as the process of undoing the convolution effects on $r[n]$ (refer to Section 6.3). In this case, seismic deconvolution compresses the basic seismic wavelet $w[n]$ in the recorded seismic trace $g[n]$ and, hence, results in increasing the temporal resolution. It is usually applied before the stacking process (refer to Figure 1.3) but it is commonly applied on stacked seismic data.

Deterministically, seismic wavelet compression can be done using an inverse filter as a deconvolution operator. An inverse filter, when convolved with the seismic wavelet, converts it into a unit impulse. And when it is applied to a seismic trace, it should yield $r[n]$. A more accurate inverse filter design is achieved using the Wiener optimum design problem discussed in Section 6.2. This only applies when the unknown source wavelet $w[n]$ is that of minimum-phase. The Wiener filter converts $w[n]$ into any desired shape. For example, much like the inverse filter, a Wiener filter can be designed to convert the wavelet into a unit sample impulse but in the least square sense. Note that converting $w[n]$ into $\delta[n]$ is like asking for a perfect resolution. In practice, because of: (a) the noise $n_r[n]$ accompanying seismic traces, and (b) the assumptions made about the seismic wavelet $w[n]$ and the recorded traces $g[n]$, spiking deconvolution is not always desirable.

Furthermore, multiples and reverberations are commonly seen coherent noise types accompanying, mainly, marine data. Multiples may also be seen in land seismic data. Improper attenuation of such noise types may result in false interpretation, where reflection ghosts can be seen on final migrated seismic images. Seismic deconvolution helps attenuate reverberations and short-period multiples and, hence, will assist in the process of yielding a representation of subsurface reflectivity function $r[n]$, which, ultimately, results in better seismic images for interpretation. Ideally, deconvolution should, in addition to compressing the wavelet components as we shall see in the next chapter, eliminate other unwanted energy such as multiples and reverberations. This would result in leaving only $r[n]$ in the seismic trance. Also, the resolution of the seismic traces can be controlled by designing a Wiener linear prediction filter. In this chapter, deterministic and statistical seismic deconvolution will be discussed, with various examples.

7.2 The Seismic Deconvolution Model

Recall the convolution model in Chapter 6 and, particularly, Equation (6.26). This equation is restated here for the sake of completeness:

$$g[n] = r[n] * w[n]. \tag{7.1}$$

If a filter, say $f[n]$, was defined such that the convolution of $f[n]$ with the (known) recorded seismic trace $g[n]$ yields an estimate of $r[n]$, then:

$$r[n] = f[n] * g[n]. \tag{7.2}$$

By substituting Equation (7.2) into Equation (7.1) one obtains:

$$g[n] = w[n] * f[n] * g[n], \tag{7.3}$$

where if $g[n]$ is eliminated from both sides of the equation, then the following will result:

$$\delta[n] = w[n] * f[n]. \tag{7.4}$$

By solving Equation (7.4) for $f[n]$, the following will be obtained:

$$f[n] = \delta[n] * \frac{1}{w[n]}, \tag{7.5}$$

where in the z-domain:

$$F(z) = \frac{1}{W(z)}, \tag{7.6}$$

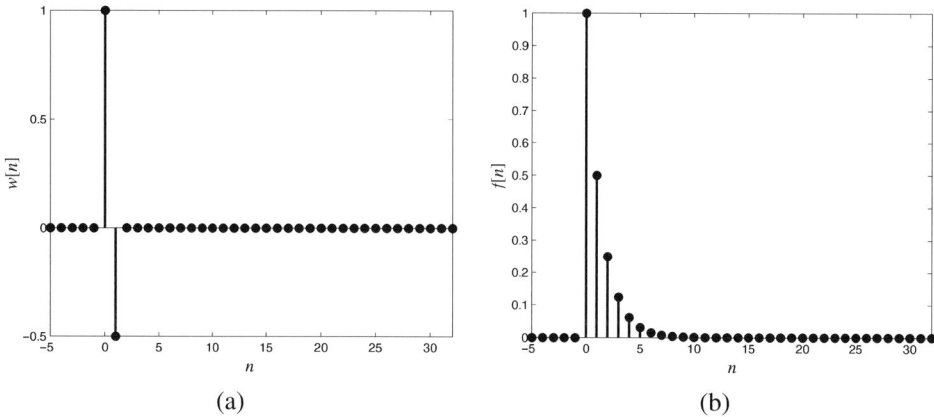

Figure 7.1 Example 7.1 for 38 samples with (a) the seismic wavelet $w[n] = \delta[n] - \frac{1}{2}\delta[n-1]$ and (b) its inverse filter $f[n](\frac{1}{2})^{n}u[n]$.

which is simply an inverse filter. Hence, the filter $f[n]$ needed to compute the Earth's impulse response $r[n]$ from $g[n]$ turns out to be the mathematical inverse of the wavelet $w[n]$. Equation (7.5) implies that $f[n]$ converts the basic wavelet $w[n]$ into a unit sample impulse at $n = 0$. Similarly, $f[n]$ converts the seismic trace $g[n]$ to a series of unit sample impulses that defines $r[n]$. This is a deterministic deconvolution approach, where it is assumed that the wavelet $w[n]$ is known. One is able to obtain $f[n]$ by using the z-transform of $w[n]$ and then applying Equation (7.6) followed by an inverse z-transform of $F(z)$. The following example illustrates this.

Example 7.1 Determine the inverse filter $f[n]$? for the seismic wavelet $w[n] = \delta[n] - \frac{1}{2}\delta[n-1]$.

Solution: Taking the z-transform of the given wavelet $w[n]$ should yield $W(z) = 1 - \frac{1}{2}z^{-1}$. And from Equation (7.6), $F(z) = \frac{1}{W(z)} = \frac{1}{1-\frac{1}{2}z^{-1}}$. Therefore, using the inverse z-transform of $F(z)$ will result in $f[n] = (\frac{1}{2})^{n}u[n]$. In this case, $f[n]$ is an exponentially decaying causal filter with infinite samples (see Figure 8.1).

From Example 7.1, $f[n]$ will have an infinite number of coefficients, although they will rapidly decay. In practice, $f[n]$ is truncated and the more included coefficients, the better results will be obtained.

Example 7.2 Based on Example 7.1, compute the resultant output of filtering the wavelet $w[n] = \delta[n] - \frac{1}{2}\delta[n-1]$, using its corresponding inverse filter $f[n]$, i.e., $f[n] * w[n]$. Also, compute the convolution with 2, 3, and 4 terms of $f[n]$ and compare your results.

Solution: The convolution in time domain is given by:

$$f[n] * w[n] = \left(\delta[n] - \frac{1}{2}\delta[n-1]\right) * \left(\left(\frac{1}{2}\right)^n u[n]\right), \qquad (7.7)$$

where in the z-transform domain,

$$F(z)W(z) = \left(\frac{1}{1 - \frac{1}{2}z^{-1}}\right)\left(1 - \frac{1}{2}z^{-1}\right), |z| > \frac{1}{2} \qquad (7.8)$$

$$= 1.$$

In this case, the inverse z-transform of $F(z)W(z)$ is $\delta[n]$. Now, when considering only the first two terms of $f[n]$, then $f_2[n] = \delta[n] + \frac{1}{2}\delta[n-1]$, the resulting convolution between $w[n]$ and $f_2[n]$ can be shown to be equal to $\delta[n] - \frac{1}{4}\delta[n-2]$. Following the same methodology, for three terms, $f_3[n] = \delta[n] + \frac{1}{2}\delta[n-1] + \frac{1}{4}\delta[n-2]$ and the $f_3[n] * w[n] = \delta[n] - \frac{1}{8}\delta[n-3]$. Finally, for four terms, one can show that $f_3[n] * w[n] = \delta[n] - \frac{1}{16}\delta[n-4]$. So as $f[n]$ is approximated with more terms, the effect of the wavelet will be attenuated. Figure 7.2 illustrates this.

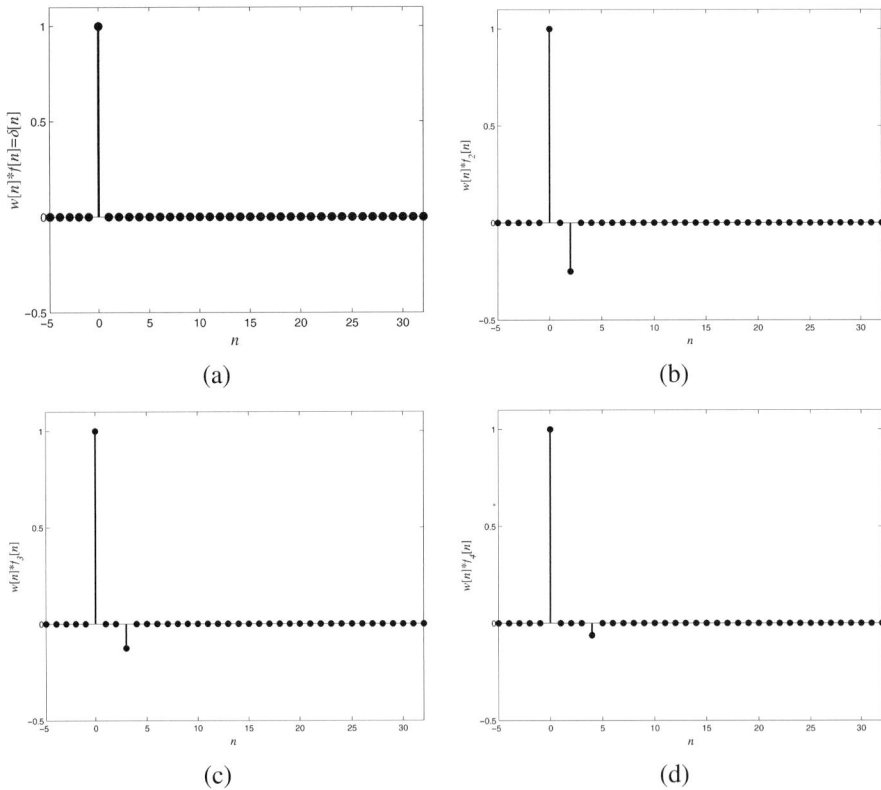

Figure 7.2 Example 7.2 for 38 samples, where (a) is $w[n] * f[n]$, (b) $w[n] * f_2[n]$, (c) $w[n] * f_3[n]$ and (d) $w[n] * f_4[n]$. As more terms are included in the approximation of $f[n]$, the better deconvolution will be obtained.

7.3 Seismic Deconvolution Based on Wiener Optimum Filtering

One can use, instead, Wiener optimum filtering to obtain the inverse filter, since in reality $w[n]$ is usually unknown and is approximated/estimated. Given an input wavelet $w[n]$ (or its estimate), one is required to determine an N-length causal FIR filter such that the error between the actual output and the desired output $d[n] = \delta[n]$ is minimum in the least square sense. The Wiener filter applies to a large class of problems in which any desired output $d[n]$ can be considered, which include:

1. A unit sample impulse known as a zero-lag spike $\delta[n]$.
2. A delayed unit sample impulse (k-lag spike) $\delta[n - k]$.
3. A time-advanced form of the input signal.
4. A zero-phase wavelet.
5. Any desired arbitrary shape.

Note that spiking deconvolution and linear prediction (which address points 1–3) will be discussed in the following two subsections. The remainder will be discussed in Chapter 8.

7.3.1 Spiking Deconvolution

When a zero-lag spike $\delta[n]$ is used as $d[n]$, the Wiener filtering process is called *Spiking Deconvolution*. Assume that the autocorrelation function $R_{xx}[n]$ of the seismic trace $g[n]$ is a scaled version of the input wavelet $w[n]$. Also, assume that $r[n]$ is random and, hence, its autocorrelation is a unit sample function. Then, when $d[n] = \delta[n]$ is crosscorrelated with $g[n]$, $\mathbf{R}_{xd}$ in Equation (6.13) becomes:

$$\mathbf{R}_{xd} = \begin{pmatrix} g[0] \\ 0 \\ \vdots \\ 0 \end{pmatrix}. \tag{7.9}$$

Hence, the normal equations in the matrix form becomes:

$$\begin{pmatrix} R_{xx}[0] & R_{xx}[1] & \cdots & R_{xx}[N-1] \\ R_{xx}[1] & R_{xx}[0] & \cdots & R_{xx}[N-2] \\ \vdots & \vdots & \ddots & \vdots \\ R_{xx}[N-1] & R_{xx}[N-2] & \cdots & R_{xx}[0] \end{pmatrix} \begin{pmatrix} h_{opt}[0] \\ h_{opt}[1] \\ \vdots \\ h_{opt}[N-1] \end{pmatrix} = \begin{pmatrix} g[0] \\ 0 \\ \vdots \\ 0 \end{pmatrix}. \tag{7.10}$$

Therefore, it is not required, for spiking deconvolution, to compute the crosscorrelation vector $\mathbf{R}_{xd}$. Rather, it is necessary to construct a zero vector of length

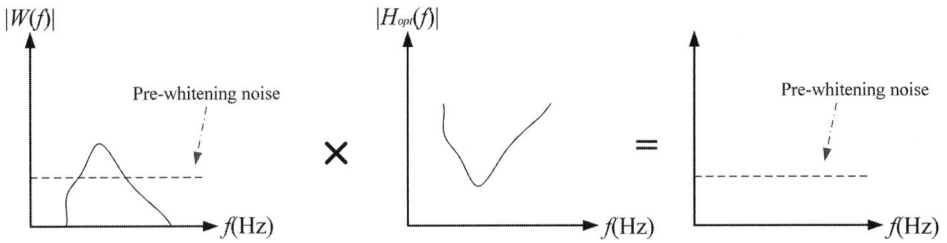

Figure 7.3 This figure illustrates the concept of spike deconvolution in the frequency domain, where the remaining pre-whitening noise in the frequency domain represents a unit sample impulse $\delta[n]$ in the time domain.

N except for the first entry to be equal to the seismic trace amplitude value at 0 sample ($g[0]$). Note that if the wavelet $w[n]$ is not minimum-phase (this will be discussed in Chapter 8), then the spiking deconvolution cannot convert it to a perfect zero-lag spike. Additionally, sometimes to ensure numerical stability of the Wiener filtering process, an artificial level of white Gaussian noise is added to the amplitude spectrum of the input seismic trace $g[n]$ before performing deconvolution. This is called *pre-whitening* (Yilmaz, 2001). Figure 7.3 graphically explains the effect of spiking deconvolution and pre-whitening. Figure 7.4a shows a seismic trace before and after performing spiking deconvolution. Clearly, from Figure 7.4b, the Power Spectral Density (PSD) of the deconvolved seismic signal has expanded its bandwidth (become more white). Also, Figure 7.5a and 7.5b shows an example from Mousa and Al-Shuhail (2011) of a real seismic data before and after applying spiking deconvolution, where the data became more spiky after applying the deconvolution. The PSD of the average seismic trace of the data in both Figure 7.5a and 7.5b shows an improvement in the bandwidth of the data after deconvolution (Figure 7.5c).

7.3.2 *The Linear Prediction Filter*

The linear prediction filter is an application of Wiener optimum filters that is used mainly to attenuate multiples that involve the surface or near-surface reflectors. Consider the input $g[n]$, where one wants to predict its value at some future time index $(n + M)$, M to be a predication lag, as seen in Figure 7.6. Based on this, $d[n] = g[n + M]$, where one can show that the crosscorrelation vector $\mathbf{R}_{xd}$ will be:

$$\mathbf{R}_{xd} = \begin{pmatrix} R_{xx}[M] \\ R_{xx}[M+1] \\ \vdots \\ R_{xx}[M+N-1] \end{pmatrix}. \tag{7.11}$$

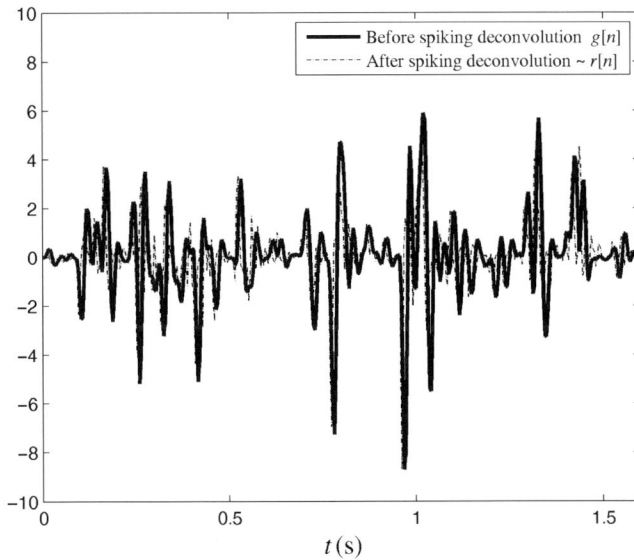

(a)

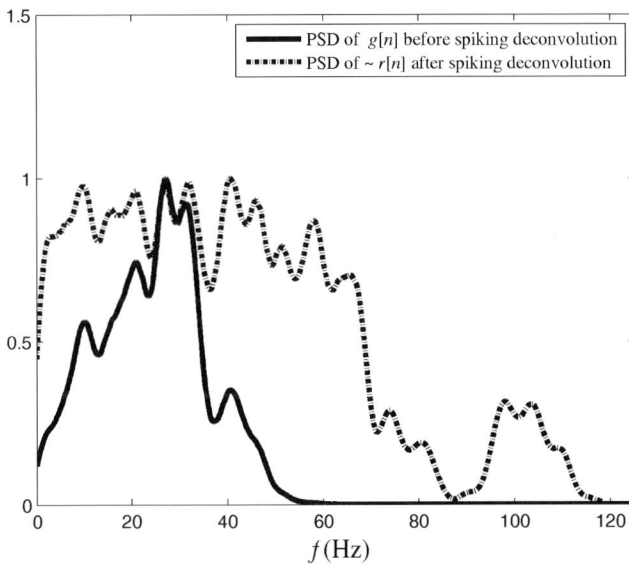

(b)

Figure 7.4 A real seismic trace signal (a) before and after spiking deconvolution and (b) its PSD. The deconvolved seismic trace signal is more spiky than the one before the deconvolution and the PSD become more spread/white.

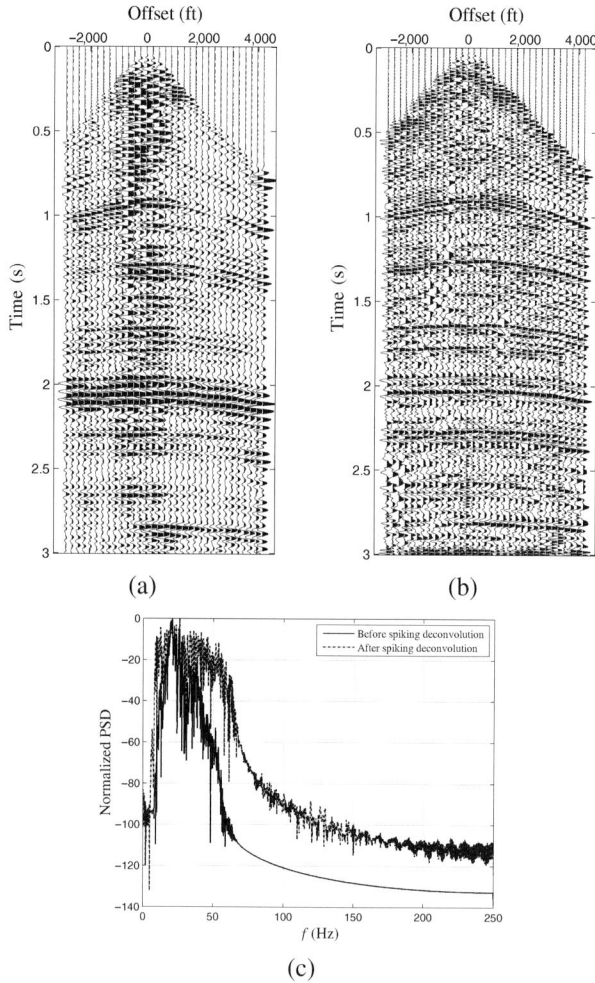

Figure 7.5 A real seismic data shot gather: (a) before and (b) after applying spiking deconvolution. (c) The PSD of the average trace for the shot gathers in (a) and (b). (courtesy of Mousa and Al-Shuhail, 2011).

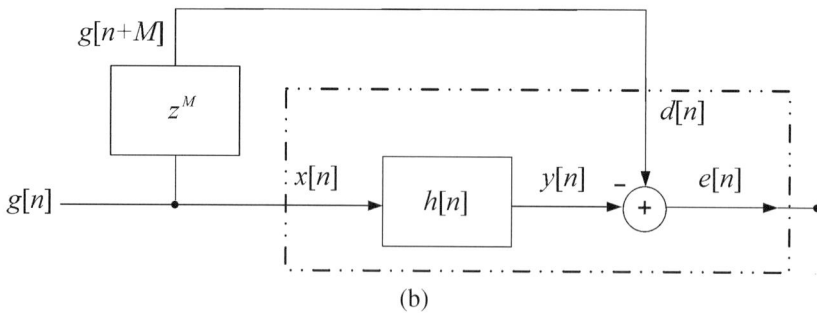

Figure 7.6 The block diagram of the linear predication deconvolution using the Wiener optimal filtering model.

In this case, the Wiener-Hope normal equations become:

$$
\begin{pmatrix}
R_{xx}[0] & R_{xx}[1] & \cdots & R_{xx}[N-1] \\
R_{xx}[1] & R_{xx}[0] & \cdots & R_{xx}[N-2] \\
\vdots & \vdots & \ddots & \vdots \\
R_{xx}[N-1] & R_{xx}[N-2] & \cdots & R_{xx}[0]
\end{pmatrix}
\begin{pmatrix}
h_{opt}[0] \\
h_{opt}[1] \\
\vdots \\
h_{opt}[N-1]
\end{pmatrix}
$$

$$
=
\begin{pmatrix}
R_{xx}[M] \\
R_{xx}[M+1] \\
\vdots \\
R_{xx}[M+N-1]
\end{pmatrix}, \quad (7.12)
$$

where it is only necessary to compute the autocorrelation sequence of the input seismic signal $g[n]$ up to $n = M + N - 1$.

Example 7.3 Consider the seismic signal sequence shown in Figure 7.7a:

$$
g[n] = 8.5\delta[n] + 4.5\delta[n-1] + 4.5\delta[n-3] + 2.25\delta[n-4].
$$

Design a linear prediction causal FIR filter of order 2 to remove the echo from the seismic signal that occurs at a prediction distance of 3 samples.

Solution: In order to obtain $h_{opt}[n]$ for a second order, i.e., $N = 3$. Note that the lag $M = 3$. Hence, $h_{opt}[n]$, based on Equation (7.12), can be computed using the following equations:

$$
\begin{pmatrix}
R_{xx}[0] & R_{xx}[1] & R_{xx}[2] \\
R_{xx}[1] & R_{xx}[0] & R_{xx}[1] \\
R_{xx}[2] & R_{xx}[1] & R_{xx}[0]
\end{pmatrix}
\begin{pmatrix}
h_{opt}[0] \\
h_{opt}[1] \\
h_{opt}[2]
\end{pmatrix}
=
\begin{pmatrix}
R_{xx}[3] \\
R_{xx}[4] \\
R_{xx}[5]
\end{pmatrix}. \quad (7.13)
$$

Hence, by using Equation (6.19) to compute R_{xx} up to $n = 5$, the filter design problem is given by:

$$
\begin{pmatrix}
117.81 & 48.37 & 38.25 \\
48.37 & 117.81 & 48.37 \\
38.25 & 48.37 & 117.81
\end{pmatrix}
\begin{pmatrix}
h_{opt}[0] \\
h_{opt}[1] \\
h_{opt}[1]
\end{pmatrix}
=
\begin{pmatrix}
48.37 \\
19.13 \\
0
\end{pmatrix}. \quad (7.14)
$$

Therefore, by solving Equation (7.14), the vector $\mathbf{h}_{opt}$ is equal to:

$$
\mathbf{h}_{opt} =
\begin{pmatrix}
0.444 \\
0.047 \\
-0.164
\end{pmatrix}, \quad (7.15)
$$

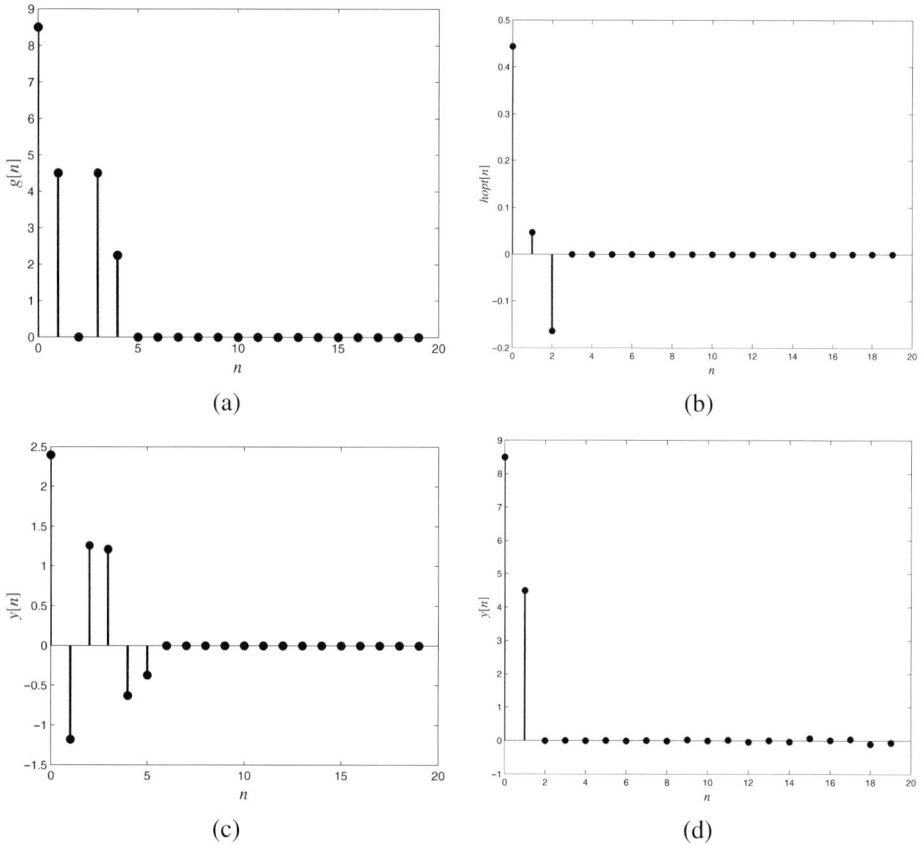

(a)

(b)

(c)

(d)

Figure 7.7 Example 7.3 for 20 samples, where (a) shows a seismic sequence with an echo. (b) The linear predication FIR filter with $N = 3$. (c) The filtered signal, where the echo still exists and other undesired samples were introduced. However, (d) when the filter length increased to be $N = 15$, the echo was greatly attenuated.

or $h_{opt}[n] = 0.444\delta[n]+0.047\delta[n-1]-0.164\delta[n-2]$ (See Figure 7.7b). The output $y[n]$, which is simply the convolution between $x[n]$ and the causal FIR filter $h_{opt}[n]$ for $N = 3$, is shown in Figure 7.7c. The filter with $N = 3$ did not completely remove the echo starting at $n = 3$. However, when $N = 15$, the echo was greatly attenuated from $g[n]$, as see in Figure 7.7d.

To show the application on real data, Figure 7.8 shows a seismic marine shot gather from Yilmaz (2001) before and after applying linear predication filtering. The data contain multiples along with reverberations. The linear prediction filters were designed with $M = 322$ and $N = 750$. Clearly, most of the multiples were attenuated.

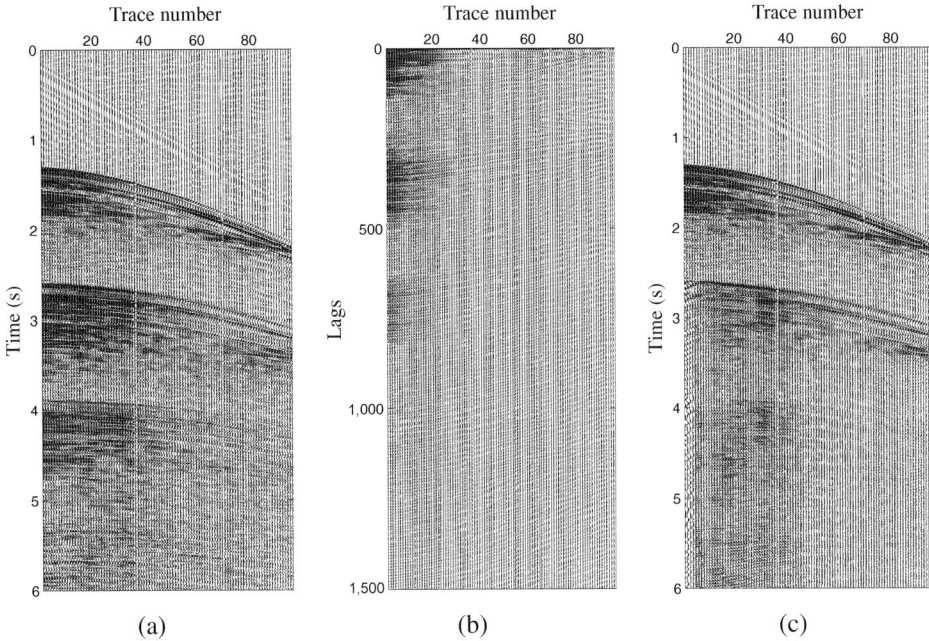

Figure 7.8 An example of real seismic data from (modified from Yilmaz, 2001) containing multiples. (a) The data before filtering. (b) The autocorrelogram of the data in (a). (c) The data after linear prediction filtering. Clearly, most of the multiples and reverberations were attenuated.

7.4 FX Deconvolution

FX deconvolution is an effective method, which attenuates random noise $n_r[n]$ (Canales, 2005) and emphasizes seismic linear events. It involves obtaining, first, the frequency space (FX) version of the seismic record. Then, for every frequency, a Wiener linear prediction optimal filter is designed and applied in the complex domain with respect to the spatial variable x. The method defines two main patterns: a predictable lateral pattern, which is considered the signal, while the unpredictable part is considered noise, which is subtracted eventually from the main data. Figure 7.9 shows the application of FX deconvolution to noisy synthetic seismic data. The noise level added to the data is 20% Gaussian noise (see Figure 7.9a). The FX deconvolved data are seen in Figure 7.9b with $SNR = 6.83 dB$. The filtered noise is seen in Figure 7.9c.

7.5 Summary

Seismic deconvolution is an important data processing step. It can be used to improve the vertical resolution of the data by spiking deconvolution. Also, it can

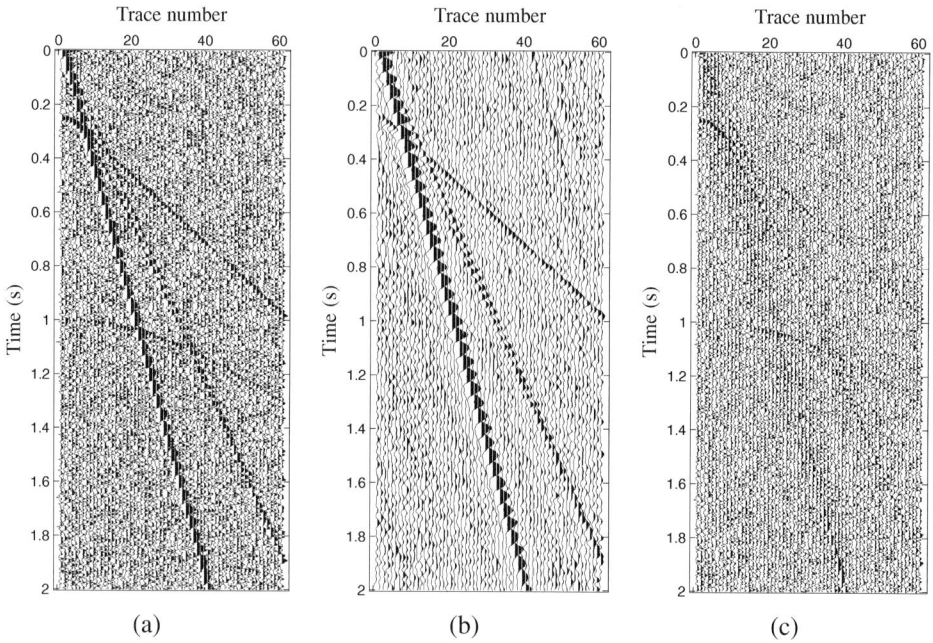

Figure 7.9 (a) An example of noisy (20% additive Gaussian noise) synthetic seismic data containing linear seismic events. (b) FX deconvolved version of (a). (c) The difference between (a) and (b).

be used to attenuate multiples using linear predication deconvolution filters. Also, it can be used to attenuate random noise and enhance seismic linear events by using linear prediction filters in the FX domain. An important assumption is that the embedded seismic wavelet must be of minimum phase in order to obtain better seismic deconvolution results.

Exercises

7.1 Compute the resultant output of filtering the following wavelets:

(a) $w[n] = \delta[n] + \frac{1}{2}\delta[n-1]$,
(b) $w[n] = -\frac{1}{2}\delta[n] + \delta[n-1]$,

using their corresponding inverse filter $f[n]$. Try 2, 3, and 4 terms of $f[n]$ and compare your results with the inverse filter with no approximation.

7.2 Let $w[n] = \delta[n] - \frac{1}{2}\delta[n-1]$.

(a) Determine the optimum filter $h_{opt}[n]$ for $N = 2$ for $d[n] = \delta[n]$.
(b) Compute the error energy.

(c) Repeat (a) and (b) but for $w[n] = -\frac{1}{2}\delta[n] + \delta[n-1]$. Compare the error energies for both filters. Comment on why one error is larger than the other.

7.3 Repeat Problem 7.2 but for $d[n] = \delta[n-1]$.

7.4 Given a recorded seismic signal $g[n] = 4\delta[n] + 2\delta[n-1] + 2\delta[n-5] + \delta[n-6]$, answer the following:

(a) Plot versus the time samples.
(b) Design a linear prediction causal FIR filter of order 2 to remove the echo from the signal that occurs at a prediction distance of 5 samples.
(c) Convolve the obtained filter with $g[n]$. Plot the result. What can you observe? What can be done to improve the resultant signal?

7.5 Let $w[n] = \delta[n] + \alpha\delta[n-1]$, where $|\alpha| < 1$. Derive the spiking deconvolution Wiener optimal filter, where the filters are assumed to be causal FIR with length equal to N.

7.6 Repeat Problem 7.5 but for $w[n] = \alpha\delta[n] + \delta[n-1]$, where $|\alpha| < 1$. Comment on your findings.

7.7 Write a MATLAB script for designing N-length linear prediction Wiener optimal FIR filters. Validate your written script by repeating Problem 7.4.

8

Seismic Wavelet Processing

8.1 Introduction

A seismic wavelet, or simply a, wavelet, is one of the basic building blocks that is used for the construction of seismic models. These models are used as a basis for various seismic signal processing algorithms. The wavelets are removed from seismic reflection data to yield the final seismic sections. The accurate estimation or measurement of seismic wavelets results in accurate seismic sections. In Chapter 6, the seismic convolutional model was studied, where any recorded seismic activity is the result of convolving a seismic wavelet $w[n]$ with the reflectivity function $r[n]$. The attenuation of such a wavelet can, for example, improve the seismic vertical (temporal) resolution by which better reflections of thin subsurface rock layers can be distinguished.

The deconvolution process that attempts to determine, control, or change the embedded wavelet shape is known as *Wavelet Processing* (Robinson and Treitel, 2000; Sheriff, 2006; Yilmaz, 2001). It involves filtering seismic data to change the shape of the underlaying wavelet in advantageous ways and can, for example, be implemented using Wiener optimum filtering. The main objective of seismic wavelet processing is to achieve some specified wavelet shape such as a zero-phase wavelet (see Section 8.2). It is mainly a two-part process:

1. The specification of a mathematical model that adequately represents the geophysical phenomenon that one is after.
2. The implementation of the numerical procedures to carry out the required filtering operation.

However, seismic wavelets are typically unknown and are assumed to be compressed in time (broadband in frequency domain) and of zero-phase. Most of the reported seismic deconvolution algorithms rely on such assumptions. The reality is that seismic wavelets are mainly dilated and of mixed-phase. Such wavelets can mislead seismic interpreters when performing seismic interpretation (Henry, 1997; Liner, 2004; Robinson and Treitel, 2008; Yilmaz, 2001). In fact, the majority of mis-tie problems between, for example, seismic data and their synthetic models are due to the remaining mixed-phase wavelets in the processed seismic data sets (Henry, 1997). This chapter aims to introduce in an illustrative manner seismic wavelets. Seismic wavelets and their various common types will be described with examples. This is followed with discussion on wavelet processing.

8.2 Seismic Wavelets

A seismic wavelet (e.g., Figure 8.1) is a waveform with the bulk of its energy confined to a finite interval on the time scale. It can be defined as an energy signal that is concentrated in a certain time interval. Therefore, an arrival time or origin time can be associated with a wavelet. The arrival time can be defined here as the reference point of the wavelet and is usually designated as time zero. Two properties thus characterize a seismic wavelet signal. The stability property of a wavelet is the first, where a wavelet is a transient phenomenon, since it has a finite energy. The arrival time property is the second property that characterizes a wavelet, since the wavelet has a definite arrival time. Three main types of wavelets exist, depending on their arrival time:

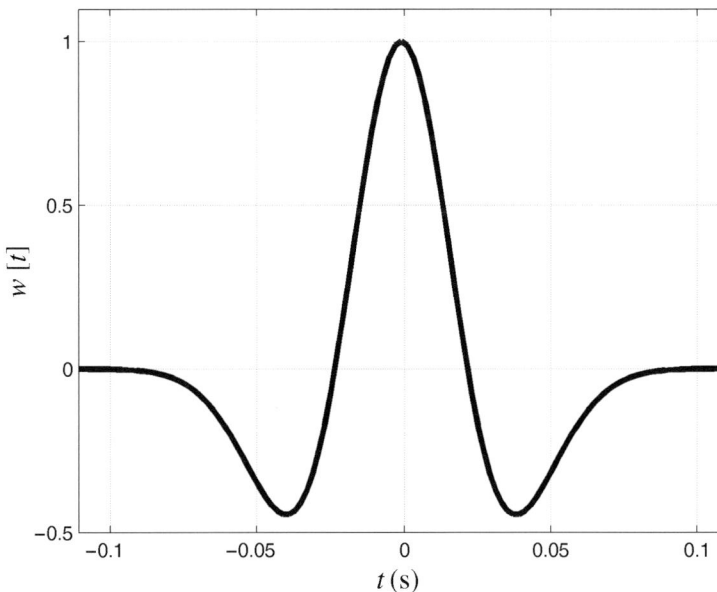

Figure 8.1 An example of a finite-time (length) seismic wavelet with two side-lobes, a central peak at zero time, and a maximum amplitude of unity.

1. Causal or one-sided wavelets: which must be:

$$w[n] = 0, \text{ for } n < 0. \tag{8.1}$$

In practice, wavelets originating from physical sources in real time must be causal.

2. Anti-causal or purely non-causal wavelets are wavelets that are:

$$w[n] = 0, \text{ for } n \geq 0. \tag{8.2}$$

3. Non-causal or two-sided wavelets: Nonzero wavelet values occur both before and after its origin time. These wavelets can be created on computers.

In general, a non-causal wavelet can be written as the following form:

$$w[n] = \sum_{k=-\infty}^{\infty} w[k]\delta[n-k]. \tag{8.3}$$

A causal wavelet can be written as:

$$w_c[n] = \sum_{k=0}^{\infty} w[k]\delta[n-k], \tag{8.4}$$

and the anti-causal wavelet will then be:

$$w_{ac}[n] = \sum_{k=-\infty}^{-1} w[k]\delta[n-k]. \tag{8.5}$$

Finally, one might want to consider finite-length wavelets, which have the following form:

$$w_f[n] = \sum_{k=-M}^{N} w[k]\delta[n-k], \tag{8.6}$$

where the length of $w_f[n]$ is equal to $M + N + 1$. Additionally, wavelets can be classified into:

1. Minimum-delay (phase).
2. Maximum-delay (phase).
3. Mix-delay (phase).

The concept of minimum- or maximum-delay wavelets only applies to causal wavelets. The concept of maximum-delay wavelets only applies to causal wavelets of finite-length. Minimum-delay wavelets have their energy concentrated near their arrival time, whereas mix-delay wavelets have their energy distributed away from their arrival time. Finally, maximum-delay wavelets are the reverse of finite-length minimum-delay wavelets. These wavelets will be further discussed in the next subsection.

8.2.1 Two-Length Wavelets or Minimum-Delay Wavelets

A two-length wavelet, known as a dipole wavelet, can be a minimum-delay, a maximum-delay, or an equi-pole wavelet. Mathematically, a dipole can be written as:

$$w[n] = w[0]\delta[n] + w[1]\delta[n-1]. \tag{8.7}$$

Also, its reverse:

$$w_r[n] = w[1]^*\delta[n] + w[0]^*\delta[n-1], \tag{8.8}$$

where $*$ denotes the complex conjugate. Taking the DTFT of Equation (8.7) will result in the following equation:

$$W(e^{j\omega}) = w[0] + w[1]e^{-j\omega} \tag{8.9}$$

$$= |W(e^{j\omega})|e^{j\angle W(e^{j\omega})}, \tag{8.10}$$

where the magnitude spectrum will be:

$$|W(e^{j\omega})| = \sqrt{w[0]^2 + w[1]w[2]\cos(\omega) + w[1]^2}, \tag{8.11}$$

and its phase spectrum is:

$$\angle W(e^{j\omega}) = \arctan\left(\frac{-w[1]\sin(\omega)}{w[0] + w[1]\cos(\omega)}\right). \tag{8.12}$$

A few observations can be deduced from Equations (8.11 and 8.12) as follows:

- The magnitude spectrum of the two-length wavelet $w[0]\delta[n] + w[1]\delta[n-1]$ and $w[1]\delta[n] + w[0]\delta[n-1]$ are identical.
- If $w[0] > w[1]$, then the phase spectrum of the wavelet $w[0]\delta[n] + w[1]\delta[n-1]$ is everywhere less than that of the phase spectrum of the wavelet $w[1]\delta[n] + w[0]\delta[n-1]$. That is,

$$\arctan\left(\frac{-w[1]\sin(\omega)}{w[0] + w[1]\cos(\omega)}\right) \leq \arctan\left(\frac{-w[0]\sin(\omega)}{w[1] + w[0]\cos(\omega)}\right), \forall\omega. \tag{8.13}$$

Therefore, the wavelet in Equation (8.7) is a minimum-delay (or minimum-phase) wavelet when it has the larger coefficient in magnitude at $n = 0$. On the other hand, Equation (8.7) is a maximum-delay (maximum-phase) wavelet when it has the smaller coefficient in magnitude at $n = 0$. Note that for a dipole wavelet, if $w[0] = w[1]$, then the wavelet is called equi-pole.

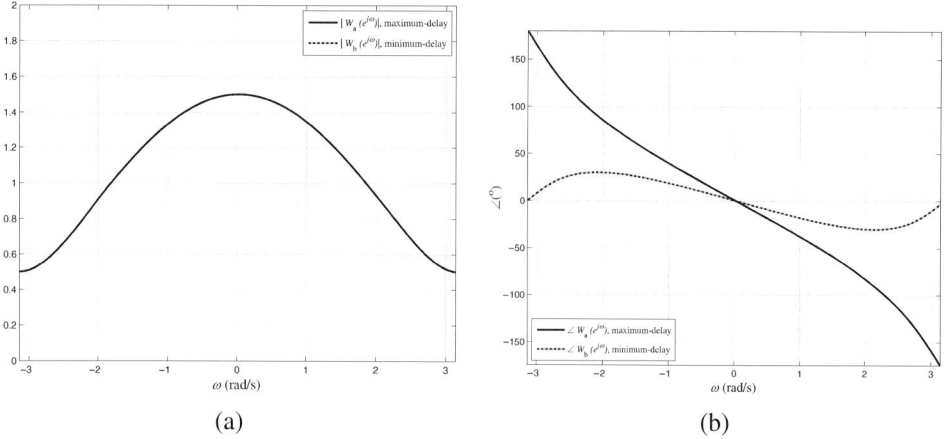

(a)   (b)

Figure 8.2 Example 8.1 showing: (a) the magnitude spectrum and (b) the phase spectrum of the wavelets $w_a[n]$ and $w_b[n]$.

Example 8.1 Determine which of the following is a minimum-delay or maximum-delay wavelet:

(a) $w_a[n] = 0.5\delta[n] + \delta[n-1]$.
(b) $w_b[n] = \delta[n] + 0.5\delta[n-1]$.

Solution: The wavelet in (a) is maximum-delay since $w[0] = 0.5 < w[1] = 1$, while the wavelet in (b) is minimum-delay since $w[0] = 1 > w[1] = 0.5$. Figure 8.2 shows the spectra of both wavelets. Both have identical magnitudes, while $w_b[n]$ has a lower phase than that of $w_a[n]$.

However, why are two-length (dipole) wavelets important? Consider the following N-length causal wavelet:

$$w[n] = \sum_{k=0}^{N-1} w[k]\delta[n-k]. \tag{8.14}$$

Taking the z-transform of Equation (8.14) will result in:

$$W(z) = \sum_{k=0}^{N-1} w[k]z^{-k} \tag{8.15}$$

$$= w[0] + w[1]z^{-1} + w[2]z^{-2} + \cdots + w[N-1]z^{-(N-1)},$$

which is a polynomial of degree $N-1$. According to the fundamental theory of algebra (Bronshtein et al., 2007), any polynomial of degree $N-1$ can be factored into $(N-1)$-polynomials of degree 1. Therefore, Equation (8.15) can be written as:

$$W(z) = \Pi_{i=0}^{N-1}(z - w[i]) \tag{8.16}$$
$$= (z - w[0])(z - w[1])(z - w[2])\dots(z - w[N-1]),$$

which is a product of $N - 1$ z-transformed dipoles. An N-length causal wavelet in Equation (8.14) is said to be of minimum-delay if and only if each of its constituent dipoles is of minimum-delay. The phase spectrum of a product of N signals/wavelets is additive and, hence, a minimum-delay (minimum-phase) signal/wavelet is obtained if all the individual signals/wavelet dipoles are of minimum-delay.

Example 8.2 Is the following 3-length wavelet $w[n] = 6\delta[n] + 5\delta[n-1] + \delta[n-2]$ a minimum-delay or maximum-delay wavelet?

Solution: The z-transform of the given wavelet is $W(z) = 6 + 5z^{-1} + z^{-2}$, which can written as $W(z) = (3 + z^{-1})(2 + z^{-1})$. In this case, $w[n]$ is composed of two minimum-delay dipole wavelets, namely, $3\delta[n] + \delta[n-1]$ and $2\delta[n] + \delta[n-1]$. Therefore, $w[n]$ is a minimum-delay seismic wavelet.

Furthermore, the cumulative energy function denoted by E_c provides a mean to identify the delay property of a given seismic wavelet. It is computed using:

$$E_c = \sum_{k \geq 0} E_k, \tag{8.17}$$

where E_k is called the partial energy, k is the time delay sample, and $E_0 = |w[0]|^2$, $E_1 = E_0 + |w[1]|^2$, $E_2 = E_0 + E_1 + |w[2]|^2$, and so on. If E_c is concentrated at the beginning and is greatly increasing, then the seismic wavelet is of minimum-delay. On the other hand, if E_c was concentrated at the end and is slowly increasing, then the seismic wavelet is of maximum-delay or mixed-delay.

Example 8.3 Compute and plot the cumulative energy function E_c for the following seismic wavelets:

1. $w_1[n] = 3\delta[n] + \frac{7}{2}\delta[n-1] + \delta[n-2]$.
2. $w_2[n] = 2\delta[n] + 4\delta[n-1] + \frac{3}{2}\delta[n-2]$.
3. $w_3[n] = \delta[n] + \frac{7}{2}\delta[n-1] + 3\delta[n-2]$.

Classify them as minimum-, maximum-, or mixed-delay wavelets. Also, plot their magnitude and phase spectra. Comment on the results.

Solution: For $w_1[n]$, the partial energy is $E_0 = 9$, $E_1 = E_0 + 12.25$, and $E_c = E_2 = E_0 + E_1 + 1 = 22.25$. Also, for $w_2[n]$, the partial energy is $E_0 = 4$, $E_1 = E_0 + 16$, and $E_c = E_2 = E_0 + E_1 + 2.25 = 22.25$. Finally, for $w_3[n]$, the partial energy is $E_0 = 1$, $E_1 = E_0 + 12.5$, and $E_c = E_2 = E_0 + E_1 + 9 = 22.25$. The wavelets are shown in Figures 8.3a–c. Also, the cumulative energy functions of the three wavelets are shown in Figure 8.3d. Clearly, $w_1[n]$ is of minimum-delay since its E_c is concentrated at the beginning and is greatly increasing. Also, $w_3[n]$ is of maximum-delay, since its E_c was concentrated at the end and is slowly increasing. The E_c of $w_2[n]$ is in between, so it can be considered as a mixed-delay wavelet. Figure 8.3e and 8.3f show the spectra of the three wavelets, where all are of the same magnitude but with different phases. The phase response of $w_1[n]$ is of minimum-delay, $w_2[n]$ is of mixed-delay, and, finally, $w_3[n]$ is of maximum-delay.

In seismic data processing, the concept of minimum-delay wavelets is important for the following two reasons: (a) of all causal wavelets with the same amplitude spectrum, the one with minimum phase spectrum has the shortest time duration. This will make minimum-delay wavelets very important for vertical resolution, and (b) many seismic data processing algorithms such as the optimum linear-predication filter perform at their best when the wavelet is minimum-delay. Therefore, the concept of an equivalent minimum-delay wavelet to a given wavelet is discussed here. An equivalent minimum-delay wavelet is simply a wavelet that has the minimum phase spectrum but retains the same amplitude spectrum as the given original wavelet. Many ways to obtain the equivalent minimum-delay wavelet of an N-length causal wavelet exist. A simple way to do so is by using the z-transform as follows:

1. Take the z-transform of Equation (8.14).
2. Decompose the result in step 1 into $N-1$ dipole as seen in Equation (8.16).
3. Reverse all the maximum-delay dipoles so that they become of minimum-delay.
4. Multiply all the factored dipoles together but with new minimum-delay ones instead of those that were of maximum-delay.
5. Find the inverse z-transform of the result of step 4.

Note that this process requires finding $N-1$ complex roots of a polynomial of degree $N-1$. Many ways to estimate the equivalent seismic wavelet exist, see, for example, the work of Lazear (1984), Lindsey (1988), Ziolkowski and Slob (1991), and Yilmaz (2001).

Example 8.4 Find the equivalent minimum-delay wavelet to $w[n] = 4\delta[n]+8\delta[n-1] + 3\delta[n-2]$.

Solution: One can show that $W(z) = (2+z^{-1})(\frac{2}{3}+z^{-1})$, where the dipole wavelets that compose $w[n]$ are $2\delta[n] + \delta[n-1]$, of minimum-delay, and $\frac{2}{3}\delta[n] + \delta[n-1]$, of maximum-delay, respectively. Hence, the equivalent minimum-delay dipole to the maximum-delay one is $\delta[n] + \frac{2}{3}\delta[n-1]$. And, $W_e(z) = (2 + z^{-1})(1 + \frac{2}{3}z^{-1}) = 6+7z^{-1}+2z^{-2}$. Therefore, the equivalent minimum-delay wavelet to $w[n]$ is $w_e[n] = 6\delta[n] + 7\delta[n-1] + 2\delta[n-2]$.

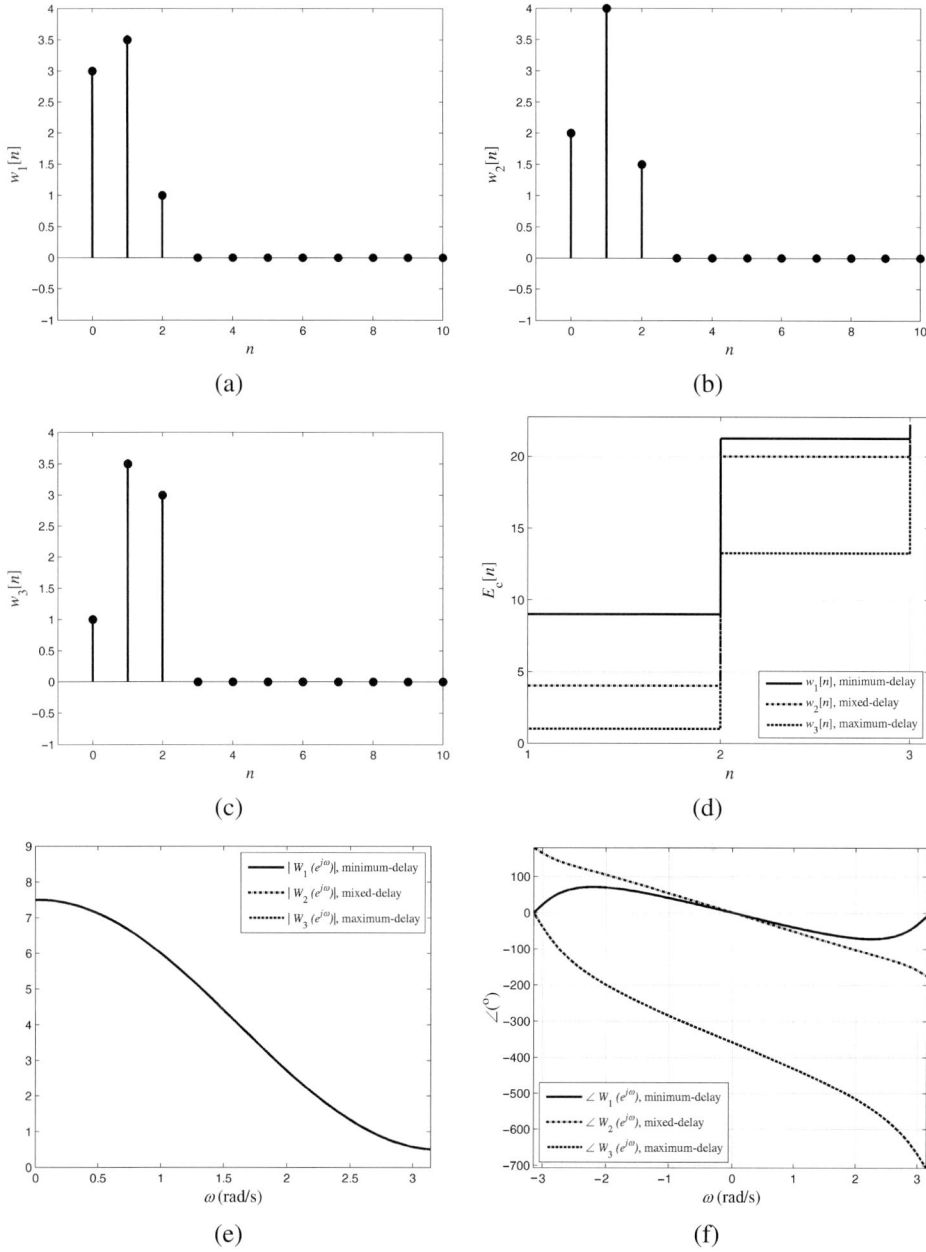

Figure 8.3 Example 8.3 showing the three wavelets in (a)–(c). Then their (d) cumulative energy. (e) The magnitude and (f) phase spectrum of the given wavelets.

8.2.2 Zero-Phase and Symmetric Wavelets

In the previous subsection, it was stated that the minimum-delay wavelet had the shortest time-duration of all causal wavelets with the same amplitude spectrum. However, they do not have the shortest duration of all wavelets with the same amplitude spectrum. Such wavelets come with zero-phase spectrum and are symmetric about their zero-time. They are, therefore, non-causal and ideally extend to infinity on both sides of time zero. Wavelets can be converted into zero-phase wavelets using an optimum filter, which is known as shaping (discussed in the next section). By doing so, the deductability of the reflected event arrival times can be maximized. Not every wavelet, however, is of zero-phase, even if it was non-causal. Consider, for example, the following symmetric wavelets:

- The rectangular wavelet, which can be given by:

$$w_{rect}[n] = \begin{cases} 1 & \text{for } \frac{-(N-1)}{2} \le n \le \frac{(N-1)}{2} \\ 0 & \text{otherwise} \end{cases}. \tag{8.18}$$

- The sinc wavelet, which can be given by:

$$w_{sinc}[n] = sinc[n]. \tag{8.19}$$

Figure 8.4 shows both $w_{rect}[n]$ and $w_{sinc}[n]$ wavelets in the time and frequency domains.

Various practical types of wavelets exist. A commonly used wavelet to generate synthetic seismic sections is the so-called *Ricker* wavelet. The Ricker wavelet is non-causal, symmetric about its time origin, and of zero-phase. In continuous time, the Ricker wavelet is given by the second derivative of Gaussian function:

$$w_{Ricker}(t) = \left(1 - \frac{12t^2}{b^2}\right) e^{\frac{6t^2}{b^2}}. \tag{8.20}$$

b is called the wavelet breadth, and it measures separation (in time) of the wavelet's two valleys, which occur on each side of the central main-lobe. Figure 8.5 shows a Ricker wavelet with various values of dominant frequencies. In the discrete-time form, the Ricker wavelet can be given by:

$$w_{Ricker}[n] = \left[1 - \frac{12n^2}{b^2}\right] e^{\frac{6n^2}{b^2}}, \forall n. \tag{8.21}$$

Figure 8.6 shows shifted Ricker seismic wavelets with various angles.

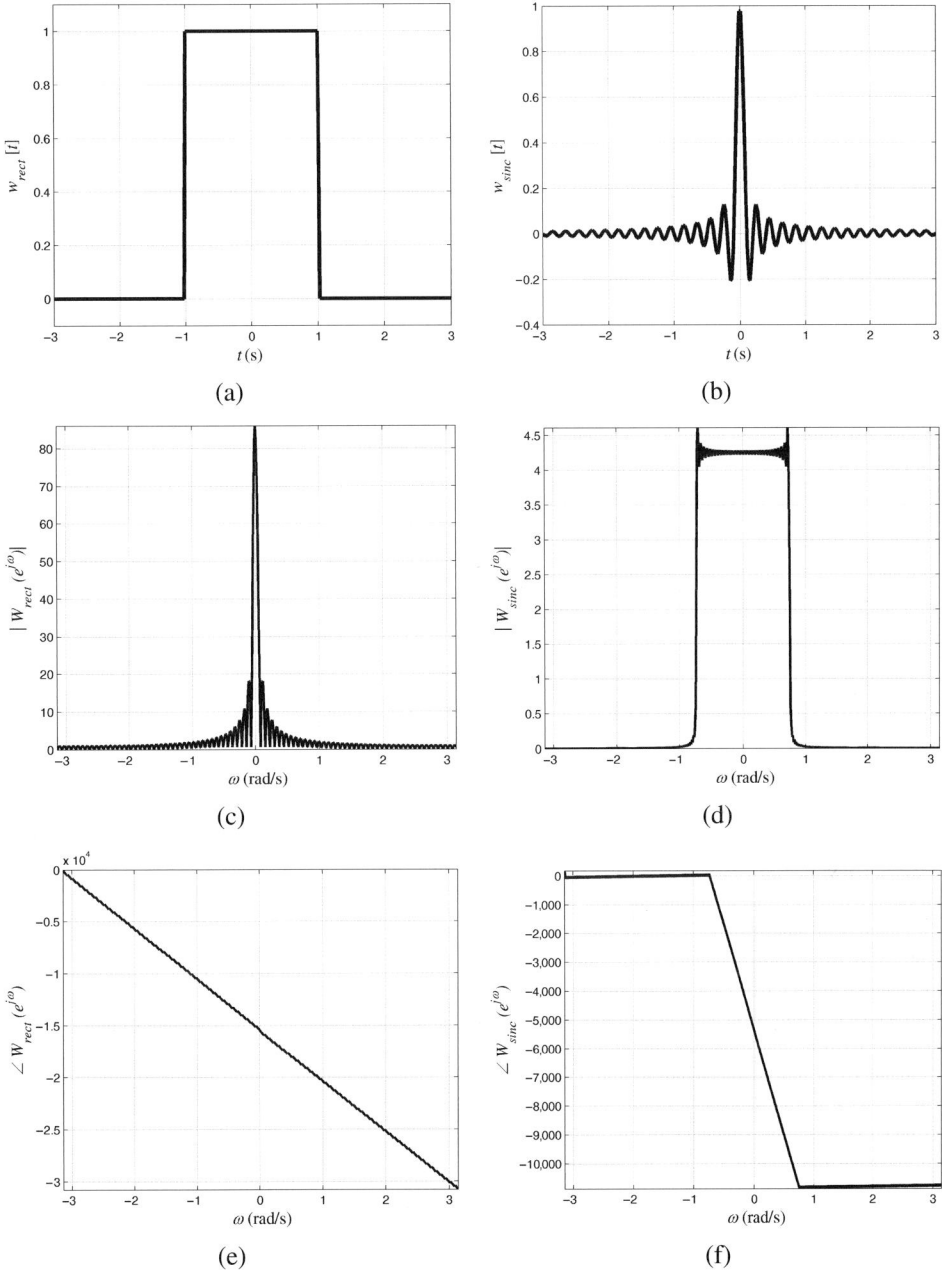

Figure 8.4 Example of (a) a rectangular wavelet with duration of 2 seconds, and (b) a Sinc wavelet with $f = 10$ Hz. Both are non-causal wavelets. (c) and (d) are the magnitude spectrum of (a) and (b), respectively. Also, (e) and (f) shows the phase spectrum of (a) and (b), respectively, which are of nonzero.

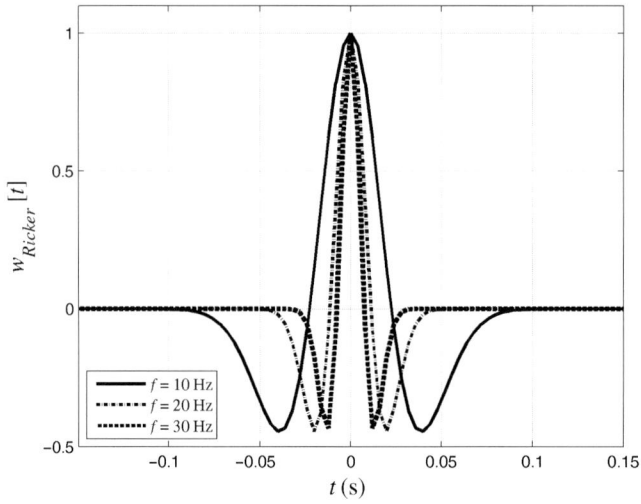

Figure 8.5 Example of Ricker wavelets of various frequencies.

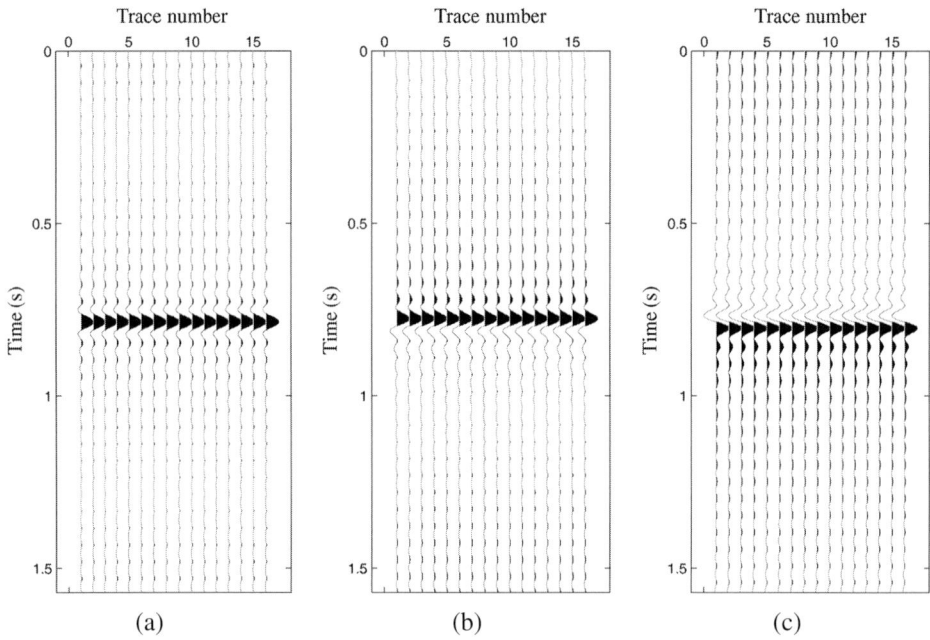

Figure 8.6 Examples of shifted Ricker wavelets with (a) 0°, (b) 45°, and (c) 90°.

8.3 Seismic Wavelet Processing

The seismic wavelet will usually be neither minimum-delay nor causal. It will often be non-symmetric and frequently quite spread out. Thus, it is not entirely desirable for a seismic interpreter. Various filtering operations, accordingly, are used to improve the quality of seismic data. These come under the heading of wavelet processing. Wavelet processing can refer to:

1. Estimating the basic embedded wavelet. Then designing a shaping filter to convert the estimated wavelet to a desired form like a zero-phase and, finally, applying the shaping filter to the seismic data.
2. Wavelet shaping in which the desired output is a zero-phase wavelet with the same amplitude spectrum as that of the input wavelet. This will only correct for the phase of the input wavelet, which is sometimes assumed to be minimum-phase.

Considering Wiener optimal filtering (see Chapters 6 and 7), for any given input wavelet, a series of desired outputs $d[n]$ can be defined as delayed unit impulse samples (spikes) $\delta[n - k]$. The least squares errors then can be plotted as a function of delay. The delay that corresponds to the least error is chosen to define the desired spike output. The actual output from the Wiener filter using this optimum delayed spike should be the most compact result. The Wiener filtering process that uses any desired arbitrary shape is called *Wavelet Shaping* and the corresponding Wiener filter is known as the *Wiener Shaping Filter*. Note that, in the Wiener filtering problem, $d[n] = \delta[n - k]$ and/or $d[n]$ is a zero-phase wavelet, and these are special cases of the more general wavelet shaping. Wavelet shaping requires the knowledge of the input wavelet to compute $\mathbf{R}_{xd}$. If it is unknown, which is the case in reality, then the minimum-phase equivalent of the input wavelet can be estimated statistically from the data (Robinson and Treitel, 2008; Yilmaz, 2001). The minimum-phase estimate is then shaped to a zero-phase wavelet.

8.4 Summary

It is important that seismic wavelets are estimated from recorded seismic data. There are various types of seismic wavelets. For a seismic interpreter, it is important that he/she be able to correctly interpret the data. Unfortunately, most of the embedded wavelets are dilated and of mixed or maximum delay. This requires careful processing to shape the wavelet into zero-phase or even into minimum-delay.

Exercises

8.1   Determine which of the following dipole wavelets is a minimum-delay or maximum-delay:

(a) $w[n] = 2j\delta[n] + 4\delta[n-1]$.
(b) $w[n] = 0.5j\delta[n] + 0.25\delta[n-1]$.
(c) $w[n] = 2.5j\delta[n] + 4.25j\delta[n-1]$.

Plot their magnitude and phase spectra. Comment on the results.

8.2   Compute and plot the cumulative energy function for the seismic wavelet $w[n] = 5\delta[n] + 2\delta[n-1] + \delta[n-2]$. Is it a minimum-, maximum-, or mixed-delay wavelet? Plot their magnitude and phase spectra. Comment on the results.

8.3   Consider the following source seismic wavelet:

$$w[n] = \frac{3}{2}\delta[n] + \frac{7}{2}\delta[n-1] + \delta[n-2].$$

(a) Show that this wavelet is not minimum-phase.
(b) Find its equivalent minimum-phase $w_e[n]$.
(c) Is the equivalent minimum-phase wavelet causal? If yes, then find its inverse $F(z)$ and $f[n]$ (only up to the 3rd term).
(d) Convolve $f[n]$ with $w_e[n]$ for the first three terms.

8.4   Consider the reflectivity sequence:

$$r[n] = 3\delta[n-1] + 2\delta[n-3] + \delta[n-5].$$

The embedded seismic wavelet is not minimum-delay and is given by:

$$w[n] = \frac{3}{2}\delta[n] + \frac{7}{2}\delta[n-1] + \delta[n-2].$$

(a) Compute and sketch the resulting seismogram $g_1[n]$ from $r[n]$ and $w[n]$.
(b) Find its equivalent minimum-delay $w_e[n]$.
(c) Compute and sketch the resulting seismic signal $g_2[n]$ from $r[n]$ and $w_e[n]$. How different is $g_1[n]$ from $g_2[n]$?

8.5   The Ricker wavelet in Equation (8.21) has $b = 5$. It is to be truncated to become of length equal to 5, i.e., $n = -2, -1, 0, 1, 2$, then obtain a causal version of this given Ricker wavelet. Is it a minimum-delay, maximum-delay, or mixed-delay wavelet? Why?

(a)   (b)

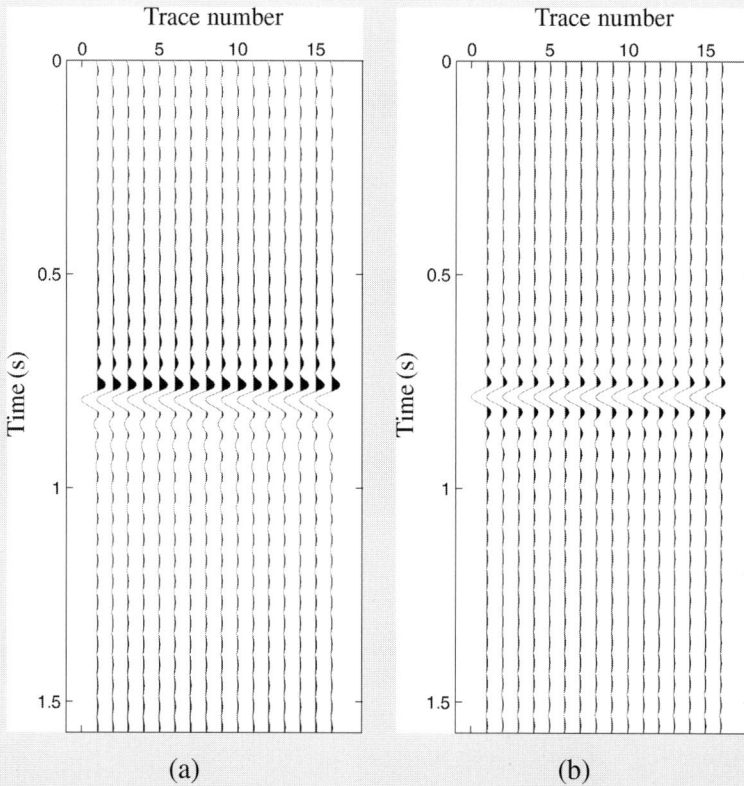

Figure 8.7 Problem 8.6.

8.6 Given the shifted Ricker seismic wavelets in Figure 8.5, identify their approximate phase angle.

8.7 Write a MATLAB script to compute the equivalent seismic wavelet. The script should also plot the spectra of the wavelet and its equivalent.

References

Al-Lehyani, A. F. 2008. *Data-Driven Mapping and Attenuation of Near-Surface Diffractors*. M.Phil. thesis, King Fahd Univeristy of Petroleum & Minerals.

Al-Shuhail, A. A., Al-Dossary, S. A., and Mousa, W. A. 2017. *Seismic Data Interpretation using Digital Image Processing for Seismic Data Analysis*. J. Wiley & Sons.

Ansari, R. 1987. Efficient IIR and FIR fan filters. *IEEE Transactions on Circuits and Systems*, **34**(August), 941–945.

Avseth, P., Mukerji, T., and Mavko, G. 2010. *Quantitative Seismic Interpretation: Applying Rock Physics Tools to Reduce Interpretation Risk*. Cambridge University Press.

Bacon, M., Simm, R., and Redshaw, T. 2003. *3-D Seismic Interpretation*. Cambridge University Press.

Baeten, G. J., Belougne, V., Combee, L., Kragh, E., Orban, J., Özbek, A., and Vermeer, P. 2000. Acquisition and processing of point receiver measurements in land seismic. *SEG 2000 Expanded Abstracts*.

Bamberger, R. H. and Smith, M. J. T. 1992. A filter bank for the directional decomposition of images: Theory and design. *IEEE Transactions on Signal Processing*, **40**(4), 882–893.

Baraniuk, R. G. 2007. Compressive sensing. *Signal Processing Magazine, IEEE*, 118–121.

Berkhout, A. J. and Verschuur, D. J. 2006. Imaging of multiple reflections. *Geophysics*, **71**(4), SI209.

Berkner, K. and Wells, R. O., Jr. 1998. Wavelet transforms and denoising algorithms. *Conference Record of the Thirty-Second Asilomar Conference on Signals, Systems and Computers*, **2**(Nov.), 1639–1643.

Beylkin, G. 1987. Discrete radon transform. *IEEE Transactions on Acoustics, Speech, and Signal Processing*, **35**(2), 162–172.

Biondi, B. L. 2006. *3D Seismic Imaging*. SEG.

Bjarnason, I. Th. and Menke, W. 1993. Application of the POCS inversion method to cross-borehole Imaging. *Geophysics*, **58**(7), 941–948.

Blacquiere, G. and Ongkiehong, L. 2000. Single sensor recording: Antialias filtering, perturbations and dynamic range. *SEG 2000 Expanded Abstracts*.

Boschetti, Fabio, Dentith, Mike D., and List, Ron D. 1996. A fractal-based algorithm for detecting first arrivals on seismic traces. *Geophysics*, **61**(4), 1095–1102.

Boussakta, S., Al-Shibami, O., Aziz, M., and Holt, A. G. 2001a. 3-D vector radix algorithm for the 3-D new Mersenne number transform. *IEE Proceedings-Vis: Image & Signal Processing*, **148**(2), 115–125.

Boussakta, S., Al-Shibami, O., and Aziz, M. 2001b. Radix 2 x 2 x 2 algorithm for the 3-D discrete Hartley transform. *IEEE Transactions on Signal Processing*, **49**(12), 3145–3156.

Bronshtein, I. N., Semendyayev, K. A., Musiol, G., and Mhlig, H. 2007. *Handbook of Mathematics*. 5th edn. Springer.

Buttkus, B. 2000. *Spectral Analysis and Filter Theory in Applied Geophysics*. Springer.

Campman, X., Behn, P., and Faber, K. 2016. Sensor density or sensor sensitivity? *The Leading Edge*, **35**(7), 578–585.

Canales, Luis L. 2005. *Random Noise Reduction*. Pages 525–527.

Candès, E. J. 2006. Compressive sampling.

Candès, Emmanuel J. 2008. The restricted isometry property and its implications for compressed sensing. *Comptes Rendus Mathematique*, **346**(9–10), 589–592.

Candès, Emmanuel, and Demanet, Laurent. 2003. Curvelets and Fourier integral operators. *Comptes Rendus Mathematique*, **336**(5), 395–398.

Candès, E. J. and Wakin, M. B. 2008. An introduction to compressive sampling. *IEEE Signal Processing Magazine*, **25**(2), 21–30.

Cao, Jingjie, Wang, Yanfei, Zhao, Jingtao, and Yang, Changchun. 2011. A review on restoration of seismic wavefields based on regularization and compressive sensing. *Inverse Problems in Science and Engineering*, **19**(5), 679–704.

Cao, Jing-Jie, Wang, Yan-Fei, and Yang, Chang-Chun. 2012. Seismic data restoration based on compressive sensing using regularization and zero norm sparse optimization. *Chinese Journal of Geophysics*, **55**(2), 239–251.

Cao, Zhihong. 2007. Analysis and application of the Radon transform. *Masters Abstracts International*.

Cao, Zhihong and Bancroft, John C. 2005. A semblance weighted Radon transform on multiple attenuation. *CSEG National Convention*, **2**(1917), 298–301.

Çetin, A. E., Gerek, O. N., and Yardimci, Y. 1997. Equiripple FIR filter design by the FFT algorithm. *Signal Processing Magazine, IEEE*, **14**(Mar), 60–64.

Chapman, C. H. 1981. Generalized Radon transforms and slant stacks. *Geophysical Journal International*, **66**(2), 445–453.

Cheng, Qiansheng, Chen, Rong, and Li, Ta-Hsin. 2001. Simultaneous wavelet estimation and deconvolution of reflection seismic signals. *IEEE Transactions on Geoscience and Remote Sensing*, **34**(2), 377–384.

Clearbout, J. F. 1985. *Imaging the Earth's Interior*. Blackwell Scientific Publications.

Coppens, F. 1985. First arrival picking on common-offset trace collections for automatic estimation of static corrections. *Geophysical Prospecting*, **32**, 1212–1231.

Criss, C. J., Kappius, R., and Cunningham, D. 2003. First arrival picking using image processing methods. *65th EAGE Conference and Exhibition Extended Abstracts*.

Curry, William and Shan, Guojian. 2010 (June). Interpolation of near offsets using multiples and prediction-error filters. Page WB153 of: *Geophysics*, vol. 75.

Donno, Daniela. 2011. Improving multiple removal using least-squares dip filters and independent component analysis. *Geophysics*, **76**(5), V91.

Donoho, D. L. 2006. Compressed sensing. *IEEE Transactions on Information Theory*, **52**(4), 1289–1306.

Dow, W. G. and Magoon, L. B. (eds). 1994. *The Petroleum System: From Source to Trap*. American Association of Petroleum Geologists (AAPG).

Dudgeon, D. E. and Mersereau, R. M. 1984. *Multidimensional Digital Signal Processing*. Prentice-Hall.

Duquet, B. and Marfurt, K. J. 1999. Filtering coherent noise during prestack depth migration. *Geophysics*, **64**(4), 1054–1066.

Duval, L. and Rosten, T. 2000. Filter bank decomposition of seismic data with application to compression and denoising. *SEG 2000 Expanded Abstracts*.

Elad, M., Starck, J. L., Querre, P., and Donoho, D. L. 2005. Simultaneous cartoon and texture image inpainting using morphological component analysis (MCA). *Applied and Computational Harmonic Analysis*, **19**(3), 340–358.

Forel, D., Benz, T., and Pennington, W. D. 2005. *Seismic Data Processing with Seismic Un*x: A 2-D Seismic Data Processing Primer*. Society of Exploration Geophysicists (SEG).

Freed, Dennis. 2008. Cable-free nodes: The next generation land seismic system. *The Leading Edge*, **27**(7), 878–881.

Garibotto, G. 1979. 2D recursive phase filters for the solution of two-dimensional wave equations. *IEEE Transactions on Acoustics, Speech, and Signal Processing*, **27**(4), 367–373.

Gazdag, J. 1978. Wave equation migration with the phase-shift method. *Geophysics*, **43**, 1342–1351.

Gholami, A. and Siahkoohi, H. R. 2009. Simultaneous constraining of model and data smoothness for regularization of geophysical inverse problems. *Geophysical Journal International*, **176**(1), 151–163.

Gholami, Ali and Sacchi, Mauricio D. 2012. A fast and automatic sparse deconvolution in the presence of outliers. *IEEE Transactions on Geoscience and Remote Sensing*, **50**(10 PART2), 4105–4116.

Ginie. 2013. GI in the wider Europe. *The Journal of Infectious Diseases*, **207**(8), NP.

Gluyas, J. and Swarbrick, R. 2014. *Petroleum Geoscience*. John Wiley & Sons.

Gulunay, Necati. 2003. Seismic trace interpolation in the Fourier transform domain. *Geophysics*, **68**(1), 355.

Haddad, K. C., Stark, H., and Galatsanos, N. P. 1999. Design of digital linear-phase FIR crossover systems for loudspeakers by the method of Vector Space projections. *IEEE Transactions on Signal Processing*, **47**(11), 3058–3066.

Haddad, K. C., Stark, H., and Galatsanos, N. P. 2000. Constrained FIR filter design by the method of vector space projections. *IEEE Transactions on Circuits and Systems*, **47**(8), 714–725.

Hale, D. 1991. Stable explicit depth extrapolation of seismic wavefields. *Geophysics*, **56**, 1770–1777.

Hart, B. S. 2011. *Introduction to Seismic Interpretation*. American Association of Petroleum Geologists (AAPG).

Hart, D. I. 1996. Reliably picking first breaks with neural networks. *58th EAGE Conference and Exhibition Extended Abstracts*, B033.

Haykin, S. 2002. *Adaptive Filter Theory*. 4th edn. Prentice Hall, Upper-Saddle River, NJ.

Hennenfent, Gilles and Herrmann, Felix. 2005. Sparseness constrained data continuation with frames: applications to missing traces and aliased signals in 2/3D. *SEG Technical Program Expanded Abstracts 2005*, Jan, 2162–2165.

Hennenfent, Gilles and Herrmann, Felix J. 2006. Seismic denoising with nonuniformly sampled curvelets. *Computing in Science and Engineering*, **8**(3), 16–25.

Hennenfent, Gilles and Herrmann, Felix J. 2008. Simply denoise: Wavefield reconstruction via jittered undersampling. *Geophysics*, **73**(3), V19–V28.

Henry, S. G. 1997. Catch the (seismic) wavelet. *AAPG Explorer*, March, 36–38.

Hermanowicz, E. and Blok, M. 2000. Iterative technique for approximate minimax design of complex digital FIR filters. *7th IEEE International Conference on Electronics, Circuits and Systems*, **1**, 83–86.

Herrmann, Felix J. 2009a. *Compressed Sensing and Sparse Recovery in Exploration Seismology*. Tech. rept.

Herrmann, F. J. 2009b. Sub-Nyquist sampling and sparsity: How to get more information from fewer samples. 2009 *SEG Annual Meeting*.

Herrmann, Felix J. and Hennenfent, Gilles. 2008. Non-parametric seismic data recovery with curvelet frames. *Geophysical Journal International*, **173**(1), 233–248.

Herrmann, Felix J. and Wason, Haneet. 2012 (Jul). Compressive Sensing in Marine Acquisition and Beyond. In: *74th EAGE Conference & Exhibition – Workshops*.

Herrmann, Felix J., Wang, Deli, and Hennenfent, Gilles. 2007a. Multiple prediction from incomplete data with the focused curvelet transform. *SEG Technical Program Expanded Abstracts*, **26**(1), 2505–2509.

Herrmann, Felix J., Böniger, Urs, and Verschuur, Dirk Jacob. 2007b. Non-linear primary-multiple separation with directional curvelet frames. *Geophysical Journal International*, **170**(2), 781–799.

Herrmann, Felix J., Erlangga, Yogi, and Lin, Tim T. Y. 2009a. Compressive simultaneous full-waveform simulation. *Geophysics*, **74**(4), A35.

Herrmann, Felix J., Erlangga, Yogi, and Lin, Tim T. Y. 2009b. Compressive-wavefield simulations. *SAMPTA'09, International Conference on Sampling Theory and Applications*, 2–5.

Herron, D. 2012. *First Steps in Seismic Interpretation*. Society of Exploration Geophysicists (SEG).

Holberg, O. 1988. Towards optimum one-way wave propagation. *Geophysical Prospecting*, **36**, 99–114.

Iske, A. and Randen, T. (eds). 2000. *Mathematical Methods and Modelling in Hydrocarbon Exploration and Production*. Springers.

Kaplan, Sam T. and Innanen, Kristopher. 2008. Adaptive separation of free-surface multiples through independent component analysis. *Geophysics*, **73**(3), V29.

Kaplan, Sam T., Naghizadeh, Mostafa, and Sacchi, Mauricio D. 2010. Data reconstruction with shot-profile least-squares migration. *Geophysics*, **75**(6), WB121.

Karam, L. J. and McClellan, J. H. 1997. Efficient design of digital filters for 2-D and 3-D depth migration. *Signal Processing, IEEE Transactions on*, **45**(4), 1036–1044.

Kayran, A. H. and King, R. A. 1983. Design of recursive and nonrecursive fan filters with complex transformation. *IEEE Transactions on Circuits and Systems*, **30**(12), 849 – 857.

Kearey, P., Brooks, M. and Hill, I. 2002. *An Introduction to Geophysical Exploration*. 3rd edn. Blackwell Science.

Keho, Timothy H. and Zhu, Weihong. 2009. Revisiting automatic first arrival picking for large 3D land surveys. *SEG Technical Program Expanded Abstracts*, **28**(1), 3198–3202.

Kreyszig, E. 1978. *Introductory Functional Analysis with Applications*. John Wiley and Sons.

Kuchment, Peter. 2006. Generalized transforms of Radon type and their applications. *Proceedings of Symposia in Applied Mathematics*, **0000**, 1–32.

Landa, Evgeny, Belfer, Igor, and Keydar, Shemer. 1999. Multiple attenuation in the parabolic tau-p domain using wavefront characteristics of multiple generating primaries. *Geophysics*, **64**(6), 1806–1815.

Latif, A. and Mousa, Wail A. 2017. An Efficient under-sampled high-resolution Radon Transform for Exploration Seismic Data Processing. *IEEE Transactions on Geosciences and Remote Sensing*, **55**(2), 1010–1024.

Lazear, Gregory D. 1984. An examination of the exponential decay method of mixed-phase wavelet estimation. *Geophysics*, **49**(12), 2094–2099.

Lehmann, Thomas M. 1999. Survey: Interpolation methods in medical image processing. *IEEE Transactions on Medical Imaging*, **18**(11), 1049–1075.

Levi, A. and Stark, H. 1983. Signal restoration from phase by projections onto convex sets. In: *Proceedings of IEEE International Conference on Acoustics, Speech, and Signal Processing (ICASSP)*.

Levinson, N. 1947. The Wiener RMS error criterion in filter design and prediction. *Journal of Mathematical Physics*, **25**, 261278.

Lin, Tim T. Y. and Herrmann, Felix J. 2007. Compressed wavefield extrapolation. *Geophysics*, **72**(5), SM77.

Lin, Tim T. Y., Herrmann, Felix J., and Columbia, British. 2009. Unified compressive sensing framework for simultaneous acquisition with primary estimation SEG Houston 2009 International Exposition and Annual Meeting SEG Houston 2009 International Exposition and Annual Meeting. *2009 SEG Annual Meeting*, 3113–3117.

Lindsey, J. P. 1988. Measuring wavelet phase from seismic data. *The Leading Edge*, **7**(7), 10–16.

Liner, C. 2004. *Fundamentals of Geophysical Interpretation*. Society of Exploration Geophysicists (SEG).

Liner, C. L. 1999a. *Elements of 3-D Seismology*. PennWell.

Liner, C. L. 1999b. *Elements of 3-D Seismology*. PennWell.

Linville, A. F. and Meek, R. A. 1995. A procedure for optimally removing localized coherent noise. *Geophysics*, **60**(1), 191–203.

Liu, Bin and Sacchi, Mauricio D. 2004. Minimum weighted norm interpolation of seismic records. *Geophysics*, **69**(6), 1560.

Lu, W. and Antoniou, A. 1992. *Two-Dimensional Digital Filters*. 1st edn. Marcel Dekker Publisher.

Lu, Wenkai. 2006. Adaptive multiple subtraction using independent component analysis. *Geophysics*, **71**(5), S179.

Luo, Yinhe, Xia, Jianghai, Miller, Richard D., Xu, Yixian, Liu, Jiangping, and Liu, Qingsheng. 2009. Rayleigh-wave mode separation by high-resolution linear radon transform. *Geophysical Journal International*, **179**(1), 254–264.

MacKenzie, D. 2009. Compressed sensing makes every pixel count. *What's Happening in the Mathematical Sciences*, 114–127.

Madisetti, V. K. and Williams, D. B. (eds.). 1998. *The Digital Signal Processing Handbook*. CRC Press and IEEE Press.

Malvern, L. E. 1969. *Introduction to the Mechanics of a Continuous Medium*. Prentice-Hall.

Mansour, Hassan, Wason, Haneet, Lin, Tim T. Y., and Herrmann, Felix J. 2012. Randomized marine acquisition with compressive sampling matrices. *Geophysical Prospecting*, **60**(4), 648–662.

Marks, R. J. 1991. *Introduction to Shannon Sampling and Interpolation Theory*. Spinger-Verlag.

Matus, Frantisek and Flusser. Jan. 1993. Image representations via a finite Radon transform. *IEEE Transactions on Pattern Analysis and Machine Intelligence*, **15**(10), 996–1006.

McCowan, D. W., Stoffa, P. L., and Diebold, J. B. 1984. Fan filters for data with variable spatial sampling. *IEEE Transactions on Acoustics, Speech, and Signal Processing*, **32**(6), 1154–1159.

Miall, A. D. (ed). 2008. *The Sedimentary Basins of the United States and Canada*. Elsevier.

Miao, X. and Cheadle, S. 1998. Noise attenuation with wavelet transform. *SEG 1998 Expanded Abstracts*.

Milkereit, Bernd. 1989. Stacking charts: An effective way of handling survey, quality control and data processing information. *Canadun Journal Of Exploratlon Geopyhsics*, **25**(1), 28–35.

Minaeian, V., Javaherian, A., and Moslemi, A. 2009. Multiple Attenuation by FX Parabolic Radon Transform. *1st International Petroleum Conference & Exhibition*, 4–6.

Molyneux, J. B. and Schmitt, D. R. 1999. First-break timing: Arrival onset times by direct correlation. *Geophysics*, **64**(5), 1492–1501.

Moon, T. K. and Stirling, W. C. 2000. *Mathematical Methods and Algorithms for Signal Processings*. Prentice Hall.

Mousa, W. A. 2006. *Design and Implementation of Complex-Valued FIR Digital Filters with Application to Migration of Seismic Data*. Ph.D. thesis, University of Leeds.

Mousa, W. A. 2012a. Frequency-space wavefield extrapolation using infinite impulse response digital filters: Is it feasible? *Geophysical Prospecting*, April.

Mousa, W. A. 2012b. Iterative design of one-dimensional efficient seismic L lt;sub gt;p lt;/sub gt; infinite impulse response f-x digital filters. *IET Signal Processing*, **6**(6), 541–545.

Mousa, W. A. 2014. Imaging of the SEG/EAGE salt model seismic data using sparse $f - x$ finite-impulse-response wavefield extrapolation filters. *IEEE Transactions on Geoscience and Remote Sensing*, **52**(5), 2700–2714.

Mousa, W. A. and Al-Shuhail, A. A. 2011. *Processing of Seismic Reflection Data Using MATLAB*. Morgan & Claypool.

Mousa, Wail A. and Al-Shuhail, Abdullatif A. 2012. Enhancement of first arrivals using the tau-[sans-serif p] transform on energy-ratio seismic shot records. *Geophysics*, **77**(3), V101–V111.

Mousa, Wail A., Al-Shuhail, Abdullatif A., and Al-Lehyani, Ayman. 2011. A new technique for first-arrival picking of refracted seismic data based on digital image segmentation. *Geophysics*, **76**(5), V79–V89.

Mousa, W. A., Baan, M. Van Der, Boussakta, S., and McLernon, D .C. 2006. Designing stable operators for explicit depth extrapolation of 2-D wavefields using projections onto convex sets. *Submitted to the SEG Journal of Geophysics*.

Mulk, M. Z., Obata, K., and Hirano, K. 1983. Design of fan filters. *IEEE Transaction on Acoustics, Speech, and Signal Processing*, **31**(6), 1427–1434.

Murat, M. E. and Rudman, A. J. 1992. Automatic first-breaks picking: A neural network approach. *Geophysical Prospecting*, **40**(6), 1365–2478.

Naseer, M. M. and Mousa, W. A. 2015. Linear complementarity problem: A novel approach to design finite-impulse response wavefield extrapolation filters. *Geophysics*, **80**(2), S55–S63.

Needell, Deanna and Vershynin, Roman. 2009. Uniform uncertainty principle and signal recovery via regularized orthogonal matching pursuit. *Foundations of Computational Mathematics*, **9**(3), 317–334.

Ng, Mark and Perz, Mike. 2003. High Resolution Radon Transform in the T-X Domain. *cseg.ca*, 1–4.

Ng, Mark and Perz, Mike. 2004. High resolution Radon transform in the tx domain using "intelligent" prioritization of the Gauss-Seidel estimation sequence. In: *2004 SEG Annual Meeting*.

Nguyen, T. and Castagna, J. 2002. Stable explicit wavefield extrapolation using recursive operators. *SEG 2002 Expanded Abstracts*.

Oh, S., Marks, R. J., and Atlas, L. E. 1994. Kernel synthesis for generalized time-frequency distributions using the method of alternating projections onto convex sets. *IEEE Transactions on Signal Processing*, **42**(7), 1653–1661.

Onajite, E. 2013. *Seismic Data Analysis Techniques in Hydrocarbon Exploration*. Elsevier.

O'Neil, P. V. 1995. *Advanced Engineering Mathematics*. 4th edn. Brooks/Cole Publishing Company.

Oppenheim, A.V. and Schafer, R.W. 1989. *Discrete-Time Signal Processing*. Englewood Cliffs, NJ: Prentice Hall.

Oskoui-Fard, P. and Stark, H. 1988. Tomographic image reconstruction using the theory of convex projections. *IEEE Transactions on Medical Imaging*, **7**(1), 45–58.

Özbek, A. 2000a. Adaptive beamforming with generalized linear constrains. *Geophysics Extended Abstracts*.

Özbek, A. 2000b. Multichannel adaptive interface canceling. *Geophysics Extended Abstracts*.

Özbek, A., Hoteit, L., and Dumitru, G. 2004. 3-D filter design on a hexagonal grid with application to point-receiver land acquisition. *SEG 2004 Expanded Abstracts*.

Peacock, K. L. and Treitel, Sven. 1969. Predictive deconvolution: Theory and practice. *Geophysics*, **34**(2), 155–169.

Pecholcs, Peter I., Al-Saad, Riyadh, Al-Sannaa, Muneer, Quigley, John, Bagaini, Claudio, Zarkhidze, Alexander, May, Roger, Guellili, Mohamed, Sinanaj, Sokol, and Membrouk, Mohamed. 2012. *A Broadband Full Azimuth Land Seismic Case Study from Saudi Arabia Using a 100,000 Channel Recording System at 6 terabytes per day: Acquisition and Processing Lessons Learned*. Pages 1–5.

Porsani, Milton J. and rn Ursin, Bjø. 2007. Direct multichannel predictive deconvolution. *Geophysics*, **72**(2), H11–H27.

Proakis, J. G. and Manolakis, K. D. 2006. *Digital Signal Processing*. 4th edn. Prentice-Hall, Inc., Upper Saddle River, NJ.

Recht, B., Kumar, R., Aravkin, A.Y. Y, Mansour, H., Recht, B., and Herrmann, F. J. 2013. Tu-04-13 Seismic Data Interpolation and Denoising Using SVD-free Low-rank Matrix Factorization. Pages 10–13 of: *75th EAGE Conference & Exhibition incorporating SPE EUROPEC 2013 London, UK, 10–13 June 2013*.

Robinson, E. A. 1967. *Predictive Decomposition of Time Series with Applications to Seismic Exploration*. Ph.D. thesis, Massachusetts Institute of Technology.

Robinson, E. A. 1998. Model deriven predictive deconvolution. *Geophysics*, **63**, 713–722.

Robinson, E. A. and Clark, R. D. 1987. The wave equation. *The Leading Edge*, **6**(7), 14–17.

Robinson, E. A. and Treitel, S. 2000. *Geophysical Signal Analysis*. SEG.

Robinson, E. A. and Treitel, S. 2008. *Digital Imaging & Deconvolution: The ABCs of Seismic Exploration and Processing*. SEG.

Sabbione, Juan I. and Velis, Danilo. 2010. Automatic first-breaks picking: New strategies and algorithms. *Geophysics*, **75**(4), V67–V76.

Sacchi, M. D., Verschuur, D. J., and P. M. Zwartjes. 2004. Data reconstruction by generalized deconvolution. *SEG Int'l Exposition and 74th Annual Meeting, Denver, Colorado, 10–15 October 2004*, 1989–1992.

Sacchi, Mauricio D. and Ulrych, Tadeusz J. 1995. Highresolution velocity gathers and offset space reconstruction. *Geophysics*, **60**(4), 1169.

Sacchi, Mauricio D., Porsani, Milton, Federal, Universidade, and Bahia, Da. 1999. Fast high resolution parabolic Radon transform. *SEG Technical Program Expanded Abstracts 1999*, 1477–1480.

Sacchi, Mauricio D. 1999. A Tour of High Resolution Transforms. *cseg.ca*, 665–668.

Sanchis, Charlotte and Hanssen, Alfred. 2011. Multiple-input adaptive seismic noise canceller for the attenuation of nonstationary coherent noise. *Geophysics*, **76**(6), V139–V150.

Sarajrvi, M. *Inversion of the Linear and Parabolic Radon Transform*. M.Sc.

Sastry, C., Hennenfent, G., and Herrmann, F. 2007. Signal reconstruction from incomplete and misplaced measurements. *European Association of Geoscientists & Engineers*, 11–14.

Sayed, Ali H. 2008. *Adaptive Filters*. John Wiley and Sons, Inc., Hoboken, NJ, USA:

Scales, J. A. 1997. *Theory of Seismic Imaging*. Samizdat Press.

Search, Home, Journals, Collections, Contact, About, Iopscience, My, Problems, Inverse, and Address, I P. 2001. Inversion of the attenuated Radon transform. *Inverse Problems*, **113**(1), 113.

Selley, R. C. and Sonnenberg, S. A. 2014. *Elements of Petroleum Geology*. Academic Press.

Sezan, M. 1992. An overview of convex projections theory and its application to image recovery problems. *Ultramicroscopy*, **40**, 55–67.

Shannon, C. E. 1949. Communication in the presence of noise. *Proceedings of the Institute of Radio Engineers*, **37**(1), 10–21.

Sheriff, R. E. 2006. *Encyclopedic Dictionary of Applied Geophysics*. 4 edn. Society of Exploration Geophysics.

Sheriff, R. E. and Geldart, L. P. 1995. *Exploration Seismology*. 2nd edn. Cambridge Univeristy Press.

Shoup, Robert C., Sacrey, Deborah K., Sternbach, Charles A., and Nagy, Richard L. (eds). 2003. *Heritage of the Petroleum Geologist*. American Association of Petroleum Geologists (AAPG).

Spanias, A. S., Jonsson, S. B., and Stearns, S. D. 1991. Transform methods for seismic data compression. *IEEE Transactions on Geosciences and Remote Sensing*, **29**(3), 407–416.

Stark, H. and Yang, Y. 1998. *Vector Space Projections: A Numerical Approach to Signal and Image Processing, Neural Nets, and Optics*. 1st edn. John Wiley and Sons Publisher.

Sydsaeter, K. and Hammond, P. 2001. *Essential Mathematics for Economic Analysis*. Prentice Hall.

Tang, Wen, Ma, Jianwei, and Herrmann, Felix J. 2009. Optimized compressed sensing for curvelet-based seismic data reconstruction. *Preprint*, 1–28.

Telford, W. M., Geldart, L. P., and Sheriff, R. E. 1990. *Applied Geophysics*. 2nd edn. Cambridge Univeristy Press.

Thorbecke, J. 1997. *Common Focus Point Technology*. Ph.D. thesis, Delft University of Technology.

Thorson, Jeffrey R. 1985. Velocity-stack and slant-stack stochastic inversion. *Geophysics*, **50**(12), 2727.

Toft, P. 1996. *The Radon Transform: Theory and Implementation*. Ph.D. thesis, Technical University of Denmark.

Tong, Siyou, Wang, Ruimin, Liu, Huaishan, Zhang, Jin, and Bu, Changcheng. 2009. High resolution radon transform and its applications in multiple suppression of seismic data in deep-sea. *Proceedings of the 2009 2nd International Congress on Image and Signal Processing, CISP'09*, **2**(1), 1–4.

Trad, Daniel. 2003. Interpolation and multiple attenuation with migration operators. *Geophysics*, **68**(6), 2043.

Trad, Daniel, Ulrych, Tadeusz, and Sacchi, Mauricio. 2003. Latest views of the sparse Radon transform. *Geophysics*, **68**(1), 386–399.

Trad, Daniel O., Ulrych, Tadeusz J., and Sacchi, Mauricio D. 2002. Accurate interpolation with high-resolution time-variant Radon transforms. *Geophysics*, **67**(2), 644.

Treitel, S., Shanks, J. L., and Fraster, C. W. 1967. Some aspects of fan filtering. *Geophysics*, **XXXII**, 789–800.

Trorey, A. W. 1970. A simple theory for seismic diffraction. *Geophysics*, **35**, 762–784.

Ulrych, T. J., Sacchi, M. D., and Graul, J. M. 1999. Signal and noise separation: Art and science. *Geophysics*, **64**(Sept.- Oct.), 1648–1656.

Vermeer, G. J. O. (ed.). 2012. *3D Seismic Survey Design*. 2nd edn. Society of Exploration Geophysics.

Verschuur, D. J. 1997. Estimation of multiple scattering by iterative inversion, Part II: Practical aspects and examples. *Geophysics*, **62**(5), 1596.

Verschuur, Eric Dirak Jacob. 1991. *Surface-Related Multiple Elimination: An Inversion Approach*. Ph.D. thesis, Delft University of Technology.

Womack, J. E. and Cruz, J. R. 1994. Seismic data filtering using a Gabor representation. *IEEE Transactions on Geosciences and Remote Sensing*, **32**(2), 467–472.

Wong, J., Han, L., Bancroft, J. C., and Stewart, R. R. 2009. *Automatic Time-Picking of First Arrivals on Noisy Microseismic Data*. Tech. rept. CREWES Research Report, University of Calgary.

Yang, Yi, Ma, Jianwei, and Osher, Stanley. 2013. Seismic data reconstruction via matrix completion. *Inverse Problems and Imaging*, **7**(4), 1–16.

Yilmaz, Ö. (ed). 2001. *Seismic Data Analysis: Processing, Inversion, and Interpretation of Seismic Data*. 2nd edn. Society of Exploration Geophysicists.

Zhang, H., Thurber, C., and Rowen, C. 2003. Automatic P-wave arrival detection and picking with multiscale wavelet analysis for single-component recordings. *Bulletin of the Seismological Society of America*, **93**(5), 1904–1912.

Zhang, Rongfeng and Ulrych, Tadeusz J. 2003. Physical wavelet frame denoising. *Geophysics*, **68**(1), 225–231.

Zhou, Binzhong. 1994. Linear and parabolic τ-p transforms revisited. *Geophysics*, **59**(7), 1133.

Zibulevsky, M. and Pearlmutter, B. 2001. Blind source separation by sparse decomposition in a signal dictionary. *Neural Computation*, **13**(4), 863–882.

Ziemer, R. E., Tranter, W. H., and Fannin, D. R. 1998a. *Signals & Systems: Continuous and Discrete*. 2nd edn. Prentice-Hall.

Ziemer, R. E., Tranter, W. H., and Fannin, D. R. 1998b. *Signals and Systems: Continuous and Discrete*. 4th edn. Prentice Hall.

Ziolkowski, Anton and Bokhorst, Karel. 1993. Determination of the signature of a dynamite source using source scaling, Part 2: Experiment. *Geophysics*, **58**(8), 1183–1194.

Ziolkowski, Anton, and Slob, Evert. 1991. Can we perform statistical deconvolution by polynomial factorization? *Geophysics*, **56**(9), 1423–1431.

Index

k-lag spike, 289
l_1-norm, 209
2-D DFT, 163

A/D, 23, 206
A/D convertor, 280
Aliasing, 202
aliasing, 193
Analog-to-digital convertor, 23
arrival time, 299
autocorrelation, 130, 275

Bilinear transformation, 248
Butterworth, 251

Chebyshave, 251
CMP, 18, 29
common mid-point, 29
Compressive Sensing
    Incoherence, 210
    Measurement matrix, 210
    RIP, 210
    Sensing matrix, 210
Compressive sensing, 208
    Interpolation, 211
Convolution
    Circular, 160
    Linear, 162
Convolution sum, 127
crosscorrelation, 130, 275

Deconvolution, 298
    deterministic, 286
    FX, 295
    spiking, 289
    statistical, 286
deconvolution, 30
DFT, 155
Diffractions, 101

diffractions, 279
Discrete Fourier transform, 155
Discrete-time Fourier transform, 150
display
    variable area, 27
    variable density, 27
    wiggle, 27
displaying seismic, 27
Distinct Poles, 145
DTFT, 150

Earth layers, 42
elasticity, 42–44, 46, 47
Elliptic, 251
Energy
    Cumulative, 303
exploration geophysics, 10

FFT, 238, 240
Filter
    Analog, 248
    anti-alias, 23
    Anti-aliasing, 206
    FIR, 289
    inverse, 285
filtering, 29
Filters
    fan, 261
Finite Impulse Response, 274
FIR, 274
force, 43, 47–49
Frequency, 150
frequency-wavenumber, 163

gain application, 31
Gather
    Common Midpoint, 20
    Common Receiver, 20
    Common Shot, 20
Gaussian noise, 278

Geological
    stratigraphy, 39
    structure, 39
    surveys, 8
geophone, 12
ground-roll, 224

least square, 274
least square sense, 289
Levinson algorithm, 277
Linear
    Time Invariant, 17
linear equations, 276
LSI, 123, 221
LTI, 17

migration, 30
mode conversion, 67
modulii, 45, 46, 48, 59, 63, 75
modulus, 46–48, 60, 61, 65, 111
Moveout, 217
Multiple Poles, 147
Multiples, 214
    Long-path, 102
    Short-path, 103
multiples, 279, 286
muting, 36

NMO, 29, 36
Noise
    ambient, 280
    coherent, 105
    random, 106
    wind, 280
noise, 25, 43, 273, 285
    coherent, 25, 29
    ground roll, 25
    incoherent, 25
    random, 29
normal equations, 276
normal move-out, 29
Nyquist frequency, 196, 201
Nyquist wavenumber, 201

offset, 29
    zero, 29

Partial fraction expansion, 144
Projections onto Convex Sets, 238

Q factor, 68
quality factor, 68

Radon Transform, 172
    $\tau - q$, 213, 217
    $\tau - q$, 182
    $\tau - p$, 179
    Linear, 179
    Parabolic, 182

Random noise, 280
random process, 274
reflection
    coefficient, 24
    time, 24
Reflections, 214
reflectivity
    function, 24
refractions, 280
Region of Convergence, 132
reservoirs
    anticline, 6
    faulted, 6
    lenses-type, 8
    reef-type, 8
    salt-dome, 6
    unconformity, 8
Resolution
    vertical, 298
resolution
    horizontal, 30
    vertical, 30
Reverberations, 103
reverberations, 286

Sampling, 155
Sampling frequency, 194
Sampling theorem, 196
Seismic
    acquisition, 12
    conventional acquisition, 20
    convolution model, 25, 280
    deconvolution, 281
    detectors, 15
    dynamite, 15
    event, 27
    geophones, 15
    image, 29
    interpretation, 14
    land, 14
    linear predication, 281
    Migration, 257
    models, 298
    noise cancellation, 281
    primary reflections, 279
    processing, 12
    recording system, 22
    reflection survey, 11
    single-sensor acquisition, 20
    sources, 15
    trace, 23, 27
    vibroseis, 15
Seismic Events
    Hyperbolic, 166, 181
    Linear, 164, 179, 185
    Parabolic, 186
    Random, 169

seismic waves, 42, 43, 48, 49, 54, 58, 59, 64, 65, 67, 69, 81, 97, 111
seismogram, 24
shot gather, 30
signal-to-noise ratio, 29
Signals
   Aperiodic, 121
   Dipole, 118
   Energy, 118
   Exponential, 118
   Periodic, 121
   Power, 119
   Sinusoid, 118
   Spike, 117
   Unit ramp, 118
   Unit sample, 117
   Unit step, 118
SNR, 29
sorting, 29
Stack
   Stacking chart, 19
Stacking, 36
stacking, 29, 30
statics, 31
strain, 42–49, 60, 75, 111, 238, 240–245
stress, 42–49, 75, 111
surface waves, 279
System
   Discrete-time, 122
System function, 139
Systems
   All-Pole, 140
   All-zero, 140
   BIBO stable, 126
   Causal, 125
   Convolution, 127
   Difference equation, 126
   FIR, 131
   IIR, 131
   Linear, 123
   LSI, 125
   Non-recursive, 131
   Pole-zero, 141
   Poles, 140
   Recursive, 131
   Shift-invariant, 125
   Zeors, 140

Toeplitz matrix, 277
trace
   editing, 31
Transfer function, 139
transmission losses, 67
traveltime, 24

Undersampling, 206

Velocity, 261
velocity analysis, 29

wave equation, 48–51, 222, 255
wave propagation, 42, 43, 49, 59, 60, 65, 67, 75, 78, 208
wavelet
   source, 24
   zero-phase, 289
Wavelets
   anti-causal, 300
   causal, 300
   dipole, 301
   equi-pole, 301
   equivalent, 304
   finite-length, 300
   maximum-delay, 300
   minimum-delay, 300
   non-causal, 300
   Rectangular, 306
   Ricker, 217, 306
   Sinc, 306
wavenumber, 52, 54, 71, 150, 164, 165, 172, 202, 204, 206, 221, 222, 256–259
Wiener
   optimum filters, 281
Wiener-Hopf, 276
Window
   Blackman, 232
   Hamming, 230
   Hanning, 230
   Kaiser, 232
   Rectangular, 227

zero padding, 158
zero-lag spike, 289